Received April 25, 1998

RMS Associates
109 Shasta Rd.
Plymouth Meeting, PA 19462

S0-BYE-501

THE COMPLETE
RF TECHNICIAN'S
HANDBOOK
Second Edition

THE COMPLETE
RF TECHNICIAN'S
HANDBOOK
Second Edition

BY COTTER W. SAYRE

PROMPT® PUBLICATIONS

PROMPT© Publications is an imprint of Howard W. Sams & Company, A Bell Atlantic Company, 2647 Waterfront Parkway, E. Dr., Indianapolis, IN 46214-2041.

International Standard Book Number: 0-7906-1147-3
Library of Congress Catalog Card Number: 96-69139

Acquisitions Editor: Candace M. Hall
Editor: Loretta L. Yates
Assistant Editors: Pat Brady, Natalie Harris
Typesetting: Loretta L. Yates
Cover Design: Kelli Ternet
Graphics Conversion: Brian Drum, Terry Varvel

PRINTED IN THE UNITED STATES OF AMERICA

9 8 7 6 5 4 3 2 1

Dedicated to my lovely wife Linda and my remarkable mother Janice.

Without their help, this book would not have been possible.

Table of Contents

CHAPTER 3
Oscillators and Frequency Synthesizers

CHAPTER 4
Amplitude Modulation

CHAPTER 5
Frequency Modulation 115

CHAPTER 6
Single-Sideband Suppressed-Carrier 137

CHAPTER 7 149
Other Communication Applications 149

CHAPTER 8

Microwaves 179

CHAPTER 9

Antennas and Transmission Lines 189

CHAPTER 10
Troubleshooting 203

CHAPTER 11
Support Circuits 277

Introduction to the Second Edition

This is *the* handbook for the RF or wireless communications beginner, student, experienced technician, or ham radio operator! Although meant for people with a prior foundation in electronics, this book furnishes the reader with valuable information on the fundamental and advanced concepts important to the study and application of RF wireless communications.

In this Second Edition of *The Complete RF Technician's Handbook* all chapters have been expanded, enhanced, and/or updated, including the addition of forty-five new explanatory drawings, along with an enlarged and comprehensive glossary with every need-to-know radio communications term clearly defined.

All too often electronic books are loaded with theory, but contain precious little practical knowledge. As well, many of these books refer to test equipment and circuits that have not been used by industry in over twenty years. *The Complete RF Technician's Handbook*, however, includes only those circuits that you will find in the majority of modern RF devices. Additionally, current test equipment and their applications are covered in order to help the technician become familiar with their basic functions.

Troubleshooting of RF, digital, and audio circuits down to the component level is heavily stressed, as this is what most competent technicians will be required to do.

Since digital circuits and modulation techniques have become such an important part of many electronic devices, digital methods and their real-world circuits are discussed as they apply to wireless communications.

RF voltage and power amplifiers, in addition to their classes of operation, configurations, biasing, coupling techniques, frequency limitations, noise and distortion are all included, as are the major types of LC, crystal and RC oscillators. Up-to-date frequency synthesis procedures are included as well.

Of course, all of the RF modulation and detection methods are explained in detail — AM, FM, PM, and SSB — as are their associated circuits; such as automatic gain control, squelch, selective calling, frequency multipliers, speech processing, mixers, power supplies, digital signal processing, and automatic frequency control circuits. Various multiplexing methods and data, satellite, spread spectrum, cellular, and microwave communication technologies are discussed.

The Complete RF Technician's Handbook will supply the working technician or student with a solid grounding in the latest methods and circuits employed in today's RF communications gear, as well as the ability to test and troubleshoot transmitters, transceivers, and receivers with absolute confidence.

CHAPTER 1
The Basics

Even though the reader is probably quite familiar with the basics of RF electronics, a quick review of hazy long-ago studied principles will help in fully understanding the rest of the chapters on more advanced topics. For the electronics student or technician not very familiar with the RF basics, this chapter is a good primer on the subject.

1-1. Reactance, Impedance and Phase Angle

Understanding the principles of reactance, impedance, and phase angle are vital for the understanding of all alternating current circuits. *Reactance* is the opposition to sinusoidal alternating current by inductance and capacitance. A circuit with steady DC current can not have any capacitive or inductive reactance (direct current also produces a magnetic field around a conductor, but this field is not changing). *Impedance* is the combined opposition to both the resistive (introduced by any circuit resistance) and reactive components (physical capacitors and inductors, or stray capacitance and inductance) of a sinusoidal alternating current circuit.

Phase angle is the difference in degrees between two sine voltages, or two sine currents, or a combination of both. Usually only of concern when working with certain amplifiers, phase splitters, angle (FM) modulation methods, and resistive and reactive components in an AC circuit.

1-1A. Reactance

There are two types of reactances; inductive and capacitive. Inductive reactance is the opposition to alternating sine-wave current that takes place in a coil or inductive circuit and increases with an increase in frequency.

Inductance resists any changes in current: If the current in an inductor attempts to increase, a temporary counter-EMF, produced by the expanding magnetic lines of force that are created by the increasing current, will oppose any changes in the original voltage value. This keeps the current at its original level. The current does eventually rise, nonetheless, because the counter-EMF will no longer be induced as soon as there are no rising or falling current variations (the current has reached its maximum or minimum level) and thus no expanding or contracting magnetic fields are produced. However, if the current does begin to decrease, the magnetic lines of force cut the conductors on their way back into the conductor, creating a voltage that opposes this change in current.

An inductor therefore stores current, because when the supply voltage is removed from across the inductor the magnetic lines of force contract, inducing a voltage in the inductor, creating an output current. Since an AC circuit consists of constantly changing current levels, and an inductor opposes any change in this current, this translates into a opposition to AC, or inductive reactance.

On the other hand, capacitive reactance is the opposition to alternating sine-wave current in a capacitor or a capacitive circuit. This opposition increases with a decrease in frequency. As well, the larger the capacitance of a capacitor or circuit, the lower its capacitive reactance.

A capacitor resists any change in voltage and will charge to any new level of voltage as supplied by the source. If the source decreases in voltage, the capacitor will supply current to the load until it reaches this new lower source voltage. The time it takes for the capacitor to discharge or charge is dependent on its capacitance and series resistance ($t=RC$). As an AC circuit is constantly changing voltage levels, and a capacitor opposes any variation in this changing voltage, this also translates into a resistance to AC, or capacitive reactance.

1-1B. Impedance

Impedance is the total opposition to alternating current from the capacitive and inductive reactances as well as the resistive component in a circuit. Because the reactance and resistance of a circuit are out of phase, these values cannot simply be arithmetically summed ($Z \neq R+X$) as in resistive circuits, but a phasor sum must be taken. First, find the reactances of each component by:

$$X_C = \frac{1}{2\pi fC}$$

and

$$X_L = 2\pi fL$$

Then, for a series circuit (*Figure 1-1*):

$$Z = \sqrt{R^2 + (X_L - X_C)^2}$$

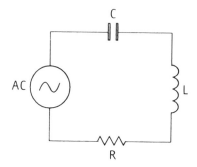

Figure 1-1. *A series LCR circuit.*

for a simple parallel reactive circuit (*Figure 1-2*), find the total circuit current by:

$$I_T = \sqrt{I_R^2 + (I_L - I_C)^2}$$

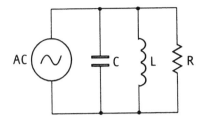

Figure 1-2. *A parallel LCR circuit.*

Then, find the total impedance by:

$$Z = \frac{V_{APPLIED}}{I_{TOTAL}}$$

Parallel circuits with multiple complex branches (*Figure 1-3*) are even more elaborate.

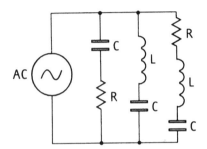

Figure 1-3. *A complex branch parallel LCR circuit.*

With real-life circuits this can begin to get very involved, as a dozen or more reactive and resistive components can make up a normal network in series and parallel combinations. Mathematics employing *complex numbers*, which will not be addressed here, are the normal method adopted for these calculations. Typically, however, simple direct measurements of currents and voltages can be taken, and Ohm's law would then be utilized to find the total impedance of the circuit:

$$Z_{TOTAL} = \frac{V_{APPLIED}}{I_{TOTAL}}$$

1-1C. Phase Angle

The phase difference between reactive and resistive components mentioned above refers to an inductor's voltage leading its current by 90° and a capacitor's current leading its voltage by 90° (*Figure 1-4*).

Resistors do not have this phase, or time, shift between their current and voltage: When the voltage is maximum across a pure resistance, its current is also maximum. Thus, a resistance has a phase angle of 0°. A purely reactive circuit has a phase angle of ±90°, the sign depending on whether it is capacitive (-90°), or inductive (+90°). A mixed reactive and resistive circuit has a phase angle somewhere between 0° and ±90°, depending on the LCR component values.

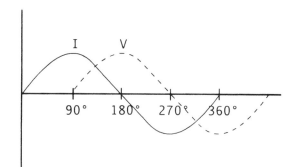

Figure 1-4. *The phase shift between a capacitor's current and voltage.*

1-2. Decibels

Decibels (*dB*) have a *logarithmic* response like that of the human ear, with 0 dB just at the threshold of human hearing. The ear responds (is more sensitive) to a change in sound intensity at lower levels than at higher levels. An increase of 4W to 5W sounds much louder than a change from 20W to 21W, yet they are both an increase of 1W. It is therefore the power ratios that matter (4W to 5W is an increase of 25% in power, while 20W to 21W is an increase of only 5%).

Not only is the ratio between two values used in dB measurements, but so is the $\log_{10}$ of this ratio. This was adopted to squeeze the extremes of small and large measurement values: Doubling the power output of an amplifier from 50W to 100W is a 3 dB increase; quadrupling the power output from 50W to 200W is only a 6 dB increase from the original 50W. Even though this quadrupling in power has occurred it is only *twice* as loud to the human ear, while an increase from 50W to 400W is only a 9 dB increase from the original 50W; the power has increased eight times, but is only three times as loud to the ear.

Doubling the *power* is a 3 dB increase, while doubling the *voltage* or *current* is an increase of 6 dB. This is due to the square of the voltage conforming to the power. When a number is squared, its logarithm is doubled, creating voltage ratios that are double the dB amounts for the equivalent power ratios.

Different power ratios and the corresponding change in dB values are given in *Table 1-1*. The formula for these decibel power calculations is:

$$dB = 10 \left(LOG_{10} \frac{P_2}{P_1} \right)$$

Or, when both input voltage and output voltage or input current and output current are across or through the same impedance, then you can use these formulas:

$$dB = 20 \left(LOG_{10} \frac{V_2}{V_1} \right)$$

and

$$dB = 20(LOG_{10} \frac{I_2}{I_1})$$

If Z_{IN} does *not* equal Z_{OUT}, then apply the formula:

$$dB = 20 \ LOG \frac{V_2 / \sqrt{Z_2}}{V_1 / \sqrt{Z_1}}$$

Power Ratio	dB
100	20
10	10
2	3
1.26	1
1	0
½	-3
1/10	-10
1/100	-20

Table 1-1. Comparison of a change in power levels to a change in decibles.

A loss in dB, such as an insertion loss through a filter, is indicated by a negative value, while an increase in dB, such as that supplied by an amplifier, is indicated by a positive number. Either way, decibels are simply added and subtracted to find their total value.

Another common decibel value is the dBc. dBc stands for *decibels below the carrier*. This measures the relative signal strength, in decibels, between the peak of a carrier and the peak of its sidebands (or its spurious responses).

1-2A. Logarithms

It may help to know what logarithms actually are. The log of a number is the exponent of the power of 10 that equals that number. For example, 163 equals $10^{2.212}$. Thus the log_{10} of 163 is 2.212.

1-2B. Reference levels

Instead of using different random values for the denominator, assorted dB *reference levels* have been generally adopted. This allows comparisons to a commonly known reference level:

0 dB: 6 mW referenced into 500Ω. $dB = 10\,LOG_{10}\dfrac{P_{OUT}}{6mW}$

0 dBm: 1 mW referenced into 50Ω. A standard measurement in almost all RF work. Equals 0.224 V$_{RMS}$ across 50Ω.

$$dBm = 10\,LOG_{10}\dfrac{P_{OUT}}{1mW}\,.$$

·0 dBmV: 1 mV referenced across 75Ω. Used for RF signal voltage, especially on 75Ω coax transmission line:

$$dBmV = 20\,LOG_{10}\dfrac{V_{OUT}}{1mV}\,.$$

0 dBu: 1 uV reference: $dBu = 20\,LOG_{10}\dfrac{V_{OUT}}{1uV}$

0 dBk: 1 KW reference: $dBk = 10\,LOG_{10}\dfrac{P_{OUT}}{1KW}$

VU Unit: 1 mW in 600Ω *or* 0.7746 V$_{RMS}$ across 600Ω. For studio measurement in radio broadcasting.

Note: When measuring with most VOMs using the dB scale, the reference level is 0.775 volts for 0 dB across 600Ω.

1-3. Sensitivity

The sensitivity of an AM receiver is usually shown on the specifications table as a certain amount of microvolts (usually around 0.5 to 1 μV) that will produce a signal-to-noise (S/N) ratio of 10 dB. *SINAD* is another such specification and is used for FM receivers. It refers to the ratio for

$$\frac{SIGNAL + NOISE + DISTORTION}{NOISE + DISTORTION}$$

or an FM receiver's sensitivity, in μVs, with a signal-to-noise-to-distortion divided by noise-and-distortion of 12 dB at the output to the final audio amplifier.

There are two types of significant internally-generated receiver noise that affect the S/N ratio, both of which produce an actual randomly fluctuating and wide frequency-ranging voltage. *White noise*, caused by a component's atoms and electrons randomly flitting around a component due to heat; and *shot noise*, caused by electrons entering the plate, collector, or drain of a tube, transistor, or FET, and especially by the random transit of electrons across a semiconductor junction. The ratio between the noise and a received signal's strength can be expressed in decibels as:

$$\text{S/N} = 10 \text{ LOG}_{10} \frac{P_{SIGNAL}}{P_{NOISE}}, \text{ or } 20 \text{ LOG}_{10} \frac{V_{SIGNAL}}{V_{NOISE}}$$

and refers to a receiver's usable sensitivity.

Above 30 MHz receiver-generated noise is generally higher than the noise received from the antenna, so the design and construction of low noise amplifiers is of special concern in the VHF and UHF regions.

1-4. Semiconductors

In the last forty years semiconductors have completely taken over rectification and amplification chores in RF from the once dominant vacuum tube — at least in the low to medium power ranges.

Semiconductors are dependable, small, sturdy, and use low supply voltages. They are used in rectification, mixing, detection, amplification, oscillation and digital circuits. ICs, and thus most modern circuits, would be impossible without semiconductors. A quick review of their internal structure is warranted.

1-4A. Diodes

A *PN junction diode* (*Figure 1-5*) consists of both N- and P-type semiconductor materials joined together. The N-type semiconductor has an excess of electrons (majority carriers) and few holes (minority carriers). This is due to the doping (adding of impurities) to the intrinsic (pure) semiconductor material by adding atoms that have five valence (outer shell) electrons as opposed to the four valence electrons of intrinsic silicon.

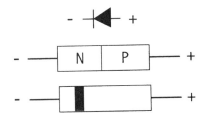

Figure 1-5. *The semiconductor diode.*

P-type semiconductors have an excess of holes and a shortage of electrons in their crystal lattice structure caused by the doping of the intrinsic semiconductor material by atoms that have three valence electrons as opposed to the four valence electrons of intrinsic silicon.

P-type semiconductor current is by hole flow through the crystal lattice. In N-type semiconductors current is by electron flow through the crystal lattice.

With no bias voltage applied to the diode (*Figure 1-6*) electrons are attracted to the P side, holes are attracted to the N side. At the PN junction a *depletion region* forms, created by the combining of these electrons and holes, producing neutral electron-hole pairs at the junction, while the depletion region (on either side of the PN junction) itself consists of charged ions. This depletion region, if the semiconductor material is silicon, has a barrier potential of 0.7 volts. The depletion region does not increase beyond 0.7V because any more majority carriers are repelled by this barrier voltage.

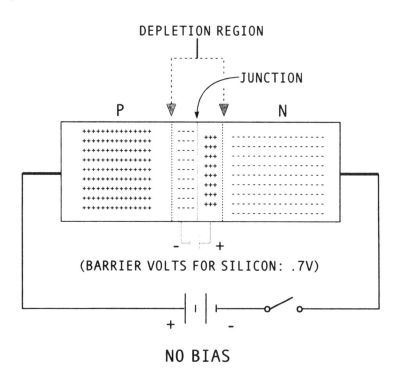

Figure 1-6. *A semiconductor diode with zero bias across its terminals.*

If an external bias voltage of sufficient potential is applied to the PN junction of the proper polarity to forward bias the junction (*Figure 1-7*), then the barrier voltage is neutralized by this bias, and electrons are able to flow. The positive terminal of the battery repels the holes and attracts the electrons, while the negative terminal of the battery repels the electrons toward the positive terminal, creating a current stream through the diode.

When a reverse bias is applied to a diode's terminals (*Figure 1-8*), the depletion region in a semiconductor diode widens as the holes are attracted to the negative side of the battery and the electrons are attracted to the positive battery terminal. With this reverse bias the diode acts as a very high resistance, allowing very little current to flow, except for some small leakage value. This widening of the depletion region can continue until the barrier potential equals that of the bias, or breakdown occurs, creating uncontrolled reverse current flow and possible damage to the diode.

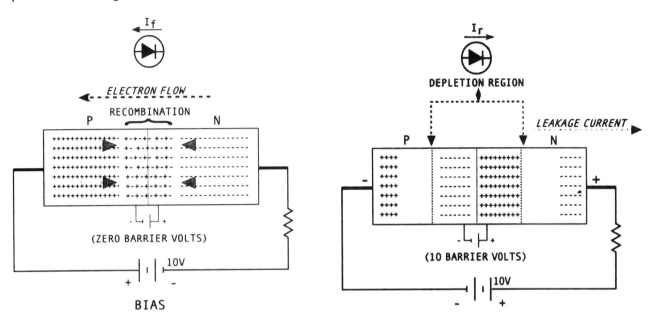

Figure 1-7. *A semiconductor diode with the proper forward bias applied.*

Figure 1-8. *A semiconductor diode with reverse bias applied.*

It must be remembered that approximately 0.7V will always be dropped across a forward-biased silicon diode, even when the current increases, as shown in the characteristic curves for an average silicon diode in *Figure 1-9*. This voltage drop across the diode will remain substantially at 0.7V, even while the diode is conducting current, because of the small value of internal (dynamic) resistance of the semiconductor materials.

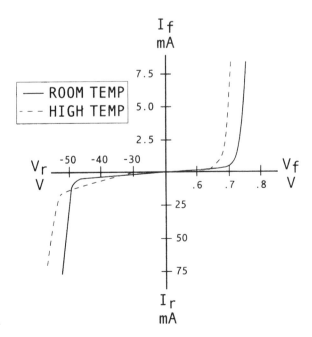

Figure 1-9. *A silicon diode's characteristic curves.*

Rectifier and *signal diodes* are used to perform rectification (converting AC to pulsating DC) and in mixer, RF detection and modulation circuits, as well as other circuits that need the ability of a device that conducts in a single direction only (*unidirectional conduction*).

Small plastic (*Figure 1-10*) and glass diode packages are used for low-current applications (up to 1A). *Power diodes* (*Figure 1-11*) are capable of surviving high forward currents (up to 1,500A), while complete *bridge rectifier diode* packages (*Figure 1-12*) are used for full-wave power supplies.

Figure 1-10. *A common small-signal diode package.*

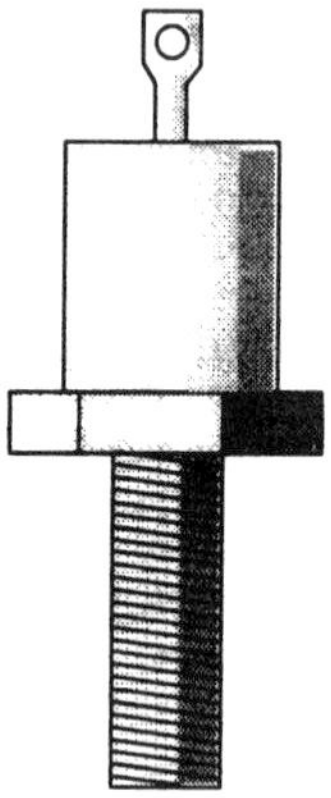

Figure 1-11. *A power diode package.*

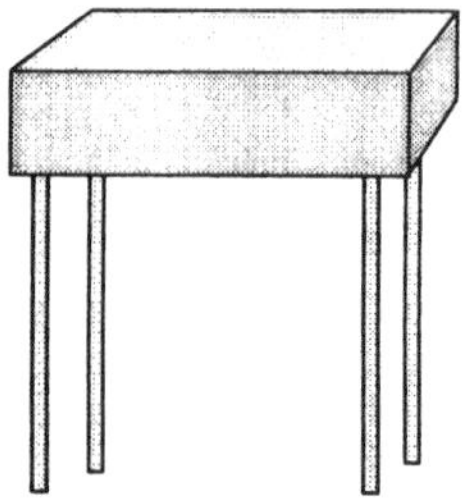

Figure 1-12. *A popular bridge rectifier package.*

Some important rectifier diode specifications:

I_R:　　　　Reverse leakage current (the small amount of current in a reverse-biased diode that increases with an increase in temperature).

$I_{F(MAX)}$:　　Maximum forward current before total destruction of a diode. Diodes can be connected in parallel to increase this current carrying capability.

PIV:　　　　Peak Inverse Voltage (maximum reverse voltage across a reverse-biased diode). Diodes can be connected in series to increase this PIV rating.

Zener diodes (Figure 1-13), utilize the Zener's ability to operate with reverse bias until *avalanche*, or reverse breakdown, occurs — without being destroyed.

As can be seen in the characteristic curves of *Figure 1-14,* for a large change in current the voltage across the Zener changes very little, making it perfect for voltage regulation and as a voltage reference.

Zener diodes with 20, 10, 5 and 1 percent tolerances are obtainable, making it possible to use low-cost 20% Zeners in non-critical circuits, while using the higher tolerance Zeners in more precise circuits.

In some circuits it must be taken into account that a standard Zener's voltage ratings fluctuate with temperature. *Voltage reference diodes,* or *temperature compensated Zener diodes,* are available that remain nearly stable with temperature variations.

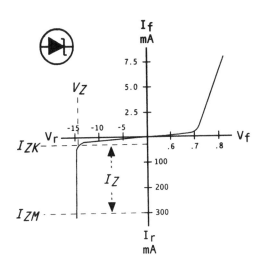

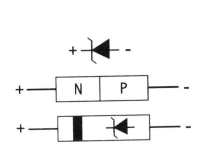

Figure 1-13. *A Zener diode.*

Figure 1-14. *A Zener diode's characteristic curves.*

Some important Zener diode specifications:

V_z: Reverse voltage across the Zener that varies little even with an increase in reverse current.

I_{ZM}:   Maximum Zener current before Zener destruction.

P_D: Recommended maximum power dissipation rating.

I_z: Zener current needed to keep Zener in V_z region.

Varactor diodes (Figure 1-15) also use reverse bias to perform their function. They are employed as *voltage variable capacitors* (VVCs). Since increasing the width of the dielectric in a capacitor decreases its capacitance, and decreasing the thickness of the dielectric increases the capacitance, this fact is used in the varactor to electrically vary its capacitance. Increasing the reverse voltage across the varactor diode widens the depletion region, which lowers capacitance by thickening the dielectric (*Figure 1-16*). Lessening the reverse voltage increases capacitance by narrowing the depletion region and thus thinning the dielectric. *Figure 1-17* displays the capacitance variations per applied reverse voltage for a standard varactor diode.

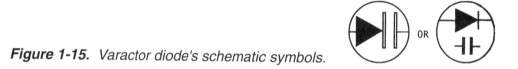

Figure 1-15. *Varactor diode's schematic symbols.*

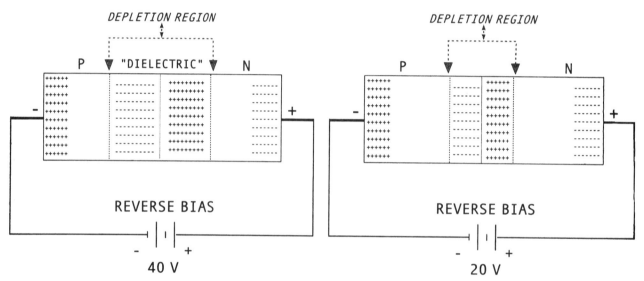

Figure 1-16. *The change in dielectric width of a varactor diode with two different reverse bias voltages.*

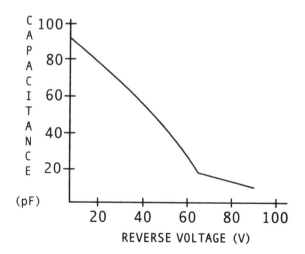

Figure 1-17. *A varactor diode's change in capacitance versus the applied reverse voltage.*

RF amplifiers that must amplify a large range of frequencies can use varactor diodes, controlled by bias voltages from a microprocessor, that vary the resonant frequency of the tanks of *non-power* RF amplifiers and radio receivers, as well as change the frequency of certain oscillators.

Varactors are available in numerous capacitance values for any application.

1-4B. Transistors

A bipolar (BJT), or junction, transistor (*Figure 1-18*) consists of either NPN or PNP doped regions. The *emitter* supplies the charges, while these charges are controlled by the *base*. The *collector* collects the charges that have not gone into the base circuit.

In the silicon NPN transistor of *Figure 1-18* we see that the emitter and the base are forward biased, while the collector is reverse-biased. The emitter's electrons are repelled by the negative terminal of the battery and are pushed into the thin base. The slender base cannot support the amount of electrons that accumulate there due to the small amount of holes available for recombination. Most of these electrons are therefore attracted by the positive potential on the collector, where they flow into and on to the positive bias supply, creating an output current.

Almost all of the current supplied by the emitter flows through the collector (around 99%), while a small amount combines with the limited number of holes in the base and then out of this base circuit.

Since $I_E = I_B + I_C$ and $I_B = I_E - I_C$, then it can be seen that the currents through a transistor are directly proportional: If emitter current doubles, then so does base and collector currents. This also means that if the base current is increased by an external bias or signal, a proportional but far greater emitter and collector current will flow, creating voltage amplification if this current is fed through a high output resistance.

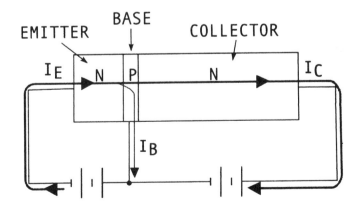

Figure 1-18. Current flow in a bipolar NPN transistor.

Since the input of a common-emitter transistor circuit is forward-biased, and therefore has a low resistance, any signal injected into the base-emitter junction is across this low input resistance. This makes bipolar transistors current controlled, as shown in their characteristic curves of *Figure 1-19*. The controlling current is managed by the DC bias and external signal voltages. The signal will then add-to or subtract-from this DC transistor bias voltage.

It should be kept in mind that transistors must always overcome their emitter-base barrier voltage, which is approximately $0.6V_{BE}$ for silicon (just as in the semiconductor diode) before appreciable current can flow. In a linear amplifier the bias circuits will set this operating point of the transistor at around the $0.6V_{BE}$ value, so that any incoming signal can swing above and below this amount. This region of active amplification is only 0.2V wide, and is the only area that a semiconductor is capable of amplifying; or between saturation (0.7V) and cutoff (0.5V). Between these two V_{BE} values the I_B, and thus the I_C, is controlled.

A transistor can be considered, for simplicity, as a current controlled resistance. A small current into the base controls the resistance of the transistor, which controls a much larger emitter-to-collector current — allowing this large current to then flow through a high load resistance, creating an amplified output voltage.

Some transistor models are a little more complex. Certain high-frequency power transistors are internally impedance matched to increase their input and output impedance, which might be as low as 0.5Ω in un-compensated, high-frequency, high-power devices.

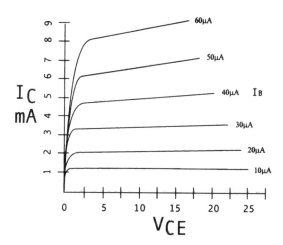

Figure 1-19. A bipolar transistor's characteristic curves.

Some important transistor ratings:

$P_{D(MAX)}$:   Maximum total power dissipation with an ambient temperature of 25°C in free air (P_D itself can be approximated by $V_{CE} \times I_C$).

BV_{CBO}:   Collector-to-base breakdown voltage, or the amount of collector voltage that will break down the collector junction.

$I_{C(MAX)}$:   Maximum allowable collector current.

$T_{J(MAX)}$:   Maximum internal junction temperature before breakdown of the semiconductor. Silicon has a maximum rating of around 200° centigrade.

f_{ae}:   b cutoff frequency, or when the frequency increases until b reaches 70.7% of its low-frequency value.

f_T: Current gain-bandwidth product, or the frequency that a transistor in a CE configuration reaches a b of unity.

I_{CEO}:   Leakage current between the emitter and the collector with the base open (I_B = 0 mA). I_{CEO} increases with an increase in the temperature, and combines with the I_C, increasing I_C.

Junction field effect transistors (JFETs) are voltage, and not current, controlled, since their input *gates* are always reverse biased, and therefore have very high input impedances as compared to bipolar transistors. JFETs are also able to accept an input of several volts to the bipolar transistor's few tenths of a volt (output clipping will begin in many BJT Class A small signal amplifiers at between 10-40 mVs input), while generating less internal noise. However, JFETs have less voltage gain and more signal distortion than transistors.

The construction of a JFET uses a *gate, source,* and *drain*. These terminals are biased so that drain-to-source voltage (V_{DS}) makes the source more negative than the drain, allowing current (I_D) to flow from source to drain through the N-channel (*Figure 1-20*).

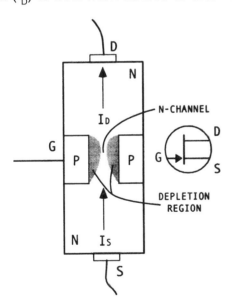

Figure 1-20. Construction of, and current flow through, a JFET; with schematic symbol.

As shown in the JFET characteristic curves of *Figure 1-21*, a JFET is a normally ON device when no bias voltage is applied to the gate. This allows maximum current flow (I_{DSS}) from source to drain. When a negative voltage is applied to the gate and source (V_{GS}), an area lacking charge carriers, called a *depletion region*, begins forming in the N-channel, and acts as an insulator. As the JFET becomes more reverse-biased and increasingly depleted of charge carriers, the N-channel continues to be narrowed by the growing depletion region, which increases the channel's resistance, lowering current output into the load resistor and thus lowering the output voltage across this resistor. As the negative voltage of V_{GS} is increased on the gate, the depletion area widens, which further increases channel resistance, lowering current flow even further. A point is finally reached where the channel is completely depleted of majority carriers, and no further decrease in current is possible. This point is called $V_{GS(OFF)}$. In other words, V_{GS} effectively controls the channel resistance of the device, and thus drain current.

V_{DS} (drain-to-source voltage) should be high enough to place the FET into a linear operating region, or above *pinch-off* (V_p). This pinch-off region is where I_D remains constant even if V_{DS} is increased — only V_{GS} can now effect I_D.

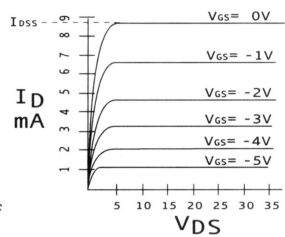

Figure 1-21. *The characteristic curves of a JFET.*

Some important JFET parameters:

Transconductance (g_m or g_{fs}, measured in Siemens or mhos): The gain of an FET, or $\dfrac{\Delta I_D}{\Delta V_{GS}}$.

V_P: Pinch-off voltage (minimum V_{DS} needed for linear operation of FET).

I_{DSS}: Maximum drain current of FET possible (with $V_{GS}=0V$).

$V_{DS(MAX)}$: Maximum allowed drain-to-source voltage.

P_D: Recommended maximum power dissipation rating.

MOSFETs (Metal-Oxide-Silicon Field-Effect Transistors), also called IGFETs (Insulated Gate FETs), use a gate that is insulated from the rest of the semiconductor material, creating a device with virtually infinite DC input resistance (bias components and high-frequency operation, however, lowers the "virtually infinite" input impedance of the MOSFET drastically. At

DC the MOSFET has very high input resistance, but at VHF and above the input impedance can be close to a BJT's).

MOSFETs function in two modes: The *depletion-mode*, which is a normally-on device, and the *enhancement-mode*, which is a normally-off device.

Depletion-mode MOSFETs (*Figure 1-22*) control drain current through the application of a negative *and* positive gate voltage, as shown in *Figure 1-23* for an N-channel device. Increasing the negative potential at the gate would eventually reach a point where no drain current could flow, due to the channel being depleted of all majority carriers. As V_{GS} (gate-to-source voltage) becomes less negative, more current begins to flow, until the V_{GS} reaches zero. At zero volts for V_{GS}, unlike the JFET, maximum current does not flow. The current, however, is still significant, since many majority carriers are present in the channel. As V_{GS} is made more positive in respect to the source, more electrons are drawn into the N-channel, increasing current through the channel and into the drain.

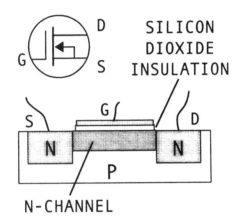

Figure 1-22. Internal construction of an N-channel depletion-mode MOSFET with schematic symbol.

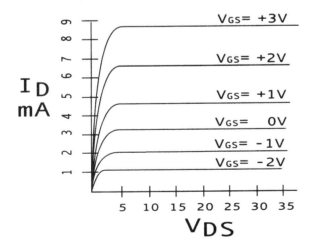

Figure 1-23. Characteristic curves of an N-channel depletion-mode MOSFET.

Depletion MOSFETs are used extensively in RF circuits due to their low noise characteristics. Dual-gate MOSFETs (*Figure 1-24*) are used in mixers and AGC controlled amplifiers, with each gate having equal control over I_D.

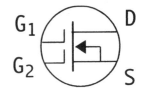

Figure 1-24. Schematic symbol of a dual-gate MOSFET.

An enhancement-mode MOSFET (*Figure 1-25*), as stated previously, is a normally-off device. When no bias is applied to the gate little source-to-drain current flows, as indicated in its characteristic curves of *Figure 1-26*. A positive voltage placed on the gate creates a channel between the source and drain (*Figure 1-27*): As electrons are drawn towards the gate, an N-channel is created in the P-type substrate, allowing electrons to flow toward the positively charged drain.

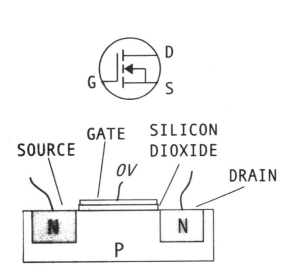

Figure 1-25. Internal construction of an enhancement-mode MOSFET with schematic symbol.

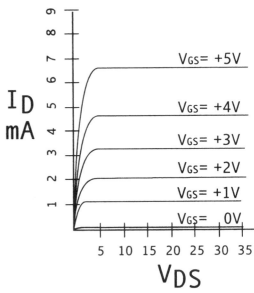

Figure 1-26. Enhancement-mode MOSFET characteristic curves.

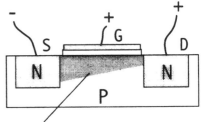

Figure 1-27. Positive gate voltage creating an N-channel in an E-MOSFET's substrate.

With enhancement-mode MOSFETs there is an approximately 1V threshold voltage before any significant current will flow. E-MOSFETs are used as voltage-controlled switches in applications using digital ICs; and are very popular for HF, VHF, and UHF DMOS and TMOS power amplifiers and drivers due to their higher power gain, higher input impedance, increased thermal stability, higher tolerance of load mismatches, and lower noise over that of the power BJT.

Another advantage in the use of any MOSFET over the BJT is that MOSFETs are designed to have a *positive temperature coefficient* at high drain currents, and thus will *decrease* their source-to-drain current as the temperature increases — unlike BJTs, which increase their emitter-to-collector currents, sometimes up to the transistor's destruction. This makes thermal runaway of the MOSFET impossible, and temperature stabilization components far less necessary, except to stabilize the FET's Q-point.

MOSFETs are, however, easily destroyed by static electricity, as any electrical spark can damage the gate insulation.

1-4C. Thyristors

Thyristors are PNPN devices that are used principally as solid-state switches, and include the *SCR*, the *TRIAC*, the *DIAC* and the *four-layer diode*. They are commonly used to switch large currents, or to control the amount of power reaching a load by allowing conduction for only part of an AC alternation, or in current protection circuits.

The SCR (*Figure 1-28*) controls, or switches, DC (or AC half-wave) current by the application of a positive control pulse at its gate (*Figure 1-29*), which causes gate current (I_G) to flow. When the SCR latches ON by this control pulse, current then flows from the cathode to the anode and through the load. Only the lowering of the anode current (I_A) below a certain value referred to as the *holding current* (I_H) will shut the SCR down, stopping current flow through the load. This value of holding current varies with the amount of gate current. Even with no gate pulse, the SCR can conduct when the anode-to-cathode voltage (V_{AK}) increases to a certain point, referred to as the *forward breakover voltage*. The SCR now displays very little resistance, and large amounts of current flows, with only a small voltage dropped across the SCR. This forward breakover voltage point is forced to occur with less V_{AK} by increasing the gate current.

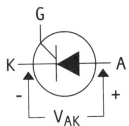

Figure 1-28. *Schematic symbol for the SCR.*

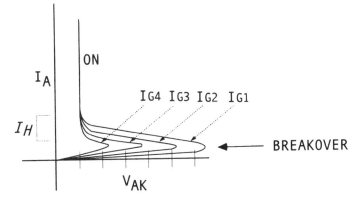

Figure 1-29. *How the application of varying gate current pulses affects the operation of the SCR: The higher the gate current the less VAK needed to switch the SCR on (IG1=Low current value; IG4=High current value).*

TRIACs (*Figure 1-30*) can be considered as back-to-back SCRs and perform as a full wave SCR for alternating current, conducting bidirectionally. Also, the gate control voltage, unlike the SCR, can be either positive or negative in polarity.

DIACs (*Figure 1-31*) are used as bidirectional triggers for SCRs, and for TRIACs to equalize the ON voltage for each AC alternation. When the voltage across the DIAC's terminals reaches a certain point (the *breakover voltage*, *Figure 1-32*) the DIAC begins to conduct current, reliably switching on the SCR's or TRIAC's gate. This breakover voltage is usually between 28 and 36 volts.

The four-layer diode (*Figure 1-33*, also referred to as the *Shockley diode*) switches ON and conducts current (I_D) when the anode-to-cathode voltage (V_D) reaches a certain level (the breakover, or *switching voltage*), as shown in its characteristic curve of *Figure 1-34*. The diode can only be switched off by lowering its I_D below a certain value called the holding current (I_H).

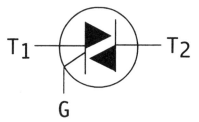

Figure 1-30. *Schematic symbol for a TRIAC.*

Figure 1-31. *Schematic symbol for a DIAC.*

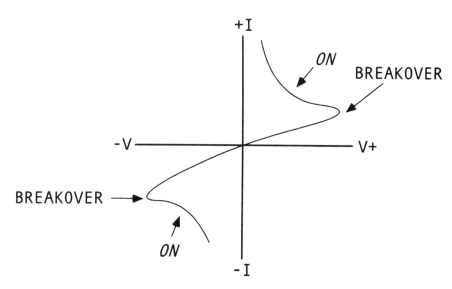

Figure 1-32. *Characteristic curve for a DIAC.*

A four-layer diode is capable of switching voltages up to a few hundred volts, but has a low power dissipation rating. Beside its normal current switching duties, the four-layer diode can also be used in relaxation oscillators and time-delay circuits.

Some ICs have been developed to trigger TRIACs, such as the *optocoupler TRIAC driver*, used in applications that need completely insulated TRIAC triggering with high electrical isolation.

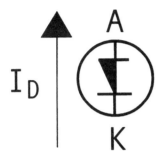

Figure 1-33. *Schematic symbol for the four-layer diode with the forward current flow.*

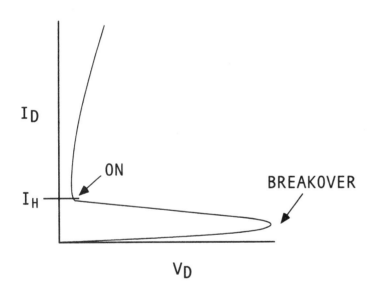

Figure 1-34. *Characteristic curve for the four-layer diode.*

1-5. Filters

There have been many filter designs and sub-variations over the years — entire books are written on the subject and still cannot fully cover the topic. Only the most common filters are covered here.

A filter is used to selectively pass or attenuate a certain band of frequencies, and can be *active* or *passive* while using LC, RC, LCR, or LR components in the common L, T, or pi configurations; or as completely resonant filters, which can be either series or parallel (tank) filters. Resonant filters enjoy widespread use in RF circuits, being used in bandpass and bandstop applications.

Naturally high-Q crystals and ceramics can also be utilized to increase filter selectivity of the passband or stopband dramatically, since the range of frequencies passed or attenuated is a function of the filter circuit's Q.

1-5A. Passive Filters

Since the radio spectrum consists of a wide range of frequencies, some way of isolating a range of desired frequencies is needed. Untuned and tuned filters solve this problem.

A simple passive LC filter takes advantage of the ability of an inductor's increasing reactance to increasing frequencies and a capacitor's decreasing reactance as the frequency increases. An inductor is usually placed in series with the output of any generator of RF to attenuate higher frequencies, while a capacitor is placed in series with the output to attenuate lower frequencies. An inductor can also be in shunt (parallel) with the output to attenuate lower frequencies, while a capacitor can be in shunt to attenuate higher frequencies.

A *low-pass filter* (*Figure 1-35*) passes only a range of lower frequencies; a *high-pass filter* (*Figure 1-36*) passes a range of higher frequencies. There are many different circuit configurations to accomplish these functions, and the filters may be cascaded to increase the sharpness of the skirts. Examples would be a class of filter referred to as an *L-type, constant-K* low-pass filter as shown in *Figure 1-37*. L-type refers to its physical configuration, while constant-K refers to a filter that not only rejects certain frequencies, but matches impedances between the generator and its load throughout its entire operational passband. Other constant-K low-pass filters are the *T-type* (*Figure 1-38*) and *pi-type* (*Figure 1-39*).

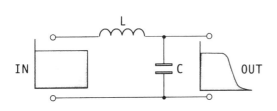

Figure 1-35. *A simple LC low-pass filter.*

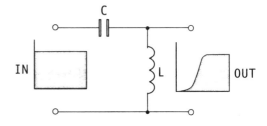

Figure 1-36. *A simple LC high-pass filter.*

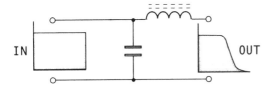

Figure 1-37. *L-type constant-K low-pass filter.*

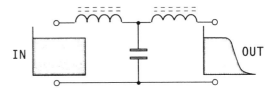

Figure 1-38. *T-type constant-K low-pass filter.*

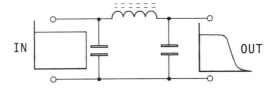

Figure 1-39. *Pi-type constant-K low-pass filter.*

Another popular classification of filter is the *m-derived type*. This is a modified constant-K that exhibits an enhanced filter response of an extremely sharp cutoff, while displaying a constant impedance throughout its operational passband.

Other filters pass only a set range of frequencies, called *bandpass filters*, while others reject a set range of frequencies, referred to as a *bandstop, band-rejection, or notch filters*. These filters are almost always of the resonant, or tuned, type, though by combining a non-resonant low-pass with a non-resonant high-pass filter, a bandpass characteristic can be obtained.

Tuned filters work on the principle of resonance. The reactance of an inductor increases with frequency, while the reactance of a capacitor decreases with an increase in frequency. Since these are completely opposing characteristics, when a capacitor and inductor are connected in *series*, a certain frequency is reached where their equal ($X_L=X_C$) but opposite reactances cancel, causing current to be maximum and impedance minimum. The only impedance to this current being the small resistance of the coil (*Figure 1-40*, with the Q of a coil being:

$$\frac{X_L}{r_e}$$

and is the reactance verses the *AC resistance*, and not the very low DC resistance, of the coil. This resistance is influenced by wire type, size, skin effect, and losses in the coil forms. Also, since a series resonant circuit will have almost zero impedance, the voltage drop across each component will be far higher than the supply voltage, causing possible arc-over between inductor windings). The resistance sets the Q, and thus the bandwidth is:

$$(BW = \frac{fr}{Q})$$

The frequency that causes $X_L=X_C$ is referred to as the *resonant frequency* (f_r) of the tuned circuit, and depends on the values of the LC components. f_r can be calculated by using:

$$fr = \frac{1}{2\pi\sqrt{LC}}$$

Figure 1-40. *The in-series resistive component of an inductor.*

Either above or below this resonant frequency the circuit behaves as any AC circuit would to an inductor and capacitor, with the reactances now no longer canceling the other, which creates a higher impedance and a decrease in current flow.

As can be observed in the above formula for the resonant frequency, the larger the values of L and C the lower the resonant frequency, while the lower the values of L and C the higher the resonant frequency.

An example of a series bandpass is shown in *Figure 1-41*. When series resonance is reached, the current is maximum and the impedance minimum, thus passing the desired frequency without attenuation to the output. The bandstop of *Figure 1-42* has the same resonant effect, but passes the undesired frequency to ground instead of passing it on to the output.

A *parallel*, or resonant, *tank* circuit has the opposite characteristic as that of a series resonant circuit: At f_r impedance is at maximum and current in the main-line is minimum. This state occurs when $X_L = X_C$.

As in series resonance, the value of the parallel resonant frequency can be calculated as:

$$fr = \frac{1}{2\pi\sqrt{LC}}$$

Just above or just below this resonant frequency, the impedance decreases and the current increases. In fact, when tuning a tank, the resonant condition can be found when current is measured as minimum and voltage is measured as maximum — the impedance will be extremely high.

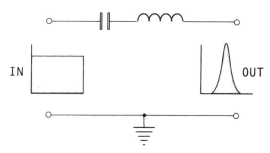

Figure 1-41. *A simple series bandpass filter.*

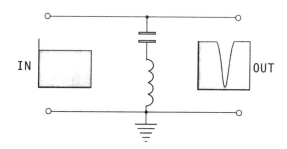

Figure 1-42. *A simple series bandstop filter.*

A certain amount of resistance is present in a tank circuit, lowering the overall resonant tank impedance and consuming energy, which must be replaced, creating a small current flow in the main line. Because the inductor and capacitor currents are 180° out of phase with each other, with a resonant tank having a very low resistance to this current, high circulating currents are present within the tank, as the currents are being exchanged between the inductor and capacitor (*Figure 1-43*). As these currents are out of phase, however, main line current flow is minimum and dependent on the pure resistance in the tank (mainly present in the inductor). Since the main line current is at some minimal value, then impedance must be at some maximal value at f_r. Because only at f_r is impedance maximum, rejection of non-resonant frequencies is accomplished, inasmuch as f_r will be the only frequency dropped across the tank.

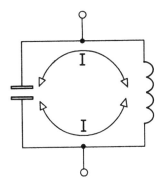

Figure 1-43. *A tank circuit at resonance, with high internal circulating currents.*

A complete sine wave need not be the input f_r to achieve this resonant effect. Because L and C have the capability to store and exchange this energy and current in a circulating fashion, complete sine waves can be furnished at the output when only pulses are inputted (See *Flywheel effect*). Only sine waves at the tank's resonant frequency are produced. Since the higher the Q of a resonant tank the narrower the BW and the higher the impedance, then the higher the output voltage possible in a high Q tank. Inversely, a shunt (parallel) damping resistor can be added to a parallel tuned circuit to lower circuit Q, thus lowering gain and impedance and increasing bandwidth.

An example of a parallel bandpass in *Figure 1-44*, which creates a high impedance to ground for the desired frequency and producing an RF output voltage across the tank, while shunting all non-resonant frequencies. The bandstop of *Figure 1-45* passes all frequencies except the resonant frequency, which is dropped across the high tank impedance, causing a lack of output at this resonant frequency.

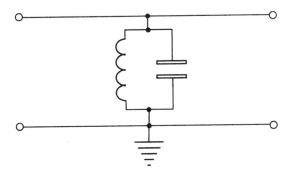

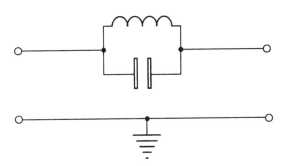

Figure 1-44. *A simple parallel bandpass filter.*

Figure 1-45. *A simple parallel bandstop filter.*

RC filters are a type of passive non-resonant filter that can be used in low and high-frequency applications, or as part of the frequency determining network of an active filter.

RC filters can be utilized to attenuate RF and pass only DC and low-frequency AC. The capacitor in *Figure 1-46* has a low reactance to high frequency, but a high reactance to low frequencies. Thus the RF is sent to ground through the capacitor and the low frequencies are not.

Another way to look at such a filter is to consider this circuit as a voltage divider, with the RF voltage dropped across the high resistance of the resistor instead of the low reactance of the capacitor, decreasing its output. Lower frequencies would cause an increase in the capacitor's reactance, dropping a larger voltage across the capacitor, and less across the resistor, as the frequency decreased, thus outputting increased lower frequency voltages.

Swapping the resistor with the capacitor will cause the opposite effect, and create a high-pass filter (*Figure 1-47*). Any low frequencies would now be dropped across the high reactance of the capacitor and not across the lower resistance of the resistor, while the higher frequencies would easily pass through the lower reactance of the capacitor and be dropped across the high resistance of the resistor, and be outputted.

Component values for both filters can be chosen to select the desired cut-off frequency.

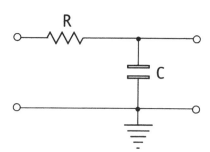

Figure 1-46. *An elementary RC low-pass filter.*

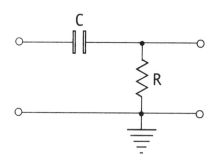

Figure 1-47. *An elementary RC high-pass filter.*

1-5B. Active Filters

Active filters are used for audio frequency filtering. The active element is typically an operational amplifier. A sharper response curve is available over that of using passive RC components only, and eliminates the need to employ inductors, which would be too large and expensive at low frequencies. Other advantages are: Lower value capacitors can be adopted; the op-amp buffers the entire filter from the characteristics of the load; and the output exhibits an insertion gain instead of a loss.

Figure 1-48 shows an active low-pass filter. Due to the lower capacitive reactance of C_1 and C_2 to higher frequencies, C_2 passes the higher frequencies to ground, while C_1 sends an increasing degenerative feedback to the non-inverting input as the frequency increases. C_2, R_1, and R_2 effectively form a simple passive low-pass filter.

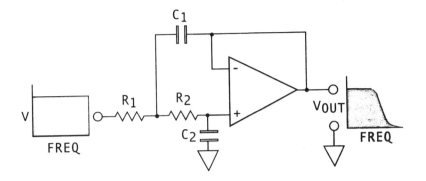

Figure 1-48. *An active low-pass filter using an operational amplifier and RC components.*

Figure 1-49 is an active high-pass filter. R_2, C_1, and C_2 are now forming a simple high-pass filter, passing the higher frequencies to the input, while the increasing capacitive reactance to the lower frequencies decrease their gain at the output.

Figure 1-50 is an active bandpass filter. The feedback network easily passes all frequencies just above and just below the passband frequencies. Since this feedback is degenerative, all but this narrow band of desired frequencies is attenuated by the negative feedback.

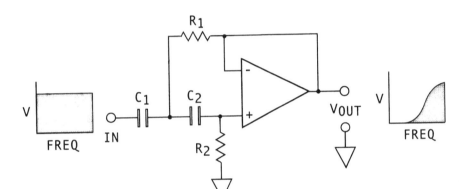

Figure 1-49. *An active high-pass filter using an operational amplifier and RC components.*

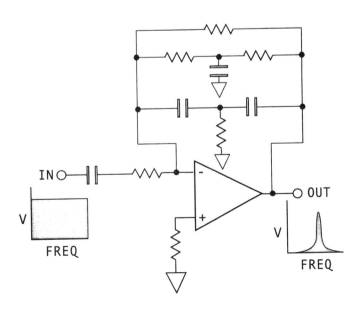

Figure 1-50. *An active bandpass filter using an operational amplifier and RC components.*

1-5C. Crystal and Ceramic Filters

Crystal or ceramic filters are used instead of RC and LC filters because of their much higher Q and subsequent narrower bandpass and much sharper response. Their frequency stability is far superior to LC circuits. Crystals perform like highly selective series and parallel resonant circuits, which have an extremely low or an extremely high impedance, respectively.

As mentioned above, all crystals have a series resonant mode, and a slightly higher parallel resonant mode (due to the capacitance of the holder (C_{PLATE}), *Figure 1-51*). Other crystal modes are also utilized besides the standard fundamental series and parallel resonant modes, such as the *harmonic*, or *overtone*, mode (as well as the "*inharmonic*", or *spurious*, modes): A crystal can be made to resonate at odd harmonics intervals of its fundamental frequency at the 3rd, 5th, 7th, 9th, and the 11th harmonic (up to 180 MHz).

Any crystal which is marked as over 29 MHz is almost certainly an overtone crystal; made to resonate at one of these odd harmonics — even though a crystal marked as low as 22 MHz may be running on its third overtone and not on its fundamental frequency.

The *inharmonic* mode uses the non-harmonically related (to the fundamental frequency) spurious responses of the crystal, of which there can be many.

Ceramic filters (CFs) are much lower in Q than crystals, but are also lower in cost and more sturdy. They both have almost the same characteristics.

R in *Figure 1-51* is the crystal's *equivalent series resistance* (*ESR*) at resonance, which is typically a pure resistance of around 25 to 250 ohms (while L and C cancel each other).

A crystal or ceramic filter uses its *piezoelectric* properties to convert mechanical vibration into electrical impulses. These vibrations occur chiefly at its natural resonant frequency: A crystal that is used in an oscillator circuit can be made to oscillate *both* mechanically and electrically, with the fundamental resonant frequency set principally by its physical proportions. However, its resonant frequency is not only dependent on the thickness of the crystal, but also in the way it is cut, the crystal substance used, and the holder capacitance.

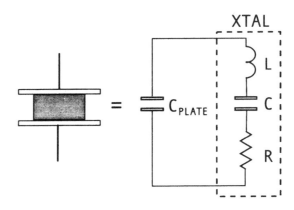

Figure 1-51. *The equivalent circuit of a crystal in its holder.*

Crystals are placed between a holder (modern crystals use silver directly deposited on their face) and inserted into a metal can (*Figure 1-52*) or into an *SMD* package, and marked with their resonant frequency.

As mentioned previously, crystals are commonly available up to a maximum of around 29 MHz on their fundamental, after which their thinness makes them impractical and highly prone to stress fracture, especially if subjected to over 1 mW of input power.

Figure 1-52. A common crystal package.

Crystals for filter applications can be utilized alone or in combination. *Crystal-lattice filters* (*Figure 1-53*) consist of several crystals, frequently in a single package, used as a very sharp-skirted bandpass filter. Each set of crystals are cut to different frequencies to achieve the proper bandwidth and selectivity desired. The matched set of Y_1 and Y_2 have a lower resonant frequency than the matched set of Y_3 and Y_4. The difference between these two resonant frequencies sets the bandwidth of the filter. This combination of crystals passes the desired signal to the output, but blocks all other frequencies. They are used extensively in SSB communications for sideband suppression.

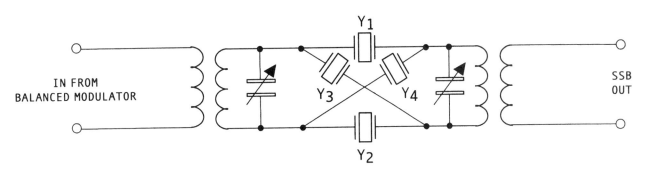

Figure 1-53. A crystal-lattice filter arrangement.

The *ceramic (or crystal) ladder filter* of *Figure 1-54* consists of a stack of two or three-terminal ceramic filters. All the crystals in this filter are cut to the same series resonant frequency, with the amount of crystals, and the value of the shunt capacitors, fixing the bandwidth of the filter. Coupling between individual crystals or ceramics is also accomplished by these capacitors.

Different ways of using and packaging single crystals have also been developed. A *monolithic crystal* employs quartz filters with several electrodes deposited on its top and bottom surface, with the space between the electrodes forming resonators. They can be used from 5 to 300 MHz and have a sharp frequency response: One crystal can obtain a two or more pole response. *Surface Acoustic Wave (SAW)* filters employ a piezoelectric crystal substrate with deposited gold electrodes. They can replace LC filters in certain wideband applications between 20 MHz to 1 GHz, and commonly experience at least a 6 dB insertion loss.

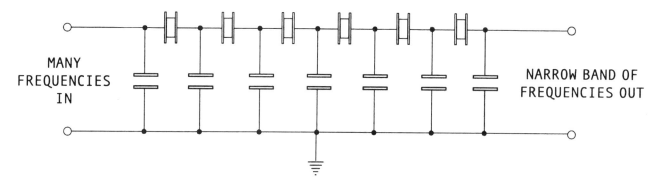

MANY FREQUENCIES IN

NARROW BAND OF FREQUENCIES OUT

Figure 1-54. A common ceramic ladder filter arrangement.

1-6. Toroids

Toroids are doughnut shaped cores made of *ferrites*, and can function as inductors (*Figure 1-55*), transformers (*Figure 1-56*), or chokes. Ferrite toroidal cores can be used from 1 kHz to 100 MHz (depending on the type of ferrite material used) in low- to medium-power circuits. They have become increasingly important in RF design.

Toroidal inductors and transformers have very little flux leakage, and are less sensitive to inductive coupling between another coil and itself. The ferrite material can be magnetized but will not conduct electrons, thus minimizing *eddy current* losses. They do not radiate RF: Unlike air-core inductors and transformers that may need shielding or changes in physical positioning to attenuate any mutual coupling between components, toroids are, because of their circular shape, almost self-shielding. They are also extremely efficient: For every magnetic field produced by the primary, almost all flux lines will cut the secondary. This is not true of the air-core transformer.

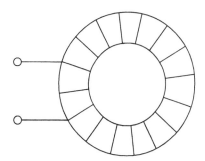

Figure 1-55. A toroid used as an inductor.

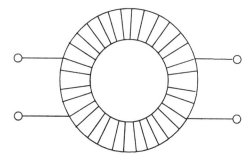

Figure 1-56. A toroid used as a transformer.

Toroids used as interstage transformers are also supplanting aircores. They are, moreover, being placed in untuned wideband finals (*Figure 1-57*) to match the impedances between the final amplifier and the transmission line (a low-pass filter is regularly added to the output to reduce harmonics, since a broadband (non-resonant) transformer or balun is incapable of attenuating these harmonics. In fact, many power amplifiers have up to eight low-pass filters arranged in banks that are switched into the RF output path when a frequency band is selected).

Toroids are, furthermore, utilized to eliminate any hum that may reach the receiver from the power line, as well as attenuating any interference escaping from the transmitter into the mains, by placing them in series with the supply power as a toroidal choke.

Toroids, as with all inductors, are capable of small variations in inductance values by separating or pressing together their windings, shorting adjacent turns, or removing turns.

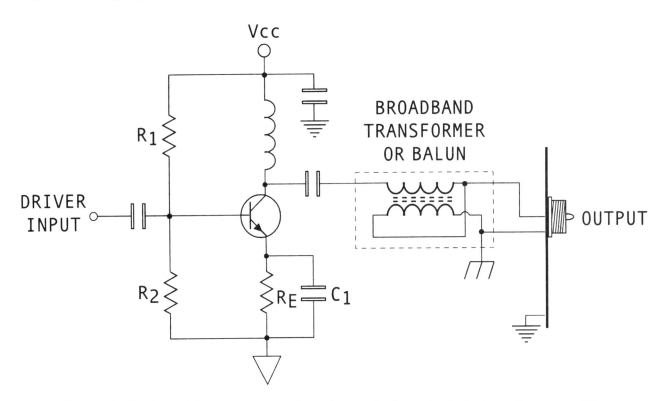

Figure 1-57. *A toroid employed as a transformer in the output of a broadband amplifier.*

Another type of toroid transformer is the *binocular broadband transformer* (also called a *broadband ferrite-core transformer*) of *Figure 1-58*, shown with its schematic representation. It is a common type of high- and low-power matching transformer for use in semiconductor RF amplifiers.

Broadband transmission-line transformers are also available, and can be used as baluns and as interstage matching transformers between RF transistor amplifiers. They consist, in various configurations, of a transmission line wrapped around a toroidal core.

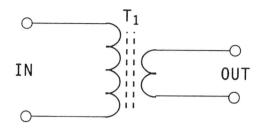

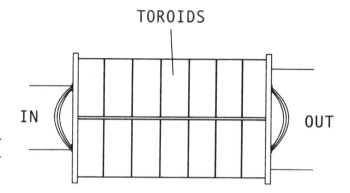

Figure 1-58. *The popular binocular broadband transformer for semiconductor amplifiers.*

CHAPTER 2
Amplifiers

Amplifiers can amplify voltage, current, or both. Since power is P=IE, power amplification is a natural product of this ability. Increasing *either* voltage or current will produce power amplification with transistors, FETs, or tubes, generating high powers into a (normally) low-resistance or low-impedance load.

All amplifiers employ a small external AC input signal to control a much larger DC output bias current. This large fluctuating direct current is then passed into an output impedance or resistance-producing component to create a fluctuating output voltage. The output component(s) can consist of either a resistor, inductor, or tuned circuit, depending on the amplifier's intended function.

There are many ways to configure an amplifier to obtain different input and output impedances, phase inversions, frequency responses, and amplification levels. As well, negative feedback can be used for thermal stability, while various bias arrangements can produce amplification at different efficiency levels. Special coupling methods may be used to match impedances, tune out reactances and filter out unwanted frequencies to other stages or loads.

2-1. Amplifier Circuit Configurations

There are three different circuit configurations possible with transistors (as well as tubes), and each has its own applications and abilities. They are referred to as *common-base* (grounded-grid), *common-collector* (grounded-plate), and *common-emitter* (grounded-cathode) amplifier circuits. This infers, correctly, that either the base (or grid), or collector (or plate), or emitter (or cathode) can be common to both the input and output of the amplifier circuit.

2-1A. Common-Base

A typical *common-base* (or common-gate for the FET) amplifier configuration is shown in *Figure 2-1*. CB circuits are used as voltage amplifiers and possess low input impedance, high output impedance, and increases in output power due to:

$$P = \frac{V^2}{R}$$

Current gain (a or h_{fb}) however, is slightly less than unity. A input signal is placed on the emitter, with the output taken from the collector circuit.

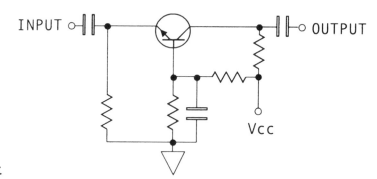

Figure 2-1. *A common-base amplifier.*

Although the CB amplifier has great temperature stability and high-frequency operational capabilities, it is rarely used. This is due to its very low input impedance, which would normally load down any amplifier attached to its input. A positive aspect of this circuit is that it does not need neutralization because of a lack of phase shift (normally) from input to output.

Common-base circuits can sometimes be found at the antenna input of a receiver because of its low Z_{IN} (50-75Ω), which matches that of many antennas, and as Class C high-frequency amplifiers, or as amplifiers in the IF of some receivers.

An example of a common-gate FET amplifier for the IF of a receiver is shown in *Figure 2-2*. C_2, C_3, R_2, and the RFC are for decoupling; C_4 is for DC blocking; C_5 is used to frequency compensate for a flatter frequency response throughout its entire passband.

A popular common-base BJT RF amplifier is shown in *Figure 2-3*. C_3 grounds the base to RF; forward base bias is provided by the divider action of R_3 plus R_4; C_1 and C_5 tune L_2 and L_3 (L_2 is tapped low to match the CBs low input impedance); L_3 and L_4 form a stepdown transformer to match the relatively high output impedance of a common-base to the lower input impedance of the next stage.

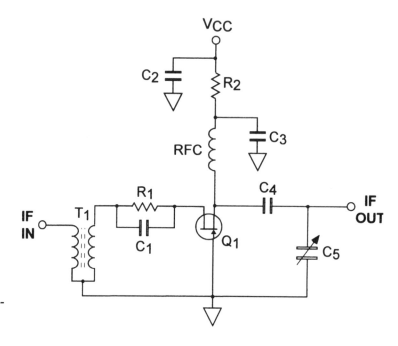

Figure 2-2. *A common-gate IF FET amplifier.*

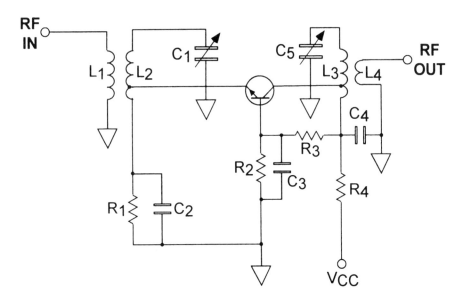

Figure 2-3. A common-base RF BJT amplifier.

2-1B. Common-Emitter

The *common-emitter amplifier* of *Figure 2-4* is by far the most popular amplifier configuration used in electronics, and has the best current and voltage gain combination of any other type. The bias arrangement shown is but one of the many ways to set up a CE amplifier (See *Amplifier biasing*).

As a signal is injected into the base the amplified output is taken from the collector circuit. This output voltage is shifted by 180° as compared to this input signal. The shift occurs because as the base goes more positive, more current flows through the transistor, which lowers its resistance and thus the voltage across the transistor. Since the output is normally taken from the voltage drop across the transistor's collector-emitter, the load resistor drops the voltage previously available to the collector's output, creating a phase shift exactly opposite to that of the input signal.

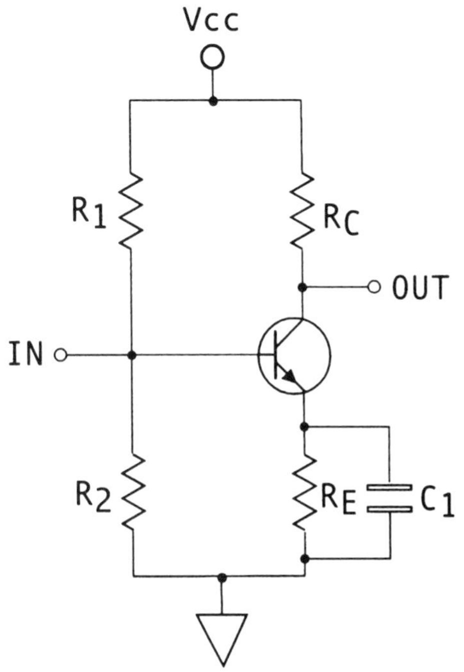

Figure 2-4. A popular type of common-emitter amplifier.

A big problem in CE amplifiers is an effect called *positive feedback*. This causes undesirable oscillations of the amplifier due to the collector-to-base internal feedback capacitance (a value between 4 and 22 pF for an average BJT) inherent in all transistors, which feeds an in-phase signal back to the base at some frequency. The transistor's internal capacitance and resistance, combined with other phase delays, can produce a strong phase shift. Only those phase delays that are close to 180 degrees (giving the common-emitter zero degrees positive feedback at its base from the normally 180 degree phase-shifted collector) can bring about amplifier instability, or oscillations. *Figure 2-5* demonstrates this phase-verses-frequency response of a single stage CE transistor amplifier.

CE configurations are capable of amplifying voltage, current, and power with a medium frequency response.

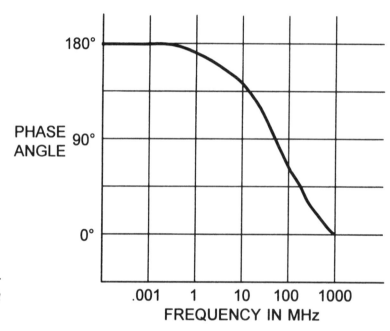

Figure 2-5. *Phase-versus-frequency response of a common-emitter amplifier.*

2-1C. Common-Collector

One type of *common-collector amplifier* bias configuration is demonstrated in *Figure 2-6*. Also referred to as an *emitter-follower*, the input signal is injected into the base, with the output taken from the emitter.

The CC amplifier supplies current gain and some power gain. It has a voltage gain, however, that is slightly less than unity.

The main advantage of this type of amplifier is its high input impedance, which makes it valuable as a buffer or impedance matching device. The CC amplifier has a low output impedance, with a good frequency response.

The CC amplifier typically lacks a collector resistor, and never uses a bypass capacitor across the emitter load resistor since any RF output voltage that would be developed across R_E would then be bypassed.

There is also no phase inversion between the input and output signal. As the input signal to the base increases, the current through the transistor increases, increasing the current through R_E, which increases the voltage drop across R_E, causing 0° phase inversion.

Table 2-1 condenses the most important characteristics of each amplifier configuration.

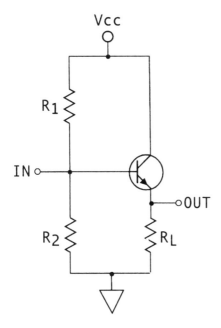

Figure 2-6. *A comon-collector amplifier.*

TYPE	Z_{IN}	Z_{OUT}	PHASE SHIFT
CE	Low	High	180°
CC	High	Low	0°
CB	Very Low	High	0°

Table 2-1. *Comparisons of amplifier parameters.*

2-2. Classes of Operation

Different classes of amplifier operation are used to accomplish different goals. The common classes of operation are Class *A, AB, B,* and *C.* Each has its own advantages and disadvantages. Every class uses circuit components to bias the active device at a different Q, or DC operating, point (*Figure 2-7*).

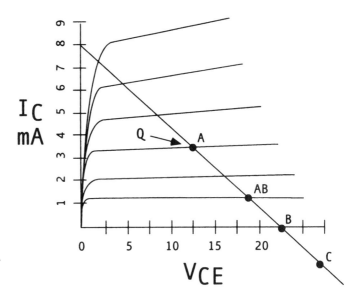

Figure 2-7. *Q-point locations for the various amplifier classes.*

2-2A. Class A

Class A operation allows a signal's amplified current to flow for 360° of the input signal (*Figure 2-8*). In other words, the output never reaches saturation (the transistor conducting as hard as it can, with $V_{CE} = 0V$) or cutoff (the transistor conducting zero current, with $V_{CE} = V_{CC}$). (In actuality, $V_{CE(SAT)}$ never really reaches 0 volts. $V_{CE(SAT)}$ can be up to two volts due to the transistor's natural internal resistances.) The biasing components must bias the transistor between saturation and cutoff, or at the Class A Q-point, allowing the signal to swing above and below this value. The output signal is therefore a linear and relatively accurate amplified representation of the input signal.

Class A single-ended amplifiers are generally used in small signal, non-power applications, such as in receiver RF and IF amplifiers, due to their very low efficiency. A large amount of constant DC supply power is needed at all times, since current is always flowing — with or without an input signal. They are capable of very low levels of distortion, especially with small values of negative feedback supplied by various circuit configurations.

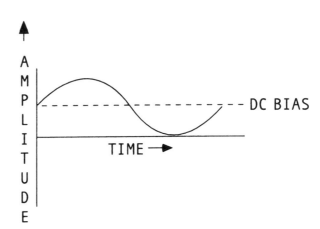

Figure 2-8. *Output waveform of a Class A amplifier.*

2-2B. Class AB

Class AB operation has a slightly higher efficiency than Class A, since output current flows for less than a complete cycle (around 300°, *Figure 2-9*). By lowering the Q-point Class AB operation is easily attained.

This type of bias cannot typically be employed as a single-ended amplifier in linear applications unless a tuned circuit is used to reconstruct the entire output signal by using the *flywheel effect* (the capability of a tank circuit to complete a sine wave at the tank's resonant frequency after a pulse has been applied). Class AB is usually used for push-pull audio amplifiers, as well as in single-ended SSB linear amplifiers outfitted with the necessary output tank, and very linear RF push-pull power amplifiers.

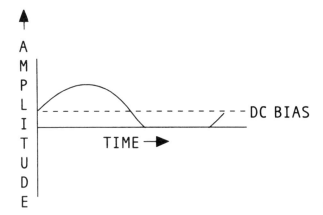

Figure 2-9. *Output waveform of a Class AB amplifier.*

2-2C. Class B

In Class B operation output current flows for about 180° of a full cycle (*Figure 2-10*). The bias is lowered to overcome the 0.6V of the base-emitter junction only. Conduction exclusively occurs when an AC signal alternation, or half cycle, forward biases the base. The other alternation reverse biases the emitter-base, causing a lack of output.

Since the Class B amplifier performs as a half-wave rectifier (amplifying only half of the incoming signal), it is typically used only in push-pull amplifier configurations (push-pull is a double-ended, or two-transistor, configuration, with each transistor amplifying each opposite alternation).

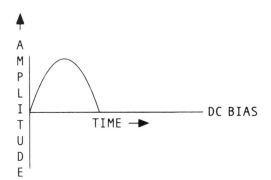

Figure 2-10. *Output waveform of a Class B amplifier.*

2-2D. Class C

Class C bias amplifies less than half of the input signal, and actually supplies only a pulse at its output port. Conduction is for 120° or less (*Figure 2-11*). The emitter-base junction is actually slightly reverse biased, or placed right at cutoff (silicon transistors, due to their 0.6V emitter-base barrier voltage, will not conduct until this voltage is overcome).

Because a pulsed output is useless for most purposes, this pulse must be reconstructed into a sine wave by the use of a tuned circuit (See *Flywheel effect*). In fact, the output tuned circuit of a Class C RF amplifier may have a peak-to-peak voltage that is twice the Vcc of the power supply, due to the flywheel effect rebuilding this missing alternation.

Class C is utilized typically in non-AM RF driver stages, in single-ended non-AM RF power output stages, in AM RF modulator stages, in mixers, or as harmonic generators and frequency multipliers.

An amplifier with a tuned output tank is not, however, automatically classified as a Class C amplifier. Amplifiers with a tuned circuit *can* be Class C, as well as Class A, AB, or B.

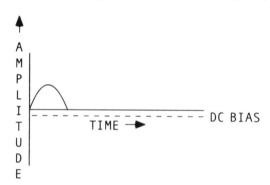

Figure 2-11. *Output waveform of a Class C amplifier.*

2-3. Amplifier Biasing

Since the common-emitter circuit is, as stated above, by far the most popular amplifier, we will go into more detail about its various biasing methods.

Biasing is used in amplifiers for two main reasons. First, two voltage supplies would be required for the emitter-collector and the emitter-base voltages to furnish the desired class of bias (A, AB, B, or C), and to supply the necessary DC input power. Biasing was developed so that these separate voltages could be obtained from a single supply.

Secondly, transistors are extremely temperature sensitive, and unless a method was found to stabilize the current increase as a transistor increased in temperature, a condition called *thermal runaway* would occur, destroying the transistor.

Many biasing schemes have been used to obtain these two results, with the most prevalent being the *base-biased emitter-feedback,* the *voltage-divider emitter-feedback*, and the *collector-feedback* bias. All three are commonly used in Class A and AB operation, while Class B and C amplifiers can also apply other methods.

2-3A. Class A and AB Biasing

As stated above, there are three common circuit configurations employed to obtain Class A or AB bias in a bipolar transistor amplifier, and their utilization depends on the desired circuit costs, stability, and other considerations. FETs will typically obtain Class A operation with another technique called *self-bias*.

The first, referred to as *base-biased emitter-feedback* (*Figure 2-12*), functions as follows: Since the base resistor, R_B, and the base-to-emitter voltage drop, V_{BE} (0.7V), and the emitter resistor, R_E, are in series with each other, and in parallel with V_{CC} (*Figure 2-13*), then as the collector current rises, due to an increase in temperature, the emitter current also increases. This increases the voltage dropped across the emitter resistor, thus reducing the voltage dropped across the base resistor (the voltage drops around a loop must equal the voltage rises). This decreased voltage across the base resistor lowers I_B, which lowers I_C.

The bypass capacitor, C_E, prevents the AC signal from degenerating the gain of the circuit by bypassing the AC around the emitter resistor. If there are *two* resistors in the CE amplifier's emitter and only one resistor is bypassed, the other one is supplying degenerative feedback to reduce oscillations, making the amplifier more stable at the expense of lowered AC gain.

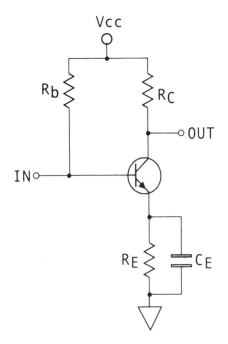

Figure 2-12. *A common-emitter amplifier with base-biased emitter-feedback biasing.*

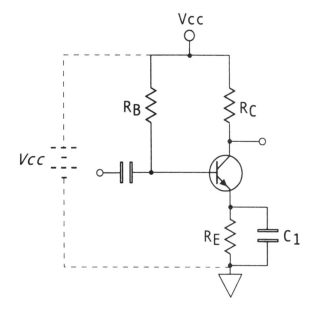

Figure 2-13. *A common-emitter amplifier showing the V_{CC} connection.*

R_B and R_C also allot the proper voltages to the collector and base, with the appropriate polarity, through one power supply. Considering that R_C, C-E (the collector-emitter), and R_E are in series and share V_{CC}'s voltage, then the collector-to-emitter voltage is now equal to V_{CC}, minus the voltage drop across R_C and R_E. Thus the collector is properly reverse biased. R_B, the emitter-base junction, and R_E are now also in series and share V_{CC}'s voltage. The voltage drop across R_B is equal to V_{CC} minus the obligatory emitter-base voltage drop of approximately 0.7V and the voltage drop across R_E. Considering that the voltage drops across the emitter-base and R_E are usually quite low, most of V_{CC}'s voltage is dropped across R_B, forward biasing the base (the smaller the value of this base resistor the more base current produced, and the more the collector current flows, since $I_B = V_{CC}-V_{BE}/R_B$). Therefore, one voltage source has now supplied the biasing needed for an NPN transistor to function as an amplifier.

Another even more effective temperature stabilization scheme than the base-biased emitter-feedback method above can be employed. It is called *voltage divider, emitter-feedback biasing* (*Figure 2-14a*), and works as such: The current through the voltage divider of R_1 and R_2 is much higher than the I_B. Consequently, any increase in temperature (which would increase base current) will not substantially vary V_{R2}, which is equal to the voltage at the base in respect to ground. A consistent voltage from base to ground is therefore maintained.

Also, If the emitter current increases due to an increase in junction temperature, then the top of R_E becomes more positive but, since the base is usually 0.7V more positive than the emitter, the base-emitter junction decreases in the voltage dropped across it, thus lowering collector current to its normal level (I_C is controlled by this V_{BE} between the base and emitter).

To further increase temperature stabilization in the above circuit for sensitive applications, *diode temperature compensation* can be used (*Figure 2-14b*). This can, as well, be utilized in many other transistor bias configurations, and either double or single diodes can be em-

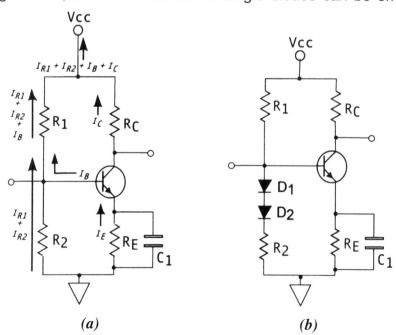

Figure 2-14. *(a) A CE amplifier with voltage divider, emitter-feedback biasing (shown with circuit current flow); (b) Same circuit with further diode temperature compensation.*

(a)   *(b)*

ployed. The diode or diodes are placed against the heatsink or transistor to somewhat accurately track the transistor's temperature changes and compensate for any temperature rise by the subsequent lowering of the diode's own internal resistance with the rising heat. This lowers the diode's forward voltage drop, decreasing the transistor's V_{BE}, and abating any temperature-induced current rise. The use of other transistors or thermistors may also be seen in circuits for amplifier temperature compensation.

A biasing scheme sometimes found in circuits that need less thermal stability than above is called *collector-feedback bias*, and is used because it employs only two resistors, and is thus low in cost to implement. In *Figure 2-15*, as the temperature increases, the transistor will of course begin to conduct more current. The base resistor, R_B, is connected directly to the collector instead of Vcc. Any increase in collector current causes more voltage to be dropped across R_C, allowing less to be dropped across R_B, thus lowering the base current and, subsequently, the collector current.

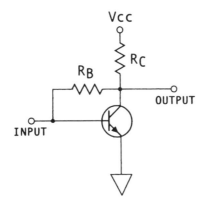

Figure 2-15. *A CE amplifier with collector-feedback bias.*

A common biasing technique to obtain Class A from an FET is a form of self-bias called *source-bias* (*Figure 2-16*). Since, unlike a transistor, no current flows in the gate circuit, the drain current is equal to the source current. Current does flow in through the source, and thus through R_S, creating a positive voltage at the top of R_S. Inasmuch as the source is common, and the gate is at 0V in respect to ground since the gate conducts no current that would create a voltage drop across R_G, the gate can now be considered *negative* in respect to this common source. The FET is thus biased at its Class A, AB, B, or C Q-point, depending on the value of the source resistor. A capacitor is usually placed across the source resistor to hold the bias voltage to a steady DC value.

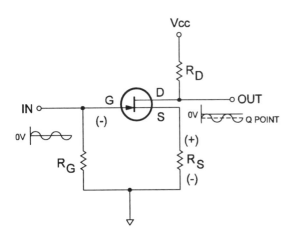

Figure 2-16. *Class A operation of an FET with source-bias.*

2-3B. Class C Biasing

Typically, the three biasing techniques commonly used for bipolar transistor in Class C operation are *signal, external,* and *self-bias.* However, most transistors can operate Class C without any bias whatsoever, such as in the circuit of *Figure 2-17.*

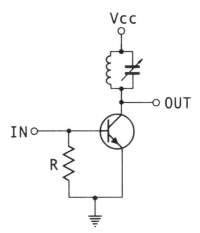

Figure 2-17. *Transistor operation at Class C without bias.*

Signal-bias (*Figure 2-18*) uses the signal to supply the negative bias needed for Class C operation. When the transistor is conducting, C is charging. C then discharges through R when the signal voltage does not have the amplitude to turn on the transistor, or when the signal reverse biases the transistor. When C discharges through the resistor, a negative potential exists at the top of the resistor, thus creating the negative bias needed for Class C operation. The RC time constant of the two components can allow for so much negative bias, if desired, that only the peaks of the incoming signal can be permitted to turn on the transistor.

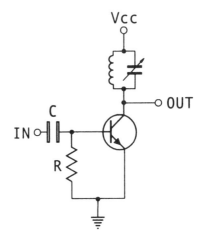

Figure 2-28. *Class C operation of a transistor with signal-bias.*

Another less utilized technique is external-bias (*Figure 2-19*), which employs a separate negative bias supply, consisting of either batteries or a power supply, to obtain the necessary negative bias. The RFC (Radio Frequency Choke) supplies a high impedance for the RF signal to be dropped across.

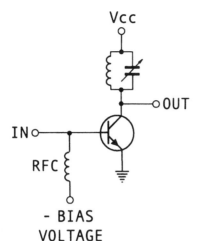

Figure 2-19. *A transistor with Class C external-bias.*

The self-bias (*Figure 2-20*) method is generally used only in applications where high input signals are available. As emitter current flows, a voltage is dropped across the emitter resistor. Since the capacitor is in parallel with the emitter resistor, it also has this same voltage across its terminals. This prevents the bias voltage from varying with the signal frequency. The top of the emitter resistor is now positive due to the direction of the current flow from emitter to collector. With the emitter now positive, and being the common element in the CE amplifier, the base (which is at DC ground potential through the RFC), is now negative in respect to the emitter, creating Class C operation.

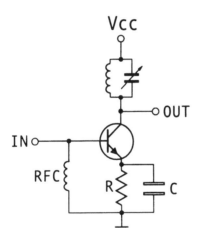

Figure 2-20. *A transistor with Class C self-bias.*

2-3C. Class B Biasing

Class B, push-pull amplifiers (*Figure 2-21*), can be operated with any bias scheme that biases the amplifiers at just above cutoff. Some Class B push-pull amplifiers are biased with *stabistors* (*Figure 2-22*), which are simply two series diodes employed to maintain an even 0.7V on each emitter-base junction, while also helping to protect against thermal runaway.

Class B can also be used single-ended into high-Q tuned RF circuits with any biasing arrangement that biases the transistor at approximately 0.7V above cutoff.

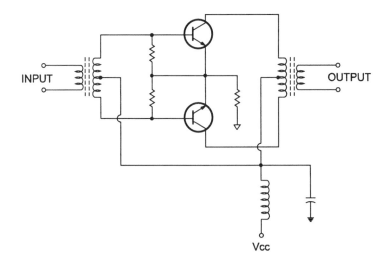

Figure 2-21. *Class B biased push-pull amplifier.*

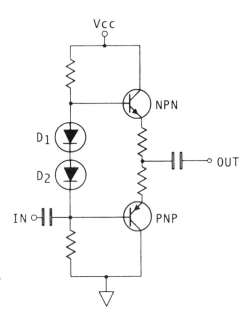

Figure 2-22. *Class B push-pull amplifier biased with stabistors.*

2-4. Amplifier Coupling

A way of coupling a signal into or out of another amplifier, or a source and its load, must be used so that the DC biasing of the stages does not adversely affect each other. In many amplifiers a good impedance match ($Z_{OUT}=Z_{IN}$) for maximum power transfer (especially for power types) and the tuning-out of any reactances between stages is also critical, as well as supplying filtering and harmonic attenuation. However, many small-signal cascaded amplifiers require a low output impedance and high input impedance to minimize current flow (and thus interstage loading) between stages, which demands mismatched impedances. This form of matching, as well as the $Z_{OUT}=Z_{IN}$ matching above, may be difficult to achieve, from a design standpoint, at higher frequencies.

The actual type of coupling depends on the sort of signal being amplified. DC, low-frequency AC, high-frequency AC, or wideband amplification all have specific coupling needs.

2-4A. Capacitor Coupling

Capacitor coupling (AKA *RC coupling*) is used in AC amplifiers only. It can amplify large variances in frequency with a good wideband response.

Figure 2-23 shows two RC coupled amplifiers. The series capacitor, C_C, blocks the DC bias to the next stage, but passes the AC signal. C_C and R_6 act as voltage dividers — since the capacitor is large enough to offer little capacitive reactance to the signal coupled to the next stage, most of the signal is dropped across the resistance of R_6 of the next stage. This adds to or subtracts from the second stage's emitter-base junction, causing its collector current to vary, producing amplification.

While RC coupling is a low cost and simple method it is, however, very difficult to match impedances from one stage to the next. It also has no attenuation effect on higher undesired frequencies, such as harmonics, from passing from one stage to the next and being amplified.

Another type of capacitive coupling can be used in some circuits and operates in conjunction with an inductor. The degree of coupling between an oscillator and its buffer (*Figure 2-24*) can be controlled by a movable tap. Coupling is increased if the tap is moved toward the bottom, decreased if moved toward the top. It is utilized to decrease loading on an oscillator stage, or to increase or decrease the output signal to the buffer.

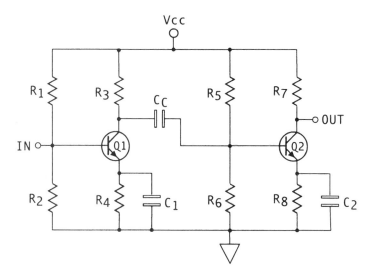

Figure 2-23. *Two CE amplifiers utilizing capacitor coupling between stages.*

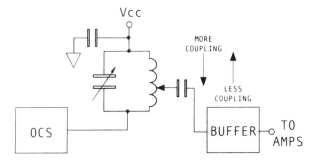

Figure 2-24. *Adjustable capacitive coupling between two stages.*

2-4B. Inductive Coupling

Inductive coupling (AKA *impedance coupling, Figure 2-25*) is also utilized for alternating current only. Similar to RC coupling, but instead of employing a collector resistor, a collector inductor is used. Since the collector inductor has very little DC resistance, insignificant DC power is consumed in the inductance, allowing for more efficient operation of the amplifier. It is only usable for a narrow band of frequencies, however, since X_L varies directly with frequency. The stage gain would thus fluctuate with any change in this X_L.

The inductor can function as a collector load for a transistor due to its reactance to the alternating collector current, which creates a varying voltage drop across the inductive load, subtracting voltage from the transistor's emitter-collector.

Do not confuse inductive coupling with *transformer coupling.*

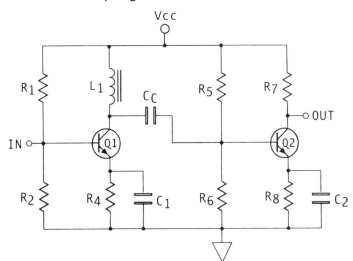

Figure 2-25. *Two CE amplifiers utilizing inductive coupling between stages.*

2-4C. Direct Coupling

Direct coupling (AKA *DC coupling*) is useful for low-frequency and DC amplification. *Figure 2-26* shows a typical DC amplifier. R_3 is used as a collector resistor for Q_1 and as a base resistor for Q_2.

A small temperature-induced current change in Q_1 is amplified by Q_2, so most DC amplifiers must be well designed, using precision components, to stabilize the circuit against these temperature variations.

DC amplifiers are perfect for ICs, since all the components are placed close together, allowing each component the same changes in temperature.

2-4D. Low-Frequency Transformer Coupling

Low-frequency (usually laminated iron-core) *transformer coupling* is used for audio AC but not DC. The DC cannot pass through the transformer from its primary to its secondary.

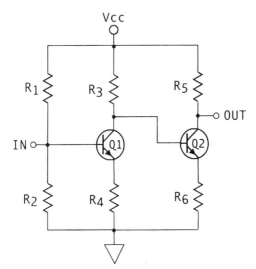

Figure 2-26. *Two CE amplifiers utilizing direct coupling between stages.*

Figure 2-27 shows two amplifiers coupled by an iron-core transformer. The transformer can be used to match the high-output impedance of Q_1 to the low-input impedance of Q_2. One end of the secondary is attached to Q_2's base, with the other end attached between its bias resistors R_5 and R_6. The signal then adds to or subtracts from the base bias, causing the base current to fluctuate, creating amplification at the output of Q_2. A transformer operated in this way is called an *interstage transformer*.

An iron-core audio transformer is heavy and expensive, so this coupling method is now rarely used for normal audio applications.

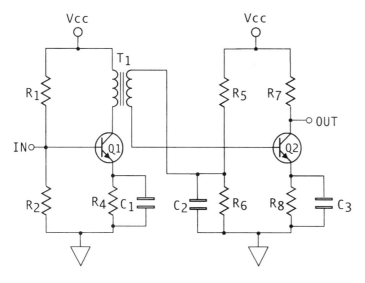

Figure 2-27. *Two CE amplifiers utilizing low-frequency transformer coupling between stages.*

2-4E. High-Frequency Transformer Coupling

High-frequency (tuned circuit, ferrite, powdered iron, and air-core) *transformer coupling* (*Figure 2-28*) is still used extensively in RF and IF amplifiers up to about 150 MHz, but less so than in the past because of its high expense as compared to other coupling methods.

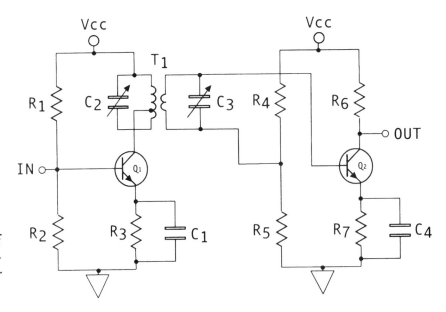

Figure 2-28. *Two CE amplifiers utilizing high-frequency transformer coupling between stages.*

Same operation as low-frequency iron-core transformer coupling above, but transformers can be air-core for low losses at high frequency, or ferrite and powdered iron cores for inductive tunability and high-inductance operation over and above that of the air-core. Transformers supply the needed impedance matching for efficient maximum power transfer between stages.

The tool used to adjust the reactance of the ferrite or powdered iron slug transformer is shown in *Figure 2-29* (some very-high-frequency coils may use *brass* slugs up to a few 100 MHz for limited tuning).

Figure 2-29. *Plastic inductor and transformer tuning tool.*

Either the primary or secondary, or both, can be tuned, forming a resonant tank circuit with a capacitor, to allow the coupling to be frequency selective. *Figure 2-28* shows *double-tuned transformer coupling*, allowing the tank circuit to pass just the desired frequencies while attenuating all others outside of its bandwidth. If only the primary of T$_1$ were tuned, then it would be referred to as *single-tuned* transformer coupling.

Toroidal transformers (*Figure 2-30*) are very popular in RF gear and have, in many applications, replaced the normal slug and air-core transformers.

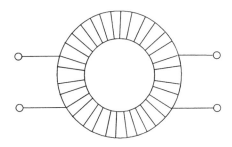

Figure 2-30. *The common toroidal transformer.*

Transfilters are small mechanical resonators that may be used as a transformer between transistor IF stages, and as a narrow-band IF filter (*Figure 2-31*).

Ceramic filters can also be used for coupling instead of ordinary transformer (*Figure 2-32*).

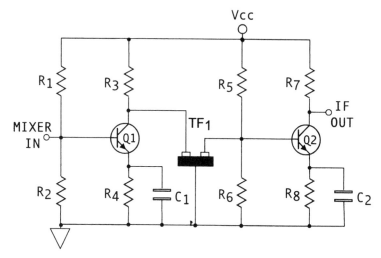

Figure 2-31. *"Transformer" coupling using a transfilter.*

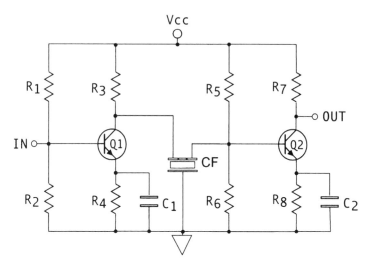

Figure 2-32. *"Transformer" coupling using a ceramic filter.*

With RF and IF interstage transformers, the degree of coupling between the primary and secondary can be a large consideration. This is usually governed by the distance between the primary and secondary windings. As the coefficient of coupling increases (*overcoupling, Figure 2-33a*), more flux lines from the primary are able to cut the secondary, creating a higher output voltage over that of *loose coupling (Figure 2-33c)*; as well as an increasing bandwidth as adjacent frequencies are also passed due to the high capacitance now present between primary and secondary. As the coefficient of coupling is decreased toward loose coupling, the amplitude as well as the bandwidth of a signal will decrease. However, loose coupling is sometimes utilized because harmonics are decreased owing to its lower capacitive coupling into the next stage, as well as supplying the narrower bandwidth needed for some applications. At *optimum coupling (Figure 2-33b)* the bandwidth is wide enough to pass a regular narrow bandwidth signal, with an amplitude that is equal to overcoupling.

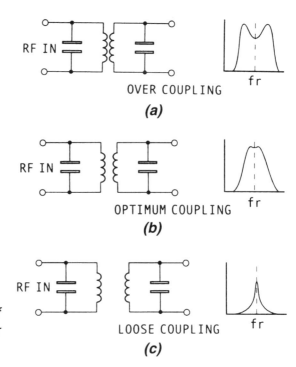

OVER COUPLING
(a)

OPTIMUM COUPLING
(b)

LOOSE COUPLING
(c)

Figure 2-33. *How the degree of coupling between RF transformer windings affects the signal.*

2-4F. Matching Networks

Various RF *matching networks* can be used between stages or their load to perform coupling, without the use of a transformer, and supply the necessary impedance matching, nullification of any reactances (making the input impedances to and from a stage into a pure resistance), and furnish a certain amount of filtering. Maximum power transfer and attenuation of harmonics between stages or their load is consequently accomplished.

Many modern RF small signal amplifiers are now designed with matching networks rather than using the more expensive transformer coupling methods, especially at the higher frequencies.

This impedance matching can be attained by simple *L-networks*, which also supply low-pass filtering to reduce harmonic output. The L-network of *Figure 2-34* matches a high output impedance amplifier to a low input impedance amplifier. Its component values are set so that at the amplifier's operating frequency the circuit appears to be a parallel resonant circuit to amplifier 1. Since a parallel resonant circuit, at resonance, has an extremely high impedance, the two stages are matched.

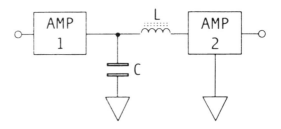

Figure 2-34. *A high-to-low impedance matching L-network.*

The L-network of *Figure 2-35* shows a low output impedance amplifier matched to a high input impedance amplifier. The inductor and the capacitor at the amplifier's design frequency appear to amplifier 1 as a series resonant circuit. A series resonant circuit, at resonance, has a very low impedance, so the two circuits are matched.

One of the most popular impedance matching networks between a low output impedance amplifier to a higher input impedance is the low-cost and small *T-network* (*Figure 2-36*). The T-network can be designed to supply any desired selectivity and impedance matching, within limits, that is not possible with the L-network. The capacitors are usually tunable for maximum output power during alignment.

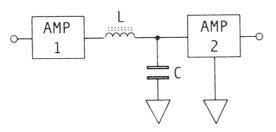

Figure 2-35. *A low-to-high impedance matching L-network.*

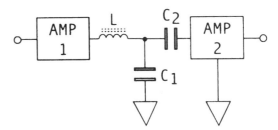

Figure 2-36. *A Z-matching T-network.*

A common tube and transistor circuit impedance matching network is the pi configuration. The pi and its equivalent circuit is shown in *Figure 2-37*. By changing the ratio between C_1 and C_2, Z_{OUT} of a power RF amp can be matched to the Z_{IN} of an antenna, as well as attenuating any harmonics and performing the needed flywheel function.

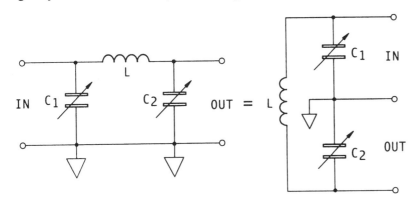

Figure 2-37. *The popular pi-matching network and its equivalent circuit.*

Although the shunt capacitor pi-network is also considered to be a low-pass filter, it has a resonant peak at a certain design frequency where the impedance of the circuit rises drastically above that of its low-pass value (*Figure 2-38*).

While C_1 is referred to as the *tuning capacitor* and tunes the plate or collector to resonance, and C_2 is called the *loading capacitor* and varies the impedance for proper power transfer into the antenna, both heavily interact with each other (if C_1 is adjusted, then C_2 must be readjusted, until a desired compromise is reached between tuning and loading).

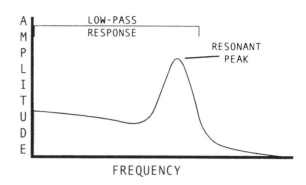

Figure 2-38. *The frequency response of a pi-network, showing the resonant peak.*

2-5. Distortion and Noise

Distortion and noise are an integral part of any amplifier, mixer, or oscillator used in a receiver or transmitter.

Distortion deforms either the carrier and any sidebands at the transmitter, causing adjacent channel interference and inaccurate reproduction of the original modulating signal at the receiver or, if it occurs within the receiver, creates undesirable, distorted output.

Distortion can be created by the production of harmonics of the complex modulated and carrier waves, or by internal mixing in any non-linear components (creating unrelated distortion products of varying frequencies), or by improper amplifier responses to the desired or undesired frequencies. Distortion may have different causes, but similar results: A waveform that is changed in shape or amplitude from that intended by the designer.

Noise also can result in anything from annoying white static to total obliteration of the received signal, originated by many internal and external causes.

2-5A. Distortion

Frequency distortion, measured as frequency response, can occur if a circuit boosts or attenuates certain frequencies better than others. Commonly a problem in audio or RF amplifiers when they are near their extreme lowest and highest frequency limits, distortion can be caused by the reactive components in a circuit acting in an undesirable fashion as bandpass, bandstop, high-pass, or low-pass filters, or by the inability of a transistor to perform at high frequencies (induced by *transit time* problems and negative effects of junction capacitance).

Amplitude distortion, often referred to as *non-linear distortion* because the output signal is not a true replica of the input signal, is created by improperly biasing an amplifier, resulting in the input signal causing either saturation or cutoff. This produces harmonics and IMDs. The harmonics and IMDs generated from the resulting *flattopping* of the signal create a raucous, coarse output that can be heard in an audio amplifier. This is sometimes referred to as *overload distortion* when the input signal overdrives the amplifier, usually creating both saturation and cutoff conditions. In a transmitter, amplitude distortion will cause interference to other channels.

Intermodulation distortion (IMD, *Figure 2-39*) is caused when frequencies, not harmonically related to the fundamental, are created by the mixing together of the carrier, any harmonics, the sidebands, the modulating audio, IMDs from other stages, etc., through non-linearities in a linear or nonlinear amplifier (such as a Class C), or in the output of any mixers. This produces an abundance of spurious responses both in and out of band.

IMDs can also originate in the finals of a transmitter when another nearby transmitter's signal and/or its harmonics reaches the power amplifier (IMDs can be caused in a city by the many thousands of FM, pager, as well as cell site, transmitters in operation). One signal will modulate the other, resulting in non-linear mixing of these signals in the power amplifier, creating many distinct sum-and-difference frequencies that are not harmonically related to the fundamental. This type of IMD can be attenuated by using a wavetrap tuned to the second transmitter's frequency. IMDs of this variety can also occur in any stage of a transmitter if the transmitter is not well shielded and grounded.

Receiver IMDs, consisting of the desired signal and a nearby transmitter's undesired signal (and/or its harmonics), which can be admitted into the front end of a receiver, can cause reception of unwanted signals and a smearing and/or obliteration of the desired frequency. A notch-filter, if the interfering signal is far enough away from the desired signal, can be effective, as can confirming that all amplifiers are biased as designed, and are not functioning in a non-linear region due to a component defect or stage misalignment.

Since intermodulation distortion is created when two or more frequencies mix in a non-linear fashion, creating many sum and difference frequency combinations (f_1+f_2, f_1-f_2, etc.), it is possible for *third-order intermodulation distortion products* to occur, which are $2f_1+f_2$; $2f_1-f_2$; $2f_2+f_1$, and/or $2f_2-f_1$. In fact, the most damaging intermodulation products can be these third-order products, because *second-order IMDs* (the $f_1 \pm f_2$ frequencies) would usually be too far from the receiver's or transmitter's bandpass to cause many problems (any IMDs created outside of the passband of the transmitter or receiver are attenuated by an amplifier's tuned circuits and the selectivity of the antenna). For example, two signals, one at 10.7 MHz and the other at 10.9 MHz, would cause false sum and difference second-order frequencies at *21.6 MHz* and *0.2 MHz*, while with third-order IMDs these same signals would be at *10.5 MHz, 11.1 MHz, 32.3 MHz*, and *32.5 MHz* (the most damaging frequencies being, of course, at 10.5 and 11.1 MHz).

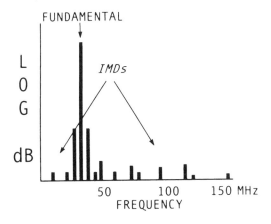

Figure 2-39. *Frequency-domain graph showing IMDs.*

In a transmitter both second and third orders may cause a signal to be broadcast on an undesired frequency; while in a receiver, a signal may be received at an undesired frequency. Higher order IMDs are created in amplifiers, but are usually too low in amplitude to create many problems.

High VSWR can also cause IMDs due to the reflected wave mixing with the transmitter's RF output and its sidebands.

When receiving in VHF and the operator begins to hear squeaks, squawks, bleeps, noises, pager tones and strange voices, then it is almost certainly IMDs caused by the desired signal mixing with undesired TV, VHF, FM, paging, and police station frequencies.

For in-shop testing purpose, using a two-tone test is the only way to generate and view IMD products efficiently.

Harmonic distortion, measured as *Total Harmonic Distortion* (THD), takes place when undesired harmonically related frequencies are produced in the amplifier and add to and distort the fundamental. Interference to receivers tuned to many megahertz away from a transmitter's carrier can be created when these harmonics are broadcast into space (*Figure 2-40*).

Since harmonics are created by amplitude distortion in an amplifier, the stronger the desired input signal the higher in amplitude are the undesired harmonics generated. An extreme case of distortion can result in the sinewave carrier turning into a rough square wave (square waves contain the fundamental frequency plus many odd harmonics).

Because no amplifier can be completely linear, a certain amount of harmonics are created in all amplifiers and they must be suppressed as much as possible — especially in a transmitter.

Harmonics, if caused by the audio sections, create harmonic multiples of the audio frequencies; and if produced in the IF and RF stages, cause harmonic multiples of the carrier wave. In a transmitter, if the audio amplifiers are overdriven, then harmonics of the audio modulation occurs, which then modulates the RF carrier, causing harmonics of the sidebands.

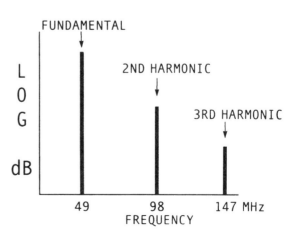

Figure 2-40. *Frequency-domain graph showing harmonics.*

2-5B. Noise

There are two major classifications of noise: *circuit generated* and *externally generated*. Both are unavoidable and limit the possible gain of any receiver's amplifiers.

Circuit generated noise causes an actual randomly-fluctuating and wide-frequency ranging voltage, created by two main causes: *white noise*, produced by a component's electrons aimlessly flitting around due to heat; and *shot noise*, caused by electrons entering the plate, collector, or drain of a tube, transistor, or FET, and especially by the random transit of electrons across a semiconductor junction.

External noise, includes static caused by atmospheric disturbances like lightning, as well as *space noise* created by solar flares and sunspots, along with *cosmic noise* induced by the stars radiating wide-ranging interfering signals. *Man-made noise* also contributes to external noise generation, and can be from any type of electromagnetic interference: from dimmer switches to car ignitions to power tools.

2-6. Voltage Amplifiers

Voltage amplifiers are used to increase low signal amplitudes to usable levels. Receiver RF and IF amplifiers are voltage amplifiers, typically Class A for linear operation and low distortion in AM and SSB.

An RF stage must ordinarily be tunable and capable of supplying relatively high gain at high frequencies, while being low in noise generation (since any noise present will be amplified by later stages), and of sufficient bandwidth to pass the desired signals and attenuate all others.

Because of the high operating frequencies, RF amplifiers are frequently *neutralized* to counterbalance any possible positive feedback and its resultant self-oscillations.

An IF amplifier must be high gain, since the sensitivity of a receiver is largely governed by the gain of the IF stages. They are fixed tuned so as to be highly selective and pass only the desired signal (some receivers employ un-tuned IF stages and make use of a crystal filter out of the mixer stage only). High gain and narrow selectivity are also easier to accomplish by making the IF lower in frequency than the incoming RF.

Voltage gain of these amplifiers can be calculated as:

$$\frac{V_{OUT}}{V_{IN}}$$

With a series of cascaded amplifiers, their gain is multiplied. For example, if each cascaded amplifier stage supplies a voltage gain of 100, and there are three stages of amplification with the same gain, then the combined value would be 100 x 100 x 100 = 100^3, or one million.

2-6A. RF and IF Amplifiers

One type of widespread RF or IF stage is shown in *Figure 2-41a*. It may be easier to identify if redrawn as in *Figure 2-41b* as the common-emitter, voltage divider, emitter feedback Class A voltage amplifier with tuned circuit RF transformer coupling. Collector bias, base-bias with emitter feedback, and inductive coupling or LC matching networks can be employed instead of the above configuration.

An AGC connection is normally applied at the base to supply gain control (See *Automatic gain control*).

The input can also be tuned in an RF amplifier, and is almost always tuned at the input of an IF amplifier. When increased selectivity is preferred, a crystal filter can also be placed in front of the IF strip, just after the mixer, to supply the desired narrow bandwidth signal.

Class A biased low noise FETs are normally used in RF amplifiers at VHF and above.

IC IF amplifiers are also quite common (*Figure 2-42*). With a few external components to supply decoupling, coupling, and resonant tank functions, an entire IF stage or section can be placed on a single chip.

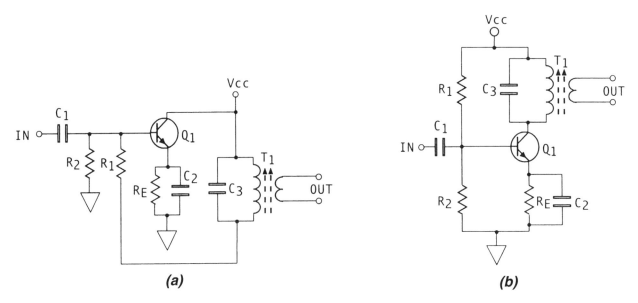

(a)   *(b)*

Figure 2-41. *An ordinary RF amplifier: (a) As shown in some schematics; (b) Redrawn.*

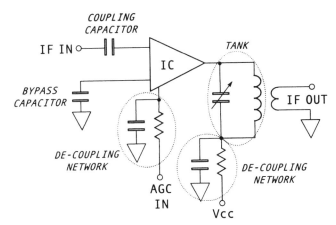

Figure 2-42. *An IF amplifier in integrated circuit form.*

Dual-gate MOSFET amplifiers, such as *Figure 2-43*, are employed as IF stages to supply high-input impedances (to reduce loading of the prior circuit), as well as the inclusion of a *control gate* to manage the gain of the stage through AGC action.

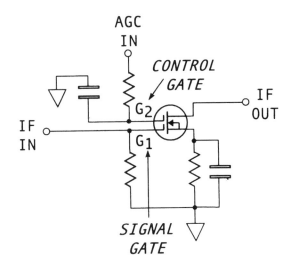

Figure 2-43. *A dual-gate MOSFET amplifier with one gate used for the input and the other for AGC.*

A dual-gate MOSFET amplifier for the RF front end is shown in *Figure 2-44*. The RF signal is inserted into G_1, while G_2 gets its positive bias from R_1. R_1 can also be used as a manually adjustable gain control.

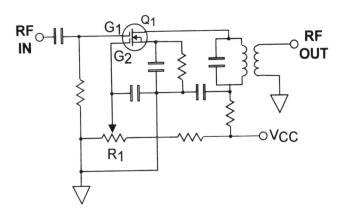

Figure 2-44. *A dual-gate RF amplifier.*

2-6B. Preselectors

A *preselector* is a tuned radio frequency pre-amp (or tunable passive filter, *Figure 2-45*) placed before the RF amplifier of a receiver to increase selectivity, sensitivity, S/N ratio, and to attenuate any image frequency.

When changing the received frequency, some preselectors must be adjusted manually by a separate control, while others are ganged with the tuning capacitor, or have electronic varactor tuning (*Figure 2-46*). This varactor tuning circuit works by changing D_1 and D_2's capacitance, and thus the resonant frequency of the tank, depending on the DC control voltage from V_{DD}, which can be factory adjusted by the trimmer R_1. C_1 acts as a filter to remove any DC fluctuations from reaching the varactors.

A simple untuned radio frequency amplifier in a receiver is sometimes referred to as a preselector.

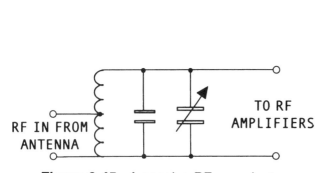

Figure 2-45. *A passive RF preselector.*

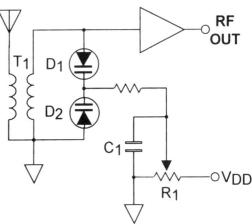

Figure 2-46. *An electronic varactor tuned preselector.*

2-6C. Cascode Amplifiers

Cascode amplifiers (*Figure 2-47*, *not* the same as a *Cascade amplifier*) are low-noise, high-gain, high-input impedance, wide bandwidth amplifiers sometimes used as a preamplifier for receivers. It consists of two transistors or MOSFETs, with the input stage acting as a common-emitter circuit feeding the second transistor, which is a common-base.

If the wide bandwidth supplied by the cascode amplifier is not sufficient, other methods can be used.

2-6D. Wideband Amplifiers

Wideband amplifiers are used in SSB, digital modulation, and video amplifiers; or in any transmitter that requires a wideband RF power output. These amplifiers are sometimes completely untuned. There are different designs exploited, depending on the needed power efficiency, power output, spurious suppression, and cost.

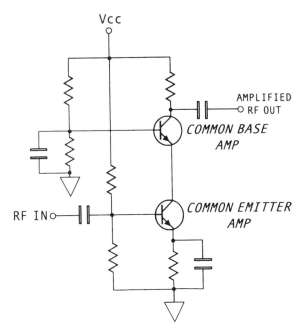

Figure 2-47. *The wideband RF cascode amplifier.*

Since

$$Q = \frac{R_P}{X_L}$$

and

$$BW = \frac{f_r}{Q}$$

a *shunt damping resistor* (R_P) can be inserted into a tank circuit to lower its Q to effectively widen an amplifier's BW (*Figure 2-48*). The lower the value of R_P, the wider the bandwidth but the lower the output voltage. In most RF communications circuits, however, a narrow bandwidth is desired to attenuate adjacent channel interference and improve the S/N ratio. TV and certain wideband communication applications use this method, as well as a technique called *stagger tuning,* to further widen the bandwidth. Stagger tuning uses two or more stages tuned to a slightly different frequency (*Figure 2-49*), and offers steep skirts and better gain than merely adding a parallel damping resistor.

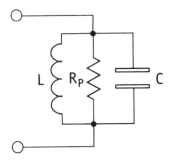

Figure 2-48. *A shunt damping resistor (R_P) used to decrease tank Q.*

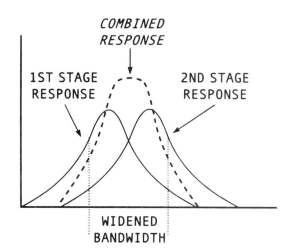

Figure 2-49. *The effects of two-stage stagger tuning on bandwidth.*

Most modern transceivers use wideband untuned RF push-pull power amplifier finals. This allows the operator to quickly change from band to band without complicated and time consuming tuning procedures.

The wideband amplifier of *Figure 2-50* uses input frequency correction to supply attenuation at 30 MHz of about 1 dB, and 12 dB at 1.6 MHz. This is for an overall gain flatness of about 1 dB to compensate for the gain variations of the RF transistors over their full 1.6 to 30 MHz wide bandwidth (amplifiers furnish more gain at their low-end frequencies than at their high-end frequencies). The input frequency correction network consists of R_1, R_2, C_2, and C_3, which are acting as a high-pass filter in series with the bases of Q_1 and Q_2. L_5 taps off degenerative feedback through R_3 and R_4 to lower gain, helping in neutralization.

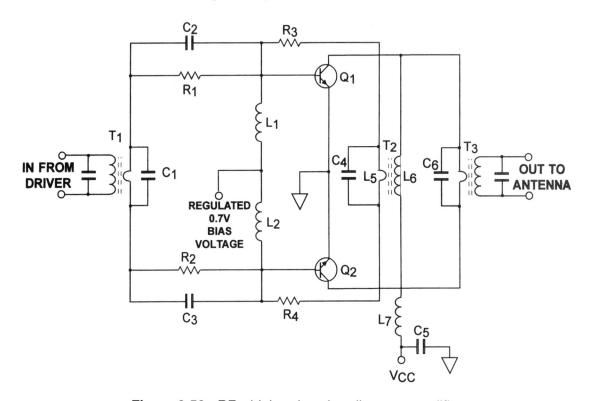

Figure 2-50. *RF wideband push-pull power amplifier.*

As well, by adjusting the low voltage bias regulator from 0.5 to 0.9 volts, the class of the amplifier can be changed from Class B to Class A for extra linear operation.

Wideband push-pull amplifiers can also supply frequency-selective negative feedback for gain flattening by using ferrite beads to attenuate the high-frequency feedback (*Figure 2-51*). Only at low frequencies is the feedback increased for more degeneration.

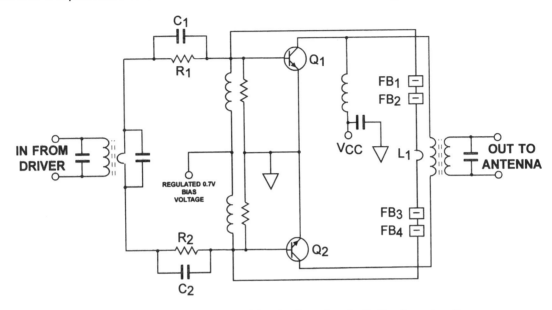

Figure 2-51. *Wideband ferrite bead gain flattening circuit.*

2-6E. Bootstrap Amplifiers

The *bootstrap circuit* (*Figure 2-52*) is an amplifier in a common-collector configuration that increases its own input impedance but, as with all CC amplifiers, has a voltage gain of less than one. A common-emitter bootstrap circuit can also be used for its increased input impedance, but voltage gain is much lower than a normal CE amplifier, and the bias (temperature) stability is compromised.

The bootstrap circuit can be used with any type of active device; BJT, JFET, or MOSFET.

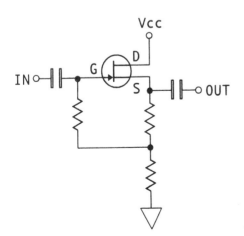

Figure 2-52. *The high input impedance bootstrap amplifier.*

2-7. Power Amplifiers

Power amplifiers (PAs) are designed to supply high power outputs into a low impedance load. The loads are low-impedance because speakers are usually 8Ω, antennas 50 to 300Ω. Power amplifiers must be able to handle large signal inputs without saturating or cutting-off (*Figure 2-53*), while controlling large amounts of output current (but generally at low values of stage gain).

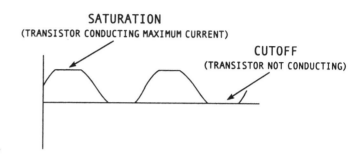

Figure 2-53. *An overdriven amplifier's waveform showing saturation and cutoff.*

There are two fundamental classes of power amplifiers to increase RF output power for transmitters, employing either a single active device (single-ended, or one lead AC grounded at the amp's input or output) or two active devices (double-ended or push-pull). These two classes can be further broken down into *linear* (Class A, Class AB and B push-pull) and *non-linear* (Class C, almost always into a tank circuit) amplifiers.

Only linear RF amplification must be used for SSB and low-level AM to maintain their delicate modulation envelopes. Class C is too non-linear to sustain these envelopes, and severe distortion will result. As well, audio power amplifiers must always be operated in one of the linear modes, or severe distortion will occur to the audio frequencies.

Linear IC audio power amplifiers (*Figure 2-54*) are also available for low-power receiver applications where low input power consumption and voltage are critical, such as in battery operation.

In the IC power amplifier example of *Figure 2-54* using, or not using, an external resistor and capacitor between pins 1 and 8 can easily and simply vary its output gain to any desired value between 20 and 200.

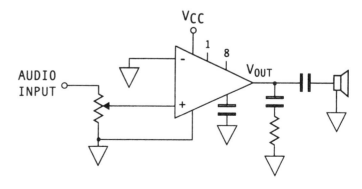

Figure 2-54. *An audio power amplifier in integrated circuit form.*

2-7A. Single-Ended Amplifiers

The Class A single-ended RF power amplifier of *Figure 2-55* can be employed in low pow-ered transmitter finals. A single-ended power amplifier can normally be identified by the very low value of bias resistors used (low-value resistors are adopted so that high amounts of power dissipation are not wasted within these components).

A special Class A single-ended amplifier biased by a temperature compensated constant-current source can deliver up to 100 watts (*Figure 2-56*).

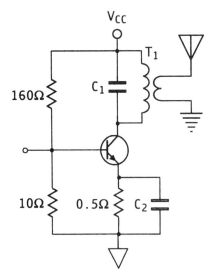

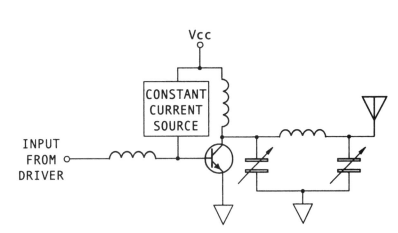

Figure 2-55. *A single-ended final for low power applications.*

Figure 2-56. *A special single-ended final capable of supplying medium output powers.*

Class C single-ended amplifiers can be used to amplify any signal that does not contain a modulation envelope, which would be distorted by its insufficient linearity. *Figure 2-57* is an ordinary Class C RF power driver stage. The transistor is operating Class C without any bias, as most unbiased transistors act as Class C amplifiers. A low-value R_E keeps the transistor stable during temperature variations by supplying degenerative feedback. If R_E were not bypassed by C_{RE} further negative feedback would occur, lowering gain. RFC and C_1 are a decoupling network for the power supply, while C_2 is a tunable capacitor that resonates with the primary of T_1 to form the tank circuit. A ferrite bead or beads can be added to the base lead to suppress any VHF and UHF parasitics as well as high-frequency harmonics.

Class C amplifiers can also be used in single-ended RF power output stages, as FSK, FM, and CW amplifiers, or as harmonic generators and frequency multipliers. They are known for their very high efficiency of operation.

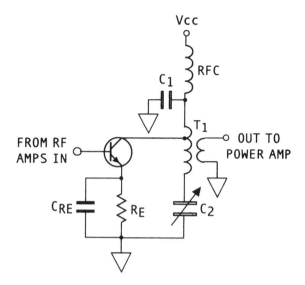

Figure 2-57. *A Class C RF driver amplifier.*

2-7B. Double-Ended Amplifiers

A common audio and RF power amplifier is the double-ended, or *push-pull*, type. Push-pull amplifiers use two transistors, with each transistor amplifying an opposite half of the incoming signal. The two amplified halves are then assembled at the output.

There are two basic ways to obtain push-pull operation. One takes the input signal and splits it into two 180° out-of-phase signals (*Figure 2-58*) and feeds this into the input of two identical active devices. The other technique uses an NPN and a PNP transistor; and no phase splitting is needed (*Figure 2-62*).

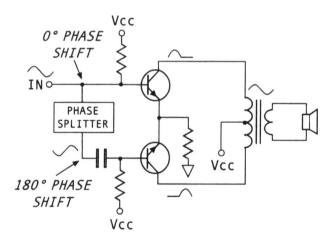

Figure 2-58. *A push-pull amplifier with phase splitter.*

There are two common ways to accomplish this phase splitting for the first type of push-pull amplifier (the two out-of-phase signals are needed so that as one transistor is increasing its current flow, the other stage is decreasing its current flow). The first is by using a passive center-tapped transformer phase splitter (*Figure 2-59*). The center-tapped transformer functions as a phase splitter because the signal between the center tap and the *top* of the transformer are 180° opposite in phase (polarity) to this center tap and the *bottom* of the transformer.

Push-pull, transformerless input power amplifiers have been developed. With the use of an active phase splitter, the circuit of *Figure 2-60* needs only a center-tapped *output* transformer, and can dispense with the expensive and bulky center-tapped input transformer.

The *paraphase* amplifier (*Figure 2-61*) is a common and simple way to accomplish this active phase splitting. By using identical values for the collector and emitter resistors, an equal-amplitude, but 180° phase-shifted, output is produced. The paraphase amplifier can thus be used to supply the necessary out-of-phase signals to the input of a push-pull amplifier.

The output transformer of *Figure 2-60* is operated to reconstruct the alternate outputs from Q_2 and Q_3. When Q_2 conducts, current will flow through the top of the coil; when Q_3 conducts, a opposite current will flow through the bottom of the coil, inducing an opposite voltage into the transformer's secondary, creating a complete low-distortion signal for the speaker coil.

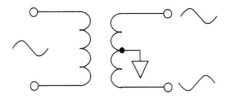

Figure 2-59. *A center-tapped transformer used as a phase splitter.*

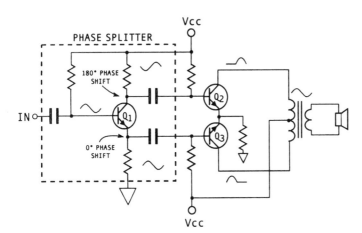

Figure 2-60. *A push-pull amplifier with active input phase splitter.*

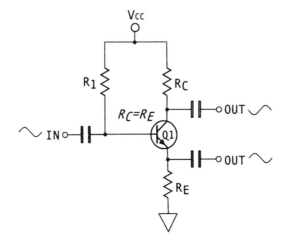

Figure 2-61. *A paraphase active phase splitter circuit.*

Keep in mind that these opposite output currents occur at different periods in time, due to the phase shift created at the input, so they do not oppose each other, but are merely "assembled" at the output transformer. The output transformer also functions as an impedance match for the output of the audio amplifier to the low impedance of the speaker coil, as well as isolating the speaker from the DC current flow.

The second type of push-pull amplifier is the completely transformerless audio push-pull amplifiers, referred to as *complementary symmetry* with *stacked-series output* (*Figure 2-62*). They are designed to eliminate the expensive and heavy (at audio frequencies) input and output transformers. By using evenly matched NPN and PNP transistors no phase inversion is necessary at the input, since they both naturally conduct at opposite times. The output is taken from the emitters in a common-collector configuration.

No voltage gain is possible with these amplifiers, but this setup does supply a high driving current into a low impedance load. As shown, D_1 and D_2 (referred to as *stabistors* when used in this configuration) keep an even 0.7V on each emitter-base junction, which consequently biases the transistors just above cutoff (Class B), while also helping to protect against thermal runaway. R_3 and R_4 also aid in thermally stabilizing the two transistors.

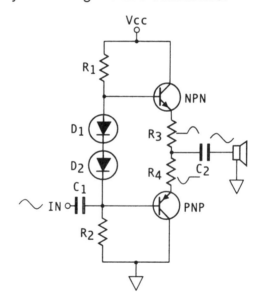

Figure 2-62. A transformerless push-pull amplifier.

Push-pull wideband Class B and AB RF amplifiers that need not be tuned by the operator are commonly available that amplify signals between 2 and 30 MHz (*Figure 2-63*) and above, and are similar in operation to the audio push-pulls. These are simply push-pull amplifiers without tuned input or output circuits.

These RF amplifiers function due to the phase shifting abilities of a center tapped transformer. The phase between the top of the tapped secondary of T_1 and the bottom of T_1 are 180° out of phase — the upper stage amplifies the positive alternation of the incoming signal, while the lower stage amplifies the negative alternation. At the primary of T_2 the two signals are added and the signal is outputted through the secondary to the antenna. T_1 is also supplying an impedance match between the output of the driver stage and the input to the push-pull power amplifier, while T_2 is supplying the output of the power amplifier with an impedance match for the input to the antenna. D_1 and R_3 bias the two transistors right at cutoff, for true Class B operation, to eliminate any crossover distortion that would result if the input signal had to overcome the 0.7V of the base-emitter junction before they could conduct.

D_1 also supplies temperature stabilization of the stage. The diode pulls current up through ground (creating a 0.7V drop), through R_3 (which also creates a voltage drop), and on to Vcc. Since the diode is typically placed on the transistors' heat sink or mounting tabs, D_1's temperature will closely track that of the transistor's: As the temperature of the transistors rise, the junction temperature of the diode will rise equally, causing a lowering of resistance across, more current through, and less of a voltage drop over the diode. This lowers the transistors' bias by a few tenths of a volt, and thus decreases their increase in current due to the increase in junction heat.

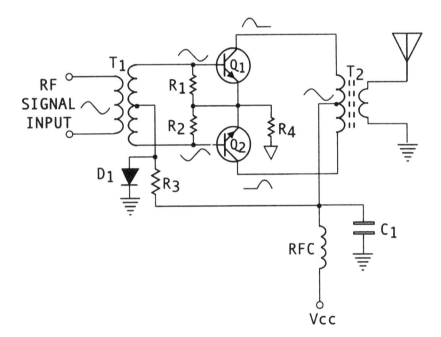

Figure 2-63. *A broadband RF push-pull amplifier.*

In the last few years, enhancement mode MOSFETs have become hugely popular in broadband RF power amplifiers, and are very common in the HF, VHF and UHF regions. *Figure 2-64* is a good example of such a circuit. Q_1 and Q_2 each have their own regulated and separate bias voltage supplies. Negative feedback is supplied by T_2 and travels through R_{20} and C_{47}, and R_{19} and C_{46} for neutralization. R_{17} and R_{18} lower the MOSFET's input impedance to further prevent oscillations. R_{15} and R_{16} supply additional stabilization at higher frequencies through C_{40} and C_{41} (lowering the input impedance yet again). C_{38}, C_{42}, C_{43}, C_{39}, C_{45}, C_{44}, R_{14}, R_{13}, and R_{17} and R_{18}, are a high-pass filter for the necessary gain equalization for its wide bandwidth of 1 to 30 MHz (to provide a flat gain across its entire operational bandwidth).

Class B push-pull is a common way for power amplifiers to be configured, due to their very high efficiency over that of Class A. Class AB is also frequently used in push-pull power amplifiers that require less distortion, but with less efficiency than Class B. Class C RF push-pull is used for signals that do not need perfect linear amplification — a tuned circuit at the output reconstructs the input signal by the *flywheel effect*.

Regular push-pull amplifiers also cancel hum from the power supply and self-generated even-order harmonics (these currents are out of phase and cancel in the output transformer); can supply large currents to the load, thus producing large output powers; and, since push-pull amplifiers have series transistors, their input and output capacitance is half that of a single device, which is good for higher frequency operation.

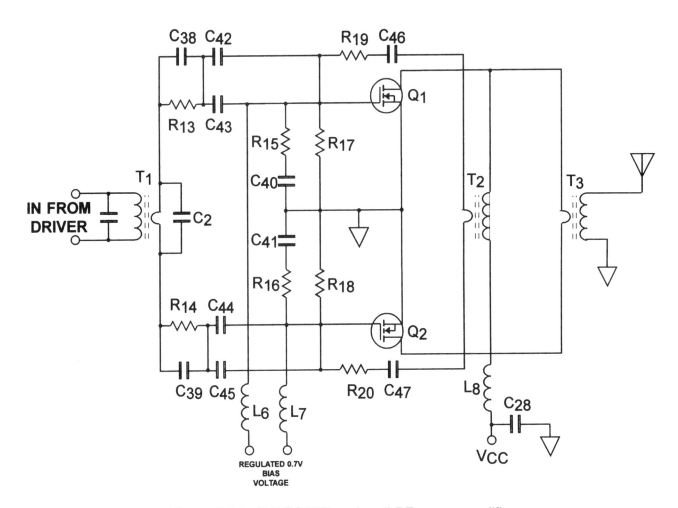

Figure 2-64. *E-MOSFET push-pull RF power amplifier.*

An RF *parallel power amplifier* (*Figure 2-65*) is used to double the radio frequency output power over that of a single-ended power amplifier. In this configuration, up to several hundred watts output power can be obtained with power transistors.

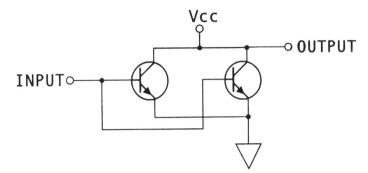

Figure 2-65. *A basic parallel amplifier circuit.*

Parallel amplifiers differ from push-pull in that each transistor is on or off during the *same* time interval, unlike push-pull, which alternately shares the power workload back and forth for equal time periods. The parallel configuration, since the output current load is shared equally amongst the transistors or tubes, as it is in any parallel circuit, doubles the power handling capabilities (P=IE) over that of a single-ended amplifier. This arrangement allows the circuit to act as if it were a single high-power transistor with two, or more, perfectly matched transistors increasing the output power-handling ability of a transmitter's final. *Figure 2-66* shows a complete biased circuit, with impedance-matching and filtering being accomplished by the pi-network at its output.

Parallel amplifiers can be biased Class A, AB, B or C. They must, however, have perfectly matched transistors — and their input and output capacitances are twice that of a single device, which can be a problem in high-frequency operation.

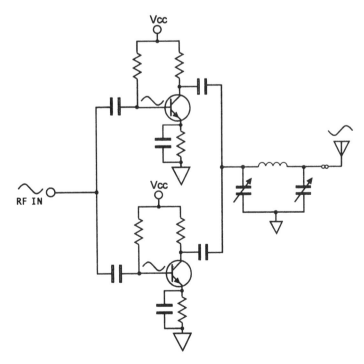

Figure 2-66. *A parallel final power amplifier showing bias components.*

2-8. DC Amplifiers

Direct current amplifiers are used to amplify DC and low-frequency AC currents or voltages. A simple single-stage amplifier (*Figure 2-67*) without coupling capacitors can be used if high gain is not needed. If more than one stage of amplification is required then AC coupling methods, which utilize transformers, inductors, or capacitors, cannot be used in a DC amplifier. The DC signal would be blocked or shunted by these components. Other techniques must be employed.

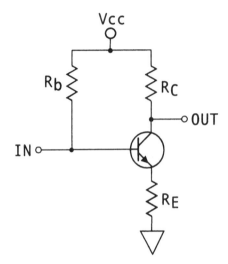

Figure 2-67. A basic single-stage DC amplifier.

2-8A. Darlington Pair Amplifiers

One type of DC amplifier is the *Darlington pair* amplifier of *Figure 2-68*, configured as a high voltage-gain common-emitter circuit. Darlington common-collector amplifiers are also popular.

The pair work as one transistor with an h_{fe} x 2, or $\beta = \beta_{Q1}$ x β_{Q2}. In practice, the current is not quite doubled due to other constraints, such as bias voltage adjustments and impedance mismatching.

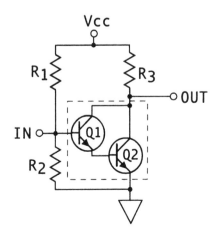

Figure 2-68. Darlington pair DC amplifier.

R_1 and R_2 are used for voltage divider bias, R_3 is the collector resistor, with Q_1 controlling the conduction of Q_2. When a DC or AC signal is placed at the input, the *base* current of Q_1 (I_{BQ1}) increases, if DC, or varies, if AC. The *emitter* current of Q_1 (I_{EQ1}) then increases or varies; which increases the *base* current of Q_2 (I_{BQ2}); which increases the *collector* current of Q_2 (I_{CQ2}) — creating amplification. Notice that no R_E is used in this particular circuit due to the degenerative action that would result in a substantially lower DC gain.

Darlingtons can be placed in the same three-terminal package as a standard bipolar transistor.

2-8B. *Differential Amplifiers*

Another popular DC amplifier is the *differential pair* amplifier (or *diff amp*), which provides great thermal stability, automatic canceling of common-mode noise, and high gain. It is now manufactured in integrated circuit form and rarely as a discrete circuit. This circuit is usually configured to amplify the voltage *difference* between the V_1 and V_2 outputs.

The diff amp is versatile in that there are two basic ways that it can be configured. For *differential output*, both inputs are fed a signal, with the output taken across V_1 and V_2 (*Figure 2-69*). Amplification will only take place if the input signals are either out of phase or of different amplitudes. Equal, or *common-mode*, signals at the inputs will not be amplified since a difference must exist for amplification to occur. R_1 and R_2 provide voltage divider bias for Q_1, R_6 and R_7 supply voltage divider bias for Q_2, while R_4 is used as the emitter resistor for Q_1 and Q_2. R_3 and R_5 are employed as collector load resistors.

R_4 is typically replaced with a constant current source for improved *common-mode rejection* (CMRR) by allowing only a set amount of current to flow in the circuit, which will not permit a common-mode signal to increase this total circuit current.

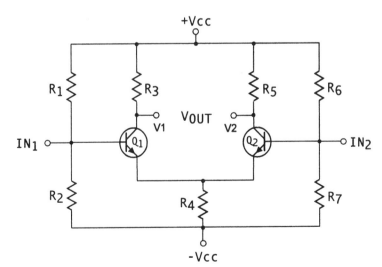

Figure 2-69. *The differential DC amplifier arrangement.*

A *single-ended input* configuration is possible for the diff amp (*Figure 2-70*) in which one input is grounded with the other being used for the input signal, while the output is taken between V_1 and V_2.

Other less common configurations, such as placing two out-of-phase (push-pull) signals at the input and taking the output from between V_1 and V_2 (*differential input*), or taking the output from *either* V_1 *or* V_2 to ground (*single-ended output*) are sometimes encountered.

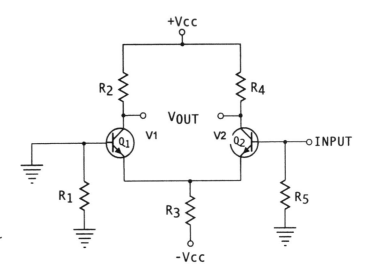

Figure 2-70. *The differential amplifier in a single-ended input configuration.*

2-8C. Operational Amplifiers

IC *op-amps* are frequently utilized for DC and low-frequency AC amplification, while a few new types are designed to operate in the low RF range. Op-amps can have very high gain, very high input impedance, and very low output impedance.

To provide the operational amplifier with both a negative and positive voltage swing, a positive-negative power supply is used to power the IC. However, op-amps do not *have* to be run on a positive/negative power supply. They can be designed to run on a single polarity supply by simply grounding the -Vcc of the op-amp while attaching +30 volts to +Vcc. There will now be a 30-volt difference between +Vcc and -Vcc, as there would be with a -15V and a +15V supply. It is strictly a matter of reference, since an op-amp does not have to be grounded at all (differential). Either way, the input must be biased between the two rails; either 0V for ±15 Vcc, or +15V for +30Vcc.

The input stage of an op-amp is a differential amplifier, followed by high-gain voltage amplifiers and a high-output current buffer amplifier (*Figure 2-71*).

Open loop gain (no negative feedback) can be very high, but closed loop configurations, used to supply the degenerative feedback, are normally employed. The feedback, while lowering the gain substantially, increases the op-amp's bandwidth, while lowering distortion and eliminating the possibility of the amplifier breaking out into oscillation.

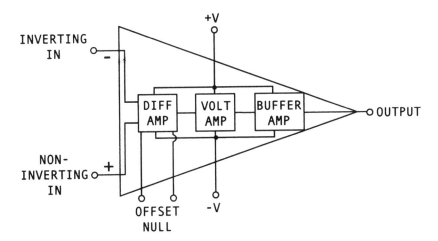

Figure 2-71. *Stages of an operational amplifier (op-amp).*

Depending on use, the op-amp is commonly configured for amplification in three ways.

First, the *inverting op-amp* (*Figure 2-72*) configuration has a gain as set by the ratio between R_2 and R_1, or:

$$A = \frac{R_2}{R_1}$$

R_2 supplies the negative feedback. The lower the value of R_2 the more feedback and the lower the gain but, as stated before, the wider the bandwidth. The entire input signal is dropped across R_1, with the input impedance essentially being the value of R_1 ($Z_{IN}=R_1$). In this configuration R_1 is usually of a very high resistance, giving it a high to medium Z_{IN}, while the output impedance is usually very low. Because the signal is placed at the inverting input, there is a 180° phase shift between the input and output signal.

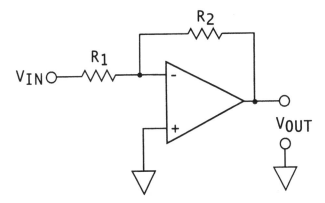

Figure 2-72. *The inverting op-amp configuration.*

Secondly, with *non-inverting* configurations (*Figure 2-73*), the input is placed at the non-inverting input, with the feedback still at the inverting input to obtain the necessary negative feedback. R_2 supplies degenerative feedback because it furnishes an in-phase signal to the inverting input, which lowers the difference between the - and + inputs, consequently lowering V_{OUT} and A_V. This configuration has a very high input impedance, a very low output impedance, and no phase shift between the input and output. The gain is equal to the ratio between R_2 and R_1 plus 1, or:

$$A = \frac{R_2}{R_1} + 1$$

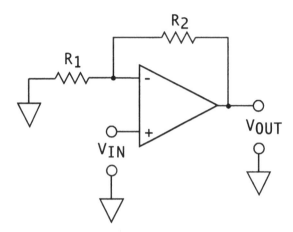

Figure 2-73. *The non-inverting op-amp configuration.*

The third amplifier configuration for an op-amp is the *voltage follower* (*Figure 2-74*), which is used as an emitter follower. In other words, it has an extremely high input resistance and an extremely low output resistance, with no phase shift between the input and output. The voltage follower has a voltage gain of one, so it is used as a buffer or impedance matcher between stages, or between a stage and its load.

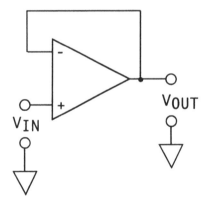

Figure 2-74. *The voltage follower op-amp configuration.*

Some important op-amp specifications:

A_v: Voltage gain of an op-amp in open loop configuration (no degenerative feedback). Specified in volts. EX: 200 V/mV (For every 1 mV input, 200 V output will attempt to occur, or a voltage gain of 200,000).

CMRR (Common-Mode Rejection Ratio): An op-amp's ability to reject in-phase, equal amplitude signals at the inverting and non-inverting terminals.

Common-mode signal: An in-phase, equal amplitude signal.

Offset-null: When an op-amp has the same potential difference between its two (diff-amp) input terminals, the output should be zero volts. If not, a proper value resistor or trimmer can be placed between its terminals to allow the op-amp's output to be adjusted to 0 V.

Slew rate: The rate that the output voltage in an op-amp can change per second without causing distortion (Ex: 1 V/μs).

2-9. Vacuum Tubes

Those who feel that tubes are no longer used in modern equipment would be incorrect. While low-power receiving tubes are not seen today, high-power triodes are employed for some power audio and low-frequency RF applications; as well as power pentodes and beam power tubes, outfitted with extra large or finned plates for high heat dissipation, found in many high-power transmitters. Vacuum tubes are used in many transmitter finals that must dissipate over several hundred watts output, not to mention most high-power microwave transmitters and all older equipment still in operation.

Tubes are normally ON devices and can be considered as basically a vacuum tube diode with a *control grid*. A *vacuum tube diode* (*Figure 2-75*), like the semiconductor diode, conducts if biased in the forward direction (with the anode, or plate, positive in respect to the cathode) and, unlike the semiconductor version, the cathode must also be heated to the proper temperature. This heating surrounds the cathode with a space charge of electrons (*thermionic emissions*), which are now free to travel towards a positive potential. If the anode becomes negative in respect to the cathode, or heating of the cathode is stopped, and conduction will cease.

The cathode can be heated indirectly by a heater element, or directly through a heated-wire cathode, using either AC or DC power.

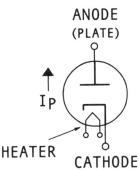

Figure 2-75. *A vacuum tube diode.*

In a *triode* (*Figure 2-76*), or three-element tube, a control grid is added to the diode. When the control grid is at the same potential as the cathode, the grid does not effect cathode-to-plate current flow, which would now be at maximum. When the control grid, constructed of fine wires, is supplied with a negative potential (in respect to the cathode), the attraction of the electrons for the plate is reduced, lowering the plate current (*Figure 2-77*). As the control grid is made more negative, a point is reached where there is zero cathode-to-plate current flow. This is referred to as *cut-off*. For Class A operation, the tube is biased somewhere between these two opposite extremes of maximum current (saturation) and cut-off (zero current). Now, any incoming signal, injected at this control grid, will add to or subtract from this negative DC bias, which then allows the tube to conduct more or less current through a suitable resistance, creating an amplified reproduction of the input signal.

The control grid should never go positive (unless biased by *grid-leak*), or current will flow in through the grid, possibly destroying it. Also, if the DC bias is lost to the grid, high currents will flow through the tube, likewise destroying the entire tube. A cathode resistor is sometimes placed in series with the cathode to limit this possible excessive current flow.

Vacuum tubes are voltage controlled, thus having a very high input impedance. This means they draw almost no current from the prior stage.

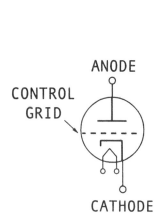

Figure 2-76. *A vacuum tube triode.*

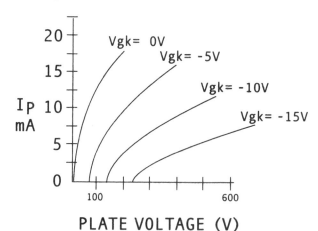

Figure 2-77. *Characteristic curves for a triode tube.*

2-9A. Tube Types

There are different types of tubes for various applications, usually dependent on frequency, cost, power, noise, and other design considerations.

The triode, which was described above, is now used mainly at audio frequencies due to the large interelectrode capacitance between all the elements, causing undesirable feedback and the resultant instability and decreased performance with higher frequencies. Low-frequency high powered RF amplifiers may still employ triodes, with *neutralization* used to help extend the limited frequency range. Although plate current is controlled by the control grid, changes in plate voltage also significantly affect the plate current (See *Figure 2-77*). Triodes are known for their low tube noise, which can be a huge asset in some audio and RF applications.

The problem of excessive interelectrode capacitance was solved by the design of the *tetrode* (*Figure 2-78*). The interelectrode capacitance is reduced significantly by the inclusion of a fourth element, the *screen grid*, between the control grid and the plate. The extra element performs as an electrostatic shield, reducing the interelectrode capacitance by three hundred times or more. This also results in a potentially higher gain than a triode. As well, the screen grid makes the plate current relatively independent of changes in plate voltage.

The tetrode is now rarely used, except in a few high-power RF amplifiers as a *beam power* tetrode, because of secondary emissions caused by the positive screen grid accelerating the electrons toward the plate to a high velocity. This increased speed causes some of the electrons to bounce off the plate, to be picked up by the positively charged screen grid. This action effectively reduces plate current, which is highly undesirable.

Neutralization may still be needed for tetrodes at high frequencies (above 50 MHz), even with the minor interelectrode capacitance present, due to the tube's high power sensitivity.

The *pentode* (*Figure 2-79*) has higher gain than a tetrode, with no neutralization typically needed at lower frequencies by virtue of the inclusion of the tetrode's screen grid. The pentode also solves the problem of secondary emissions by including a fifth element, the *suppressor grid*, between the screen grid and plate. The suppressor grid is usually placed at cathode potential (in some pentodes, the suppressor grid is internally connected directly to the cathode). Since it is at the same potential as the cathode, any secondary emissions that occur are bounced back toward the plate to add to the plate current. A pentode's characteristic curves are shown in *Figure 2-80*.

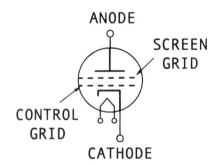

Figure 2-78. *A vacuum tube tetrode.*

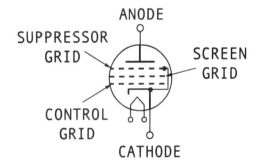

Figure 2-79. *A vacuum tube pentode.*

As with the tetrode, neutralization may still be needed at high frequencies, even with the minor interelectrode capacitance present, due to the tube's high power sensitivity.

Beam power tubes are special high-efficiency tetrode vacuum tubes with almost the same parameters as a pentode. Instead of a suppressor grid, however, they use two beam-forming electrodes at cathode potential, as well as control and screen grids that are aligned with each other. This setup repulses any secondary emissions back toward the plate and focuses the electrons into high-density beams, creating a highly efficient power tube.

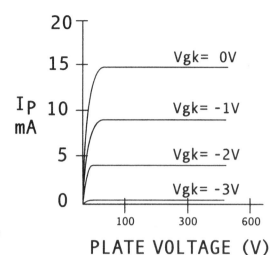

Figure 2-80. *Characteristic curves for a pentode tube.*

2-9B. Tube Amplifier Biasing

There are three major methods of biasing a vacuum tube amplifier for Class A, AB, B, or C operation.

The first, *grid-leak bias*, is a way to reliably and simply obtain any class of bias in an RF final tube circuit (*Figure 2-81*). The biasing components for grid-leak consists of R_1 and C_C. The class of bias depends on the amplitude of the input signal as well as the values of R_1 and C_C. Typically, however, grid-leak bias is utilized in Class C power output amplifiers, since grid-leak bias has a proclivity to produce a certain amount of extra distortion of the output signal — caused when the grid conducts current to supply the needed bias. Also, grid-leak cannot be used in small-signal amplifiers; and Class C grid-leak final amplifiers usually require a power driver stage located before it to produce the proper grid current.

DC grid-leak bias is created by the incoming signal going positive at the grid, causing grid current to flow (*Figure 2-82*), which charges the C_C to a negative potential on the grid, supplying the Q-point bias for the tube. With time, the charge across C_C leaks off through R_1 — with this time period dependent on the RC time constant of the two components.

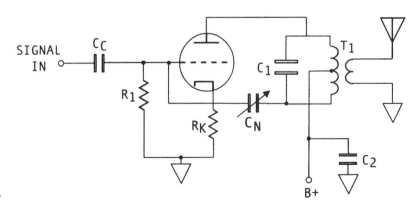

Figure 2-81. *A final tube circuit biased by grid-leak.*

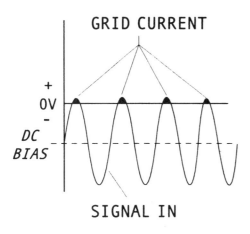

Figure 2-82. *How the input signal supplies grid current for the grid-leak capacitor.*

The grid-leak bias is self-adjusting for different signal input levels, since the bias level is dependent on the signal's amplitude and frequency. The signal is clamped at a particular positive peak input voltage level, usually at about +1V at the grid. Grid current must flow in order to keep the charge on the grid-leak capacitor, and thus the necessary negative DC bias. If the signal decreases in amplitude, the capacitor will discharge through the resistor, lowering the bias voltage until a certain point. Grid current then begins to flow as the top of the signal is naturally kept at +1V at the grid, which charges the capacitor to this new signal level. If the signal amplitude increases again, the negative bias, supplied by the charging capacitor, will also increase, keeping the positive peak of the incoming signal always at +1V at the grid.

A resistor (R_K) is sometimes placed in series with the cathode to limit current in case all grid-leak bias is lost, which would normally destroy the tube, or; many Class C grid-leak vacuum tube amplifiers may have an overload relay at their output for protection against lost grid-leak.

All classes of operation from Class A to C can be obtained by simply increasing the input signal level and adjusting the RC time constant of the grid-leak components.

Another form of bias, referred to as *cathode-bias* (*Figure 2-83*), is also used in low and medium powered Class A tube amplifiers. As current flows through the cathode resistor (R_K) the top of the resistor becomes more positive than the bottom and, since no current flows through the high-impedance grid circuit, the cathode becomes more positive than the control grid. This is the same as making the control grid more negative, thus setting its quiescent (Q) operating point. The exact Class A Q-point is set by choosing a certain value of R_K. The signal is developed across R_1, which also puts the grid at ground potential. R_K must typically be bypassed by C_K so that the constantly alternating voltage across R_K does not change the bias along with the constant change in the alternating cathode current, which would substantially lower gain through degenerative feedback.

Fixed-bias can be used to set an amplifier in Class A, AB, B, or C operation, and uses a power supply or supplies with separate bias voltages for the control grid and plate.

For high-power tubes and biasing, graphite plate tubes are available that can handle about 20 percent more power than the equivalent metal plate tubes for increased dissipation requirements. Also, most tubes are forced-air cooled by a fan, while some very high-power tubes may be water cooled.

As well, high voltage and high-power tubes have heavy filaments that need to be fed current slowly on start-up so that the resultant thermal shock is lessened — or the tube's service life can be very short. For this purpose tube circuits can employ either a variable autotransformer, which is manually turned up progressively to furnish filament current, or a special timing circuit that automatically supplies the heater voltages gradually.

Certain tubes may take up to four minutes to sufficiently warm up before plate voltage should even be applied — and some transmitters have a built-in timer circuit that will automatically supply plate voltage after an allotted time after filament turn-on.

Some high-frequency tubes have two or more leads to the cathode so as to reduce cathode-to-ground lead inductance, while others may have a plate connection through the top of the tube.

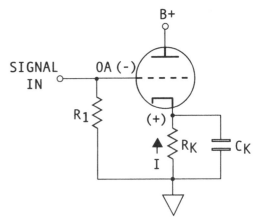

Figure 2-83. *Cathode-bias for a tube amplifier.*

2-10. Amplifier Frequency Limitations and Compensation

All amplifiers have an upper frequency limit dictated by the internal composition and physical structure of the semiconductor or vacuum tube, as well as its bias and coupling components and its configuration. There are, however, techniques that are employed to extend these frequency limitations.

2-10A. Frequency Limitations

Transit time problems, as well as other conditions in an amplifier, limit an amplifier's ability to increase the gain of a signal at higher frequencies. Transit time refers to the action caused by a frequency being so high that it begins to take longer for an electron to pass through the base of a transistor than it does for the frequency of the input signal to change polarity, which consequently builds up an excess of electrons in the base. This action drastically lowers gain.

As well, *shunt capacitances* causes lowered frequency response. *Figure 2-84* demonstrates the shunt capacitance of a standard two-stage RC coupled amplifier, while *Figure 2-85* shows a transistor's internal equivalent structure at high frequencies. C_S is the stray wiring capacitance, and shorts some of the high-frequency signal to ground, as does the collector-to-emitter capacitance (C_{CE}) and the base-to-emitter capacitance (C_{BE}). *Miller effect* capacitance (C_M) is the varying (gain dependent) collector-to-base capacitance that is 180° out of phase with the base input, causing a degenerative AC feedback from collector-to-base at certain high frequencies, thus lowering V_{IN} and, consequently, gain.

f_{co} is the high-frequency cutoff when gain in an amplifier drops to 70.7% of its low-frequency value. This f_{co} can be extended, at the expense of overall gain, when a lower value of load resistor (R_3 and/or R_7) is used in the amplifier because:

$$f_{CO} = \frac{1}{2\pi R_L C_T}$$

The symbol used in the data books to indicate when a CE configured transistor's b in a test circuit reaches 0.707 of its low-frequency value is referred to as f_{ae}, *f*b, or f_{hfe}. Another prevalent parameter for common-emitter circuits is referred to as f_t, and is the frequency at which the current gain reaches a value of unity, or 1.

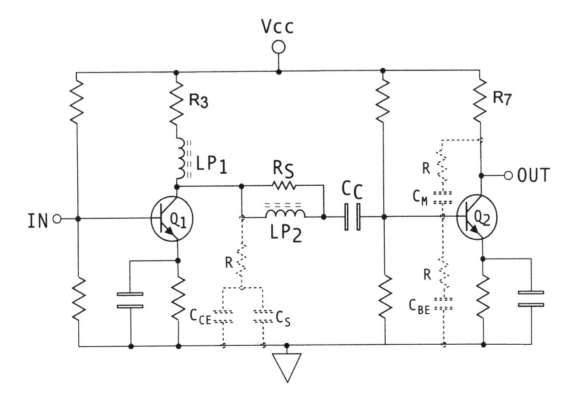

Figure 2-84. *A two-stage RC amplifier showing undesired shunt capacitance and frequency compensating components.*

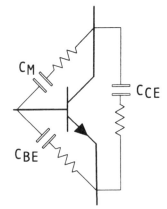

Figure 2-85. *A transistor's equivalent construction at radio frequencies.*

2-10B. Frequency Compensation

To externally compensate for a low f_{co} in a CE amplifier (See *Figure 2-84*), *peaking coils* can be utilized in the amplifier circuit by using a coil in *shunt* (LP_1), *series* (LP_2), or in *combination*, called *series-shunt*. The shunt peaking coil LP_1 almost cancels C_T (the total input and output shunt capacitances), which consists essentially of C_{CE} (collector-to-emitter capacitance), C_S (stray wiring capacitance), C_M (Miller capacitance) and C_{BE} (base-to-emitter capacitance) to ground. The shunt peaking coil is in shunt, or parallel, with these capacitances. LP_1 increases the amplifier's frequency response by resonating with C_T, causing an increase in output impedance at higher frequencies, heightening stage gain.

When employing a series peaking coil (LP_2), the output capacitance of the first stage and the input capacitance of the second stage are effectively detached and separated, allowing for a higher value of load resistor, which increases overall gain. The Q of LP_2 is usually lowered by a parallel "swamping" resistor (R_S) to prevent coil resonance, which can overcompensate for a certain range of frequencies, causing *ringing*.

When both LP_1 and LP_2 are used in combination, LP_1 offsets the output capacitance of the first stage, while LP_2 is used to counterbalance the input capacitance of the second stage. This combination can dramatically increase the frequency response across a very wide bandwidth at the output of the two coupled stages (*Figure 2-86*).

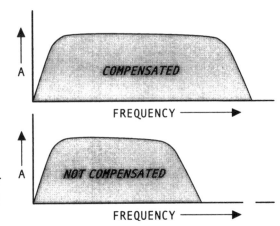

Figure 2-86. *The increase in upper frequency bandwidth possible with frequency compensation.*

Another form of frequency compensation that is quite common in many modern RF wideband (untuned) amplifiers insures that the lower frequencies are not overamplified, since the higher frequencies, as mentioned above, are normally attenuated. In the SSB wideband amplifier of *Figure 2-87*, this can take the form of negative collector-to-base feedback through a DC blocking capacitor, C_1, a series resistor, R_2 (to limit the lower-frequency feedback), and a series L_1 with its parallel resistor, R_1, which determines the feedback slope.

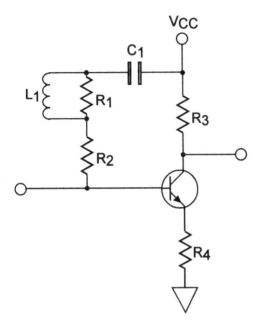

Figure 2-87. *RF wideband power amplifier.*

It is also possible to frequency compensate a wideband amplifier by a roll-off compensation network between the transistor stages (*Figure 2-88*). This can reduce low-frequency gain to give a flatter bandwidth, and is basically a high-pass filter.

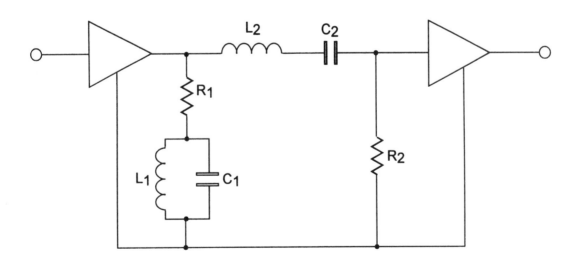

Figure 2-88. *An RF wideband amplifier with a frequency roll-off network.*

A designer can, moreover, take into account the frequency responses of the three transistor circuit configurations, the common-emitter, the common-collector, and the common-base, to increase this upper frequency limit. The common-base and common-collector configurations have a higher frequency response as compared to the common-emitter configuration. Unfortunately, the CB has a current gain that is always less than unity, while the CC always has a voltage gain that is also less than unity. By joining the CE and CC, or CE and CB amplifiers, however, increased frequency response and gain can be obtained.

Internal compensation at the time of manufacture can significantly increase the frequency response of a semiconductor. By making the base region as thin as possible, and by manufacturing the transistor with gallium arsenide, the transit time of an electron can be considerably shortened. NPN transistors also have faster transit times than a comparable PNP transistor due to the more rapid movement of electrons as majority carriers than that of holes through semiconductor materials.

CHAPTER 3
Oscillators and Frequency Synthesizers

There are many ways to generate both low- and high-frequency sinusoidal waves. In the past, the exclusive use of RC, LC, and crystal oscillators performed this function. Now, since 1970, a newer technique is being used to supply both high and low frequencies over a set design range. This method is referred to as *frequency synthesis* (FS). Oscillators are still, of course, extensively used, and have many different circuit configurations, depending on the desired frequency stability, range, adjustability, and cost.

Frequency synthesizers come in two main forms; the *phase-locked loop* (*PLL*), and the latest FS method, called *direct digital synthesis* (*DDS*).

3-1. Sine-Wave Oscillators

Since all tank circuits will ring (produce a decaying sinusoidal wave (*Figure 3-1*) at their resonant frequency when a pulse is applied to them, an oscillator can use an active device, such as a transistor, to amplify and sustain these pulses in the frequency-determining tank circuit. This tank circuit's natural resonant frequency is determined by the tank's L and C components, and can be calculated by using the formula:

$$fr = \frac{1}{2\pi\sqrt{LC}}$$

Oscillators use a small part of their output to feed a positive, or regenerative, feedback back into their own inputs to create these oscillations. In other words, the transistor constantly amplifies its own feedback. These oscillations begin by noise pulses during turn-on, after which an oscillator is basically converting DC power into AC oscillations.

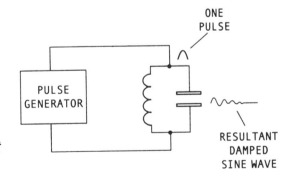

Figure 3-1. *Ringing of a resonant tank after application of a pulse.*

Since the output of most oscillators is 180° out-of-phase with the input (*Figure 3-2*), some sort of phase-shifting is needed to supply the needed in-phase, or regenerative, signal to its input (*Figure 3-3*). The reactance of inductors and capacitors perform this necessary phase shift in an LC oscillator.

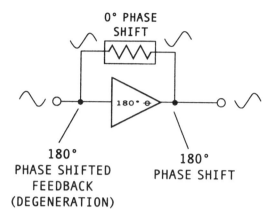

Figure 3-2. *An amplifier without a phase-shifting generally supplies degenerative feedback — no oscillations possible.*

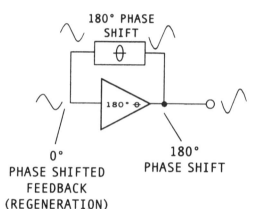

Figure 3-3. *Amplifier with phase-shifting network to supply regenerative feedback — oscillations possible.*

Both the common-emitter and the common-base configuration can be used as an oscillator, with the common-emitter being the most prevalent. The common-base oscillator is used when higher frequencies must be generated than can be handled by the common-emitter circuit. Even though there is no phase shift between a common-base's input and output, phase-shifting networks are still required, however, due to a slight phase shift (60° or less) introduced by the transistor when operated at frequencies above its *alpha cutoff* (when the gain of the transistor reaches 0.707 of its low-frequency value in its CB configuration).

Most oscillators are biased at Class C, though some are biased at Class A or AB, and typically have output powers of one watt or less.

Oscillators in the UHF region sometimes take advantage of the collector-to-base capacitance, and the resultant positive feedback, to create an oscillator without a physical phase-inversion feedback tank (*Figure 3-4*). Wire inductances and capacitances at high frequencies create a "virtual" resonant tank, causing a phase-shift and regeneration, which produces oscillation.

There are three main sine-wave oscillator classifications: The LC oscillator, the crystal oscillator, and the RC oscillator.

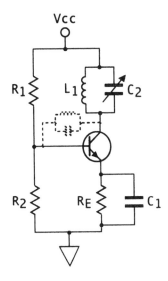

Figure 3-4. *A UHF oscillator sometimes uses a virtual tank formed by wire inductances and capacitances for phase shifting.*

3-1A. LC Oscillators

LC oscillators are generally *variable frequency oscillators* (VFOs) because they can readily be tuned to set the frequency of oscillation over a wide range by adjusting either the capacitance or, less frequently, the inductance, in the LC tank.

For any LC oscillator to be even reasonably stable it needs a good C to L ratio (for a high Q); a steady and separate power supply; constant temperature conditions; and oscillator isolation from its load. However, they still drift in frequency by 1% or more due to inherent real-world instability, which occurs when the temperature varies, the components age, or even when a hand is placed near the oscillator circuit (*hand-capacitance*). This is unacceptable for most modern transmitter and receiver uses without some form of frequency regulation (*AFC*, or *Automatic Frequency Control*).

A type of LC oscillator called a *voltage-controlled oscillator* (VCO) however, is quite common, and is utilized in frequency synthesis schemes, such as *PLLs*, and in any application where a DC control voltage is needed to change the frequency of an oscillator.

Both the *Hartley* and the *Colpitts* oscillators are the most common of the many LC designs that may be encountered.

The Hartley (*Figure 3-5*) uses a tapped coil in its tank circuit, comprised of L_1 and C_1, to invert the phase of the feedback signal to supply a positive voltage to the base through the coupling capacitor C_2, as well as set its frequency of operation. L_2 decouples the oscillator output from being injected into the power supply and also acts as a collector load. R_1 and R_2 supply the forward bias for the transistor, while C_3 allows only AC to pass to the tank, which increases oscillator stability and lowers power losses by not allowing DC to flow in the inductor. R_E and C_4 further (temperature) stabilizes the circuit as well as supplying a certain amount of self-bias.

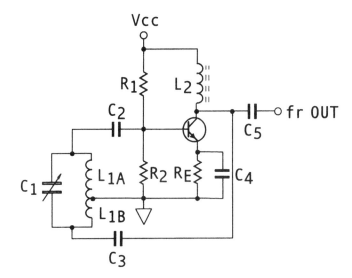

Figure 3-5. *An LC Hartley RF oscillator. Note the tapped coil.*

The Hartley's tapped coil works because the signal between the center tap and the top of the coil (*Figure 3-6*) is opposite in polarity than between the center tap and the bottom of the coil, due to current flow with respect to the grounded tap. This furnishes the necessary 180° phase shift needed for regenerative feedback, and thus oscillations. The amplitude of the positive feedback is controlled by the location of the tap on the coil.

The Colpitts (*Figure 3-7*) LC oscillator is essentially the same as the Hartley, but instead of a tapped coil sent to ground in the tank circuit, it uses twin capacitors tapped to ground. This has the same effect as the tapped coil by creating a 180° phase inversion across each capacitor, and supplying the positive feedback into the base as needed for oscillation.

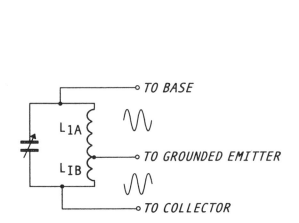

Figure 3-6. *A Hartley tapped coil with phase relationships.*

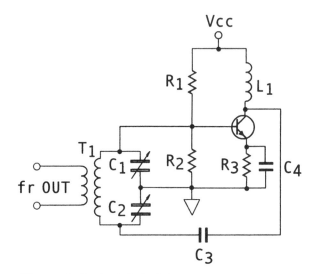

Figure 3-7. *An LC Colpitts RF oscillator. Note the tapped double capacitors.*

Since the capacitance ratio between the two capacitors changes the feedback voltage, the capacitors are always ganged to change the oscillator's frequency. A tunable inductor can also be used, with the capacitors placed at fixed values.

As stated above, VCOs are LC oscillators that can vary their output frequency by the application of a DC control voltage across *varactor diodes* (a voltage-controlled capacitor), which then varies its capacitance in the tank of an LC oscillator, changing its frequency either above or below a *rest*, or *center*, frequency. This rest frequency is set by the DC bias across the varactors, placing their capacitance at a fixed and unvarying value. By adding to or subtracting from this bias the frequency of the oscillator can be changed. In the example of *Figure 3-8*, Q_1 and its associated components function as a Hartley oscillator, while Q_2 performs as a buffer for the VCO's output (so that the oscillator is not loaded down by the low input impedance of the next stage).

VCOs are commonly used in PLLs, AFCs, and older manual electronic tuning circuits.

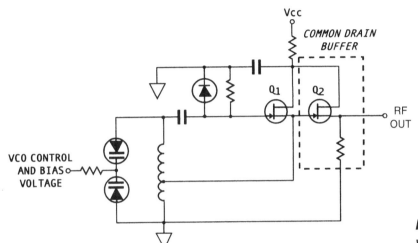

Figure 3-8. A Hartley based voltage-controlled oscillator (VCO).

3-1B. *Crystal Oscillators*

It is obvious that modern equipment cannot function properly with the excessive drift of even the best of the LC oscillators. Not only would a transmitter drift and interfere with adjacent channels, but its received signal would be unreadable as it drifted into and out of the passband of the receiver, increasing and decreasing in volume, pitch, and/or distortion. The receiver itself, if not frequency stabilized, would be hopelessly inadequate for most uses, especially in FM and SSB applications, as most signals would be unintelligible due to excessive LO drift.

One way to create a virtually drift-free oscillator output is to utilize some type of piezoelectric crystal. A crystal will produce a voltage if mechanical stresses are placed upon it, and will likewise produce mechanical expansion and contraction when an external voltage is placed across the crystal. If an alternating signal of the proper frequency is placed across the crystal, it will vibrate at its natural resonant frequency.

There are many types of crystals, but the most common is quartz because of its physical strength, heat stability, and low cost. It is the crystal material most commonly found as the frequency controlling element in a crystal oscillator. This ability of the crystal to only pass a signal at the crystal's natural resonant frequency makes it perfect for its application in oscillators.

Crystal oscillators can not readily change frequency to any appreciable degree, but may be tuned by a few hertz with a capacitive trimmer in parallel or series with the crystal. Rarely, and only in low-power applications, a crystal's frequency can be varied slightly by using an adjustable air-gap between one plate, or both plates, and the face(s) of the crystal. This type of tuning works because the spacing between the plates and the crystal, as well as the plate's pressure upon the crystal, affects its frequency.

Maximum crystal frequency can range up to 50 MHz on its fundamental frequency (most attain only 29 MHz), yet it can reach higher frequencies on harmonic operation. A *harmonic crystal oscillator* can have an output of its second, third, fourth, etc., harmonic, with the tank circuits tuned to output one of these frequencies. However, the crystal itself is actually operating at its lower fundamental resonant frequency. An *overtone crystal oscillator*, since the crystal must actually vibrate at high harmonic frequencies, will function strictly at its odd harmonic, such as its third, fifth, or seventh— with the output tank tuned to this chosen output frequency. A special overtone crystal is normally used for this type of VHF operation.

A small inductor is sometimes placed in parallel with a crystal if that crystal is part of an oscillator circuit. The inductor is there to form a parallel-resonant circuit with the crystal's holder capacitance (C_{PLATE}), and resonates out the holder capacitance. This ensures that the crystal will oscillate as marked on the can.

There are many distinct classifications of crystal oscillators, therefore we will only concentrate on the most common.

With the *Hartley crystal oscillator* (*Figure 3-9*) and the *Colpitts crystal oscillator* (*Figure 3-10*) the crystal is placed in the feedback path. Since a crystal has a very high Q (in excess of 50,000, with some as high as 100,000) and thus a very narrow BW:

$$(BW = \frac{fr}{Q})$$

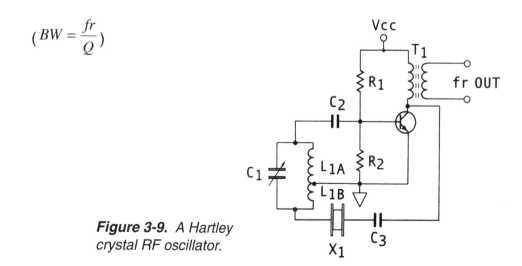

Figure 3-9. *A Hartley crystal RF oscillator.*

the series crystal will only allow the feedback current at its resonant frequency to pass onto the phase shifting LC circuits and into the oscillator's input. Any feedback that is even slightly off frequency will be severely attenuated, thus lowering feedback, and forcing the oscillator back on frequency. This makes the normal LC Hartley and Colpitts oscillators "rock steady".

The *Pierce crystal oscillator (Figure 3-11)* employs a different principle of operation. The inductor is replaced by a crystal operating in its slightly higher parallel resonant mode, which to the tank circuit appears electrically as an inductor. (Any crystal placed in its holder also has a higher *parallel* resonant frequency, as well as the lower series resonant frequency, caused by holding plate capacitance). Since impedance is at maximum at this parallel resonant frequency, the feedback across C_1, and thus into the base, is also at maximum — forcing the oscillator on frequency.

The Pierce oscillator is one of the most popular crystal oscillators for use in low-power, medium-frequency applications, and possesses a high frequency stability.

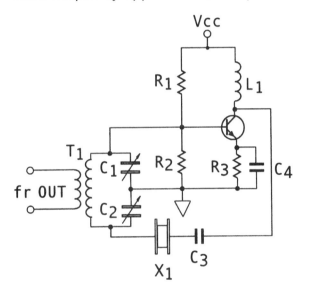

Figure 3-10. A Colpitts crystal RF oscillator.

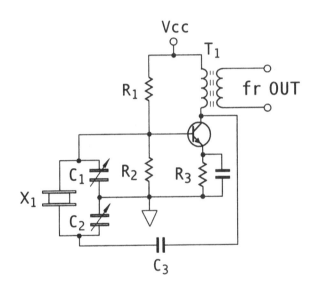

Figure 3-11. A Pierce crystal oscillator. Notice the crystal has replaced the inductor of the Colpitts oscillator.

Another popular crystal oscillator is sometimes referred to as a *TPTG (Tuned-Plate Tuned-Grid, Figure 3-12).* Also refered to as a *TITO,* or *Tuned-Input Tuned-Output*) crystal oscillator. The TPTG name is taken from an old vacuum tube LC oscillator that used twin resonant tanks; one at the input, the other at the output.

In the semiconductor TPTG oscillator, there are two tuned circuits, one in the base, consisting essentially of the crystal, and the tuned circuit at the collector, consisting of an LC output tank. R_1 and R_2 set the transistor's bias, L furnishes an elevated impedance for the crystal output into the base, while R_E provides temperature stabilization and a certain amount of self-bias, with C_1 bypassing RF around R_E for increased AC gain.

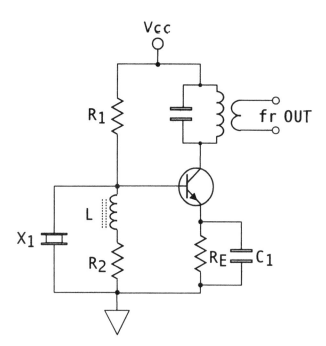

Figure 3-12. *A transistorized TPTG crystal oscillator.*

3-1C. RC Oscillators

Two types of sine-wave oscillators are commonly used as audio frequency generators: the *phase-shift* and the *Wein-bridge* oscillator. Both employ resistor-capacitor (RC) networks as the frequency setting components, inasmuch as the utilization of LC components would dictate the use of large and expensive parts at these very low frequencies (some audio oscillators may still use LC components, however).

The *phase-shift oscillator* (*Figure 3-13*) uses the phase-shifting abilities of an RC network to create the necessary positive feedback at the input to the oscillator. Since in a totally capacitive circuit the current leads the voltage by 90°, combining a capacitor with a resistor (with its zero phase shift), creates some phase shift value between 0 and 90°, depending on each component's values. Typically three 60° phase shifts can be employed, using three pairs of resistor-capacitor components to supply the full 180° phase shift needed for positive feedback. Since capacitors vary in reactance depending on the frequency of the alternating current, any frequency that does not give the proper reactance to the capacitors necessary to supply the combined 180° RC phase shift, and its resultant positive feedback, will not reach the oscillator's input, and will consequently force the oscillator back on frequency to a frequency that *does* allow an exact 180° phase shift between its output and input. This will supply the input to the oscillator with a regenerative (positive) feedback, and thus sustain continued oscillations at the proper design frequency.

The phase-shift oscillator is usually employed only in fixed-frequency applications.

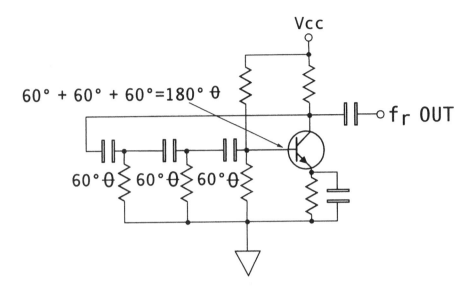

Figure 3-13. An audio
RC phase-shift oscillator.

The second type of common low-frequency oscillator is the *Wien-bridge* (*Figure 3-14*), and can be used in variable-frequency operations to output high-quality sinusoidal waves. The RC network, called a *lead-lag network*, sets the operating frequency of the oscillator by supplying both a regenerative and degenerative feedback, and ordinarily are operated in conjunction with an IC op-amp rather than any discrete amplifiers. As long as the feedback is slightly positive, oscillations will occur, but at any value where it becomes negative, no oscillations result, forcing the oscillator back on frequency.

Actual phase-shifting of the feedback voltage in a Wien-bridge oscillator does not take place — the network merely chooses the frequency at which this feedback occurs. R_1-C_1 and R_2-C_2 furnish the regenerative feedback to the positive, or non-inverting, input to the op-amp, while the degenerative feedback is created across R_3 and R_4 and placed at the negative, or inverting, input. At the oscillator's resonant frequency, the regenerative feedback across R_1-C_1 and R_2-C_2 combination exceeds the degenerative feedback voltage present across R_3-R_4, forcing the oscillator to stay on frequency. C_1 and C_2 can be ganged to vary the output frequency.

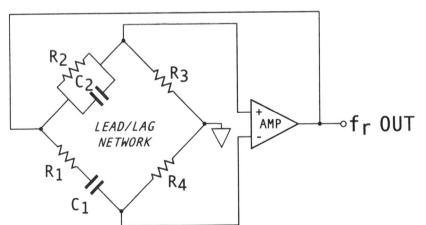

Figure 3-14. An audio RC
Wien-bridge oscillator.

3-2. Frequency Synthesis

Frequency synthesis is a technique for generating extremely accurate multiple frequencies from one (but sometimes more) crystals. It has become the dominant method for variable frequency production in receivers, transmitters, transceivers, and test equipment.

There are many ways to obtain frequency synthesis, but the most common today is by the *Phase-Locked Loop* (PLL) and a newer system called *Direct Digital Synthesis* (DDS; sometimes referred to as an *NCO*, or *Numerically Controlled Oscillator*). Both methods can vary tremendously in circuit layout from one PLL circuit to the next, or from one DDS system to the next, depending on frequency accuracy, stability, number of simultaneous frequencies needed, upper frequency limits, noise generation, tuning resolution, and bandwidth of tuning desired. Obviously then, only the basics of frequency synthesis can be covered here.

3-2A. Phase-Locked Loop

Most modern frequency synthesis is based on the phase-locked loop. *Figure 3-15* shows a simple *single loop* (one complete PLL circuit) *PLL frequency synthesizer*. A crystal oscillator frequency (the *reference frequency*) is fed into a fixed *frequency divider*, which lowers the reference frequency by a set amount and places it into the *phase-comparator*, which compares the frequency from the fixed frequency divider to that of the *adjustable frequency divider*. The adjustable frequency divider acquires its frequency from the output of the *VCO* (Voltage-Controlled Oscillator, or "voltage-to-frequency converter"). After the phase-comparator compares the two frequencies and, if they are the same, no error voltage is outputted from the phase comparator. If they differ, then a rectified DC error voltage is output to the *PLL filter*, which removes any AC variations and noise, and feeds this DC control voltage directly into the VCO (or, in some PLLs, first into a DC amplifier and then into the VCO). A varactor's bias in the VCO is affected by this DC control voltage, which forces the VCO back on frequency if it has drifted off.

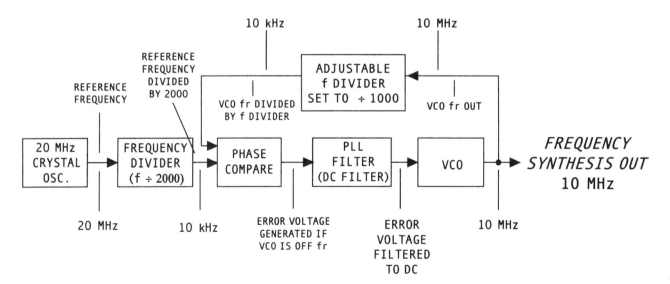

Figure 3-15. *The basic sections of a phase-locked loop frequency synthesizer.*

The internal and adjustable frequency divider of the PLL is commonly controlled by the operator through a front panel knob, with a microprocessor supplying digital control words by a serial or parallel bus. The microprocessor can also be used to decode and drive the frequency display circuits (either LEDs or LCDs), so that the operator will be visually updated on any frequency changes that may be made.

Some IC PLLs are designed to output two frequencies for frequency synthesis, one frequency as the transmitter's master oscillator, and another for input into the transmit mixer; or for a dual-conversion receiver's two LO frequencies; or one for the LO and one for the carrier of a transceiver, etc.

A *double-loop* (offset reference) *PLL*, using dual ICs, is another common type (*Figure 3-16a*) of frequency synthesis, and has the advantage over that of the single loop PLL of a much smaller frequency resolution (down to 10 Hz with a 30 MHz frequency range in the VHF region). The *offset reference* term for this PLL type refers to the frequency resolution (or *step size*) being obtained from the difference between the reference frequency and the input to the two phase detectors. This input is controlled by the first frequency dividers, with one divider being f/N (+0), and the other being f/N (+1). The output of both PLLs is fed into a mixer, filtered to obtain the sum or difference frequency, and sent out to the appropriate circuits.

The *triple-loop PLL* circuit of *Figure 3-16b* uses one PLL (PLL 1) to supply coarse tuning, a second to supply fine tuning (PLL 2), and a third as the output circuit (PLL 3; the internal VCO of PLL 3 is phase-locked to the sum frequency of PLL 1 & 2). This type of PLL circuit, while common, is difficult to manufacture and design.

Some PLL communications gear may use external VCOs, or employ simple mechanical pushbuttons or switches for channel changes directly into the PLL chip inputs.

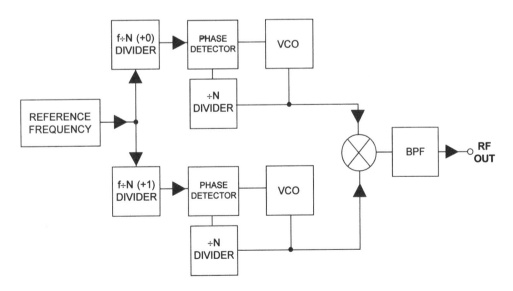

Figure 3-16a. A double-loop circuit.

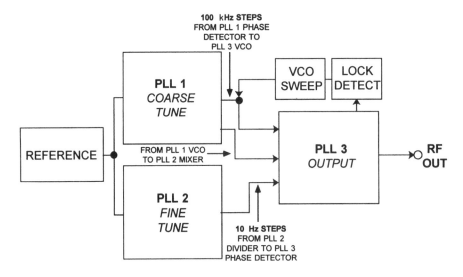

Figure 3-16b. *A triple-loop circuit.*

For PLL circuits capable of frequency synthesis at higher frequencies, the output of a crystal oscillator and the PLL can be fed into a mixer, filtered to obtain the sum or difference of the two frequencies, and resulting in a higher synthesized frequency. This is referred to as *pre-mixing* (*Figure 3-17*).

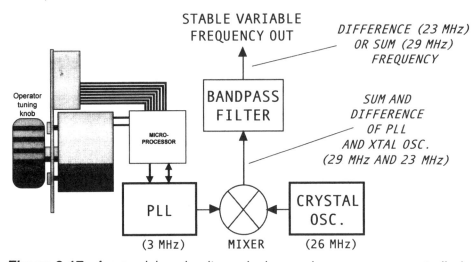

Figure 3-17. *A pre-mixing circuit employing a microprocessor-controlled PLL and a crystal oscillator.*

A very popular PLL tuning circuit is shown in *Figure 3-18*. The *shaft encoder*, which is rotated by the operator, may have two voltage outputs (A plus B), each with a square wave in quadrature phase (90-degree phase shifted) to the other. The A output is connected to the interrupt line on the microprocessor. When the A output has a falling edge, an interrupt occurs. When the microprocessor is interrupted it then looks at the B output. If B is a 1, then the knob has been rotated clockwise — and the microprocessor increments the transceiver's frequency by 1. If B is 0, then the knob has been rotated counterclockwise, and the microprocessor decrements the frequency by 1. A modern, simple, reliable, and cost effective PLL tuning circuit.

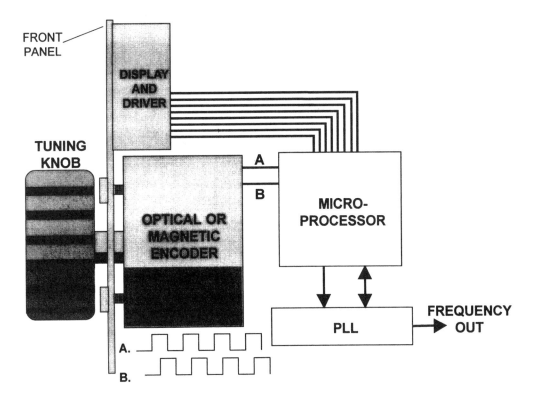

Figure 3-18. *A common PLL tuning method.*

3-2B. Direct Digital Synthesis

A very recent development in frequency synthesis is called Direct Digital Synthesis. DDS lowers the cost and intricacy of most common frequency synthesizers with a much-improved frequency resolution; giving the feel of analog fine tuning with up to 1 Hz resolution or less (tuning resolution is the fineness of the frequency steps the tuner is capable of — the lower the frequency the better). DDS usually comes in small surface-mount ICs, along with one or more support IC's.

As shown in *Figure 3-19*, one form of DDS works by utilizing a crystal oscillator as a reference frequency (the *CLOCK*), digitally calculating (in the *PHASE ACCUMULATOR*) the difference between this reference and that of the frequency the radio operator wishes to generate — with the phase accumulator also computing and outputting an appropriate address to the *WAVEFORM ROM* or RAM sine-wave look-up table. This new calculated digital sine signal information is fed to a D/A CONVERTER and then placed into a very high quality low-pass filter (LPF) — which outputs the new synthetic digitally constructed analog sine-wave.

A pure DDS system normally has limited frequency range, so DDS is usually combined with standard PLLs to increase this range to usable levels. Also, since it is a pure digital form of frequency synthesis, a type of noise is created called *quantization noise*. This simply refers to errors made when the DDS output is converted from digital to analog. The analog synthesized frequency, with its infinite amount of possible amplitudes, is actually being constructed from a multitude, but finite, of discrete digital levels — making complete accuracy during conversion impossible.

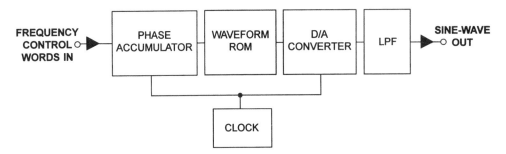

Figure 3-19. *A basic DDS circuit.*

One way to implement a practical RF DDS frequency synthesis scheme is as follows: When the operator changes frequencies the radio's on-board microprocessor outputs control words into the DDS (*Figure 3-20*) for a frequency change, which then synthesizes a low, stable synthetic analog frequency that is fed into a *summing loop* (a complex circuit that performs a type of premixing). The loop adds the low-frequency output of the DDS to a high-frequency reference, mixing the subsequent output into a much higher stable frequency. Since the frequency variations possible at the output of the summing loop are quite small, they are then inputted into a PLL (with its own crystal reference), along with coarse frequency digital control words. The output is now the desired wide-ranging, but rock steady, synthesized frequency.

There are many other ways to achieve a functional DDS frequency synthesis system, including the newer single CMOS chip implementation with on-board *Clock Reference*, *Phase Accumulator*, *Sin Rom*, and a 10-bit *DAC*. This chip is capable of low voltage (3 or 5V) and low power operation (down to 50 mW), with on-chip power-off circuitry for added power savings. DDS can also be directly modulated in-chip to produce FM, PM, AM, FSK, QPSK, etc.

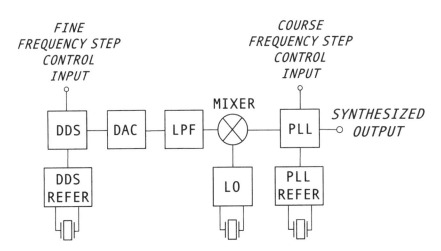

Figure 3-20. *One scheme using direct digital synthesis (DDS) to synthesize new frequencies.*

CHAPTER 4
Amplitude Modulation

Amplitude modulation (AM) is one of the earliest forms of wireless communications. It is extremely simple and low in cost to implement. Due to this low cost and simplicity, it is still extensively used — mainly in commercial and shortwave broadcast and citizens-band radio operation. Understanding AM is vital for the comprehension of newer means of radio communications.

4-1. AM Fundamentals

Modulation is the process of impressing information on a carrier wave, be it voice, digital code, images, etc. Demodulation is the process of removing this information and sending it to any number of transducers; such as a speaker or headphones for voice and music, or to a cathode ray tube for visual display, or through to digital circuits for processing or storage.

4-1A. Modulation and Demodulation

The simplest way of impressing voice or music on a carrier is by amplitude modulation. The unmodulated carrier (*Figure 4-2a*) is the radio frequency signal that transports any modulation, through space, to the receiver. The modulation is the *intelligence* (*Figure 4-1*), always at a lower frequency than that of the carrier, which is impressed on this carrier through non-linear mixing. The *amplitude* of the carrier is altered, when viewed in the time-domain, at the rate of the modulating signal's *amplitude* and *frequency* variations. Increasing the amplitude of the modulating signal increases the amplitude of the carrier (*Figures 4-1c & 4-2c*), while decreasing the modulation's amplitude decreases the carrier's amplitude (*Figures 4-1b & 4-2b*).

The modulation "rides on" the carrier to the receiver, where these amplitude variations are stripped off of the carrier and converted into audio amplitude variations (*Figure 4-3*) and amplified, then fed to an appropriate transducer. The percent of modulation (*Figures 4-2b & 4-2c*) will govern the amplitude of the final detected signal (*Figure 4-3b & 4-3c*) and the resultant audio volume at the receiver's speaker.

In Class C collector modulation (or plate modulation, in vacuum tube finals), both the positive and negative alternations of the carrier are affected equally by the modulation frequencies: The negative alternation is recreated by the output tank of the transmitter's final amplifier, generating a mirror image of the positive modulated alternation (*Figure 4-4*).

Modulation between a carrier and its baseband signal (voice, music, code, etc.) creates sidebands, viewable in the frequency domain (*Figure 4-5*), due to the non-linear mixing creating sum (USB, or *Upper Sideband*) and difference (LSB, or *Lower Sideband*) frequencies. The phase relationships between the carrier, and the LSB and USB, cause a resultant new waveform that varies in amplitude (*Figure 4-6*). When the sidebands and carrier are in phase, the amplitude of this new waveform is twice that of the unmodulated carrier. When the two

INTELLIGENCE (AUDIO)	CARRIER (AFTER MODULATION)	AFTER DEMODULATION

0V

UNMODULATED CARRIER

0V

(a)　　　　　　　　(a)　　　　　　　　(a)

LOW AMPLITUDE

50% MODULATION

HALF OF MAXIMUM AMPLITUDE

(b)　　　　　　　　(b)　　　　　　　　(b)

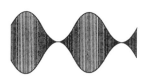

HIGH AMPLITUDE

100% MODULATION

MAXIMUM AMPLITUDE

(c)　　　　　　　　(c)　　　　　　　　(c)

MISSING INFORMATION

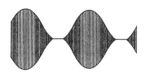

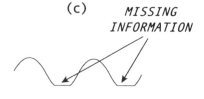

VERY HIGH AMPLITUDE

OVERMODULATION

MAXIMUM AMPLITUDE

(d)　　　　　　　　(d)　　　　　　　　(d)

Figure 4-1. *An audio modulation frequency at various amplitudes.*

Figure 4-2. *A carrier shown unmodulated and at various modulation percentages.*

Figure 4-3. *The amplitude modulated (AM) signals after demodulation.*

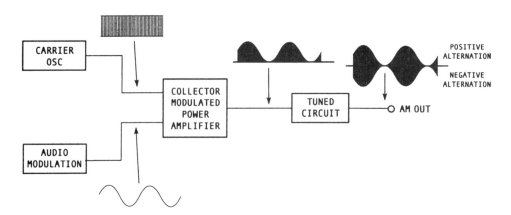

Figure 4-4. *AM generation in a Class C collector modulator.*

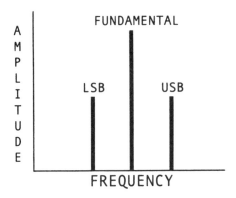

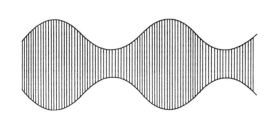

Figure 4-5. An AM carrier and its sidebands when modulated by a single tone.

Figure 4-6. A single tone amplitude modulated carrier in the time-domain.

sidebands are completely out-of-phase with the carrier, the amplitude of the new waveform is almost nil. In other words, in the time-domain, the amplitude of this new modulated carrier waveform is twice that of the unmodulated carrier at the peaks with 100% modulation, and zero in amplitude at the troughs (*Figure 4-7*).

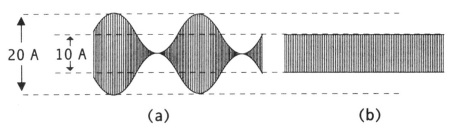

(a)   (b)

Figure 4-7. (a) A 100% amplitude modulated carrier; (b) with unmodulated carier.

When viewed in the frequency-domain on a spectrum analyzer, the carrier's frequency and amplitude does not vary, whether modulated or not (*Figure 4-8*). This is an important concept, because this proves that the carrier itself contains no information — the information is contained in the sidebands only. When amplitude modulation is viewed in this frequency-domain, it can be seen that when the modulation is changed in frequency and amplitude at the transmitter, the sidebands will change frequency and amplitude, while the carrier will remain at its original frequency and amplitude, indicating that there is no information actually contained in the carrier, but only in the sidebands (*Figure 4-9*). Each sideband will contain the exact same information and power as the other.

As mentioned, when the carrier wave is modulated by an audio wave sidebands are produced, as can be observed only in the frequency domain, on a spectrum analyzer display. These sidebands are at sum and difference frequencies of the carrier and modulating frequencies, or $f_{CARRIER} - f_{AUDIO} = LSB$ and $f_{CARRIER} + f_{AUDIO} = USB$. For example, if a 100 kHz carrier is modulated with a 5 kHz audio tone, the sideband frequencies are 100 kHz + 5 kHz or 105 kHz, and 100 kHz-5kHz or 95 kHz. The bandwidth of this resulting signal is double the audio frequency (BW = 2 X f_{AUDIO} or USB - LSB) or, in the above example, 2 X 5 kHz, or 10 kHz.

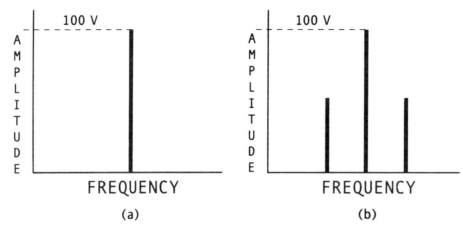

Figure 4-8. *The carrier amplitude of an AM signal remains unchanged regardless of modulation: (a) unmodulated carrier; (b) modulated carrier.*

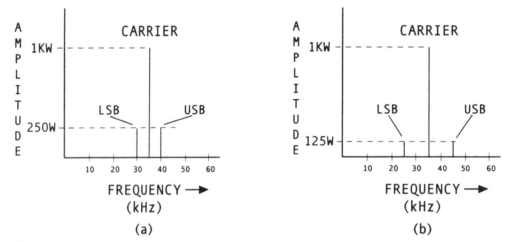

Figure 4-9. *When the modulation frequency and/or amplitude varies only the sidebands are altered: (a) 5 kHz high amplitude modulation; (b) 10 kHz low amplitude modulation.*

Since the carrier power remains unchanged no matter what the modulation percentage, the total power of an AM signal equals the sum of the carrier and sideband powers, or $P_T = P_c + P_{LSB} + P_{USB}$. Only the *total* AM power varies. When the AM carrier is modulated, antenna current increases due to the power added by these sidebands.

The percent of modulation can easily be found in the time-domain, on an oscilloscope display, by the formula:

$$\frac{V_{PEAK} - V_{MIN}}{V_{PEAK} + V_{MIN}} \ X \ 100$$

An example of a voice amplitude-modulated signal as displayed on a *spectrum analyzer* screen can be seen in *Figure 4-10*. Many sidebands are produced at varying frequencies and amplitudes, depending on the audio modulation's own amplitudes and frequencies.

When using an antenna ammeter with AM, and the AM transmitter is modulated by a sinusoidal wave at 100 percent, then antenna current will rise from its unmodulated value of, let us say, 10 amps to 12.2 amps. If modulated by music, the antenna current increase will be less. With voice, because of the lack of being sinusoidal, the antenna current may appear not to increase at all with 100% modulation.

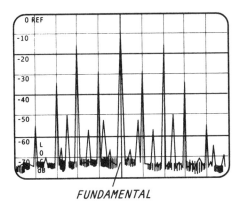

FUNDAMENTAL

Figure 4-10. *A voice signal contains many sidebands at various frequencies and amplitudes, as viewed in the frequency-domain.*

4-1B. Overmodulation

Most AM voice transmitters limit modulation frequencies to from 300 to 3000 Hz by the use of a BPF (bandpass filter), placed after the first audio (microphone) amplifier, to limit transmitted bandwidth. A *limiter* can also be used to limit the maximum audio amplitudes into the audio modulator so that *overmodulation* cannot occur.

Overmodulation causes distortion and an increase in bandwidth due to *splatter*. Splatter is harmonic generation of the original audio frequencies caused by clipping of the modulation envelope, which further modulates the carrier, causing adjacent channel interference — which is heard as a crackling noise in the receiver for many hundreds of kilohertz from the carrier frequency. Also, since part of the audio envelope is not present (See *Figures 4-2d and 4-3d*), intelligibility of the received signal is also degraded.

4-1C. AM Power Measurement

In AM the *peak envelope power* (PEP) can be used to measure the peak instantaneous power, with *100%* modulation applied, of the transmitted signal:

$$(\text{PEP} = \frac{V^2_{RMS}}{R} \text{ or } V_{RMS} \times I_{RMS} \text{ or } I^2_{RMS} \times R \text{ of the peak values at 100\% modulation}),$$

or the carrier power can be calculated using the same formulas, with zero modulation applied.

4-2. AM Receivers

AM receivers have much in common with almost all other RF receivers, in that all typically apply the superheterodyne principle, usually varying only in the form of demodulation of the incoming modulated radio signal.

The quality of a modern AM receiver — its sensitivity, selectivity, noise figure and, to a lesser extent, its dynamic range — vary between communication sets. However, all employ *basically* the same circuits, in varying numbers (whether in IC or discrete forms), to convert the RF signal into an intermediate frequency, strip the audio from the IF, and amplify the resultant AF into a speaker or headphones.

4-2A. AM Superheterodyne Circuits

Most AM receivers exploit the *superheterodyne* principle. All AM superheterodyne receivers (*Figure 4-11*) receive a high-frequency modulated signal (the RF), converting it to a lower fixed intermediate frequency (the IF). Conversion is accomplished by mixing the incoming RF with a local oscillator (LO), then filtering the mixer's output to obtain the difference frequency. This difference frequency (now called the IF) is applied to high-gain, narrow selectivity IF amplifiers. The IF is then sent to a demodulator, such as the simple *diode detector* of *Figure 4-12*, where the modulated IF is rectified (the RF is changed into a pulsating DC by cutting off the IF's negative half-cycles). The positive DC pulses of the IF itself now charge up the capacitor. Between these DC pulses the capacitor is able to discharge through the resistor, but the RC time constant is such that the charge and discharge cycle mimics the audio envelope riding on the IF — and the audio envelope is thus recovered. The *full-wave AM diode detector* of *Figure 4-13* can also be utilized. The full-wave detector produces a higher and easier-to-filter output, as well as less distortion and attenuation of the higher audio frequencies, than the single-diode type. The resultant audio frequency from any AM detector is then fed to the audio amplifiers and on to the speaker.

The audio stages of a receiver are usually Class A or linear Class AB or B push-pull, with the final audio power amplifier almost always push-pull. Also, any AM amplifier stage that contains a modulated signal, such as the RF and IF stages, cannot use non-linear stages (such as Class C), since the incoming signal contains a modulation envelope that would be severely distorted by the non-linearity.

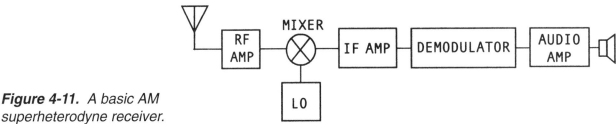

Figure 4-11. A basic AM superheterodyne receiver.

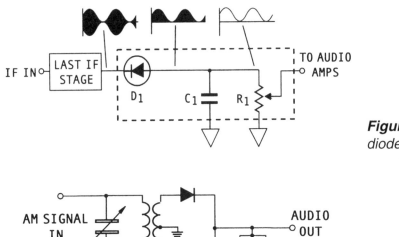

Figure 4-12. *The signal diode AM detector.*

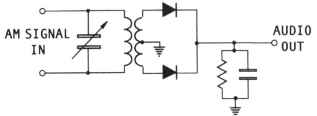

Figure 4-13. *A full-wave AM diode detector.*

The *front end* of a receiver consists of the tuned circuits from the antenna input up to and including the mixer and local oscillator, as well as any possible radio frequency amplifier placed before the mixer. Many communication receivers have two or more such RF stages. The RF amplifier stage or stages may not be present in receivers designed for reception of frequencies under 30 MHz, but their inclusion increases sensitivity, selectivity, improves the signal-to-noise ratio, the image and spurious frequency rejection, and isolates the LO from the antenna, lowering the possibility of the LO radiating, causing interference.

The volume adjustment (audio gain) is usually located just after the audio detector and before the first audio amplifier (*Figure 4-14*).

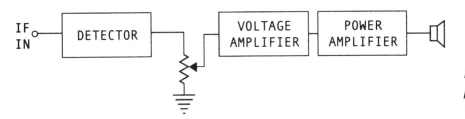

Figure 4-14. *Common placement of a receiver volume control.*

4-2B. Image Frequency

Any single conversion (one IF) superheterodyne receiver has a drawback in that any frequency received at its chosen IF will be amplified by the IF stages and outputted into the speaker as interference. The image frequency is any frequency that is at twice the IF plus the RF signal frequency ({2 · IF} + RF), or at the IF plus the LO frequency (LO + IF). In other words, it is a frequency that differs from the LO frequency by the amount of the IF, just as the desired signal does, but is higher instead of lower than the LO frequency (since the LO is usually chosen to be above the received signal by an amount equal to the IF). Subtracting the signal frequency from the LO frequency will gives us the IF (f_{IF}), which passes through the

IF amplifiers. Any frequency that is *higher* than the LO frequency by the same amount that the signal frequency (f_s) is *below* the LO frequency (f_{LO}) would give us the same frequency, and thus be passed through the IF amplifiers, causing audio interference.

To make this concept a little clearer: if we had a signal frequency of 100 MHz and a LO frequency of 101 MHz, this would give us an IF of 1 MHz. Since f_s - f_{LO} = f_{IF} then 100 MHz-101 MHz = 1 MHz. If another frequency (the image frequency, f_i) happened to be at 102 MHz, with the local oscillator at 101 MHz and the IF at 1 MHz as above, then f_i - f_{LO} = f_{IF}, or 102 MHz - 101 MHz = 1 MHz.

There are three ways to attenuate these image frequencies. One is to have a very a high-Q *preselector* (an amplifier that supplies a narrow bandwidth to pass the signal frequency only) at the input of the receiver; or to have a high intermediate frequency to remove the image frequency as far away from the desired frequencies as possible; or to utilize double or triple conversion receiver designs.

With multiple conversion receivers, the first IF is very high to keep the image far away from the signal, so that it will be out of the passband of the receiver. The second and third IFs are considerably lower in frequency to supply most of the selectivity and gain. The lower the IF the more simple, stable, sensitive, and selective the amplifiers become due to the lower stray capacitances and inductances at these lower frequencies. As well, there is no need to employ special high-frequency components or shielding.

Some double or triple conversion receivers may not use IF amps at all between the first or second conversion stages.

4-3. AM Transmitters

A transmitter, at its most basic, is an oscillator that can be turned on and off. This simple transmitter could be used to transmit information in the form of *Morse code*. Most transmitters, however, are far more complex. Even a Morse code (CW) transmitter ordinarily has supplemental circuits to supply added amplification for increased power output; tuning capabilities for transmission on more than one frequency; and filters to reduce harmonic and spurious radiation.

Any transmitter that must transmit voice or other *baseband signals* (audio, video, or pulse data modulation) must have some method of impressing, or modulating, this low-frequency baseband signal onto the higher-frequency RF carrier for transmission into space.

4-3A. Modulation

In AM, modulation is accomplished by two basic means. The first technique is referred to as *low-level modulation*. With this method the carrier is modulated at any stage before the final power amplifier, necessitating linear amplifiers after modulation to sustain the audio envelope without distortion. Feeding the carrier into the base and modulating the collector (*Figure 4-15*), the emitter (*Figure 4-16*), or even the base (*Figure 4-17*) can be used.

With the second technique, *high-level modulation*, usually only the last power amplifier stage, or both the last power stage and its driver, are modulated. *Collector modulation* is very common in both low-level and high-level modulation, and will be discussed in more detail later.

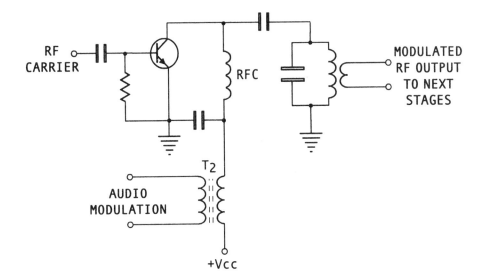

Figure 4-15. Low-level collector modulation.

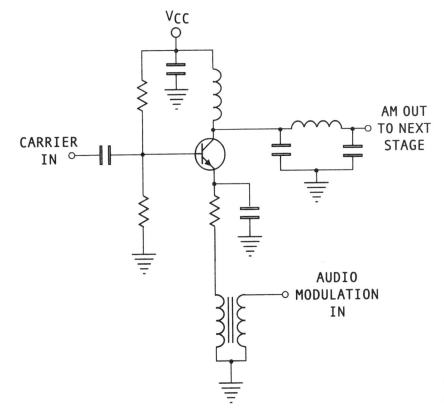

Figure 4-16. Low-level emitter modulation.

With emitter modulation the modulating signal varies the gain of the common-emitter stage, and thus that of the carrier injected into the base. For this type of modulator to function properly, the modulation amplitude must be higher than that of the carrier.

Base modulation injects both the carrier and the modulating signal into the base, with the transistor biased Class C. Since Class C will only pass half the waveform, the complete waveform is reconstructed by a low-Q tuned-output circuit. This configuration is rarely used due to its inferior linearity.

The main advantage of low-level modulation is that very little audio power is needed to create the amplitude modulated signal. Still, low-level modulation is seldom practiced because any energy savings enjoyed in using a low level of modulation power is offset by the need to operate with relatively inefficient linear amplifiers, whether single-ended Class A, or AB or B push-pull. The exception is in low-cost and low-power transmitters, or in AM generation above the VHF spectrum, which typically employ *PIN* diodes.

Indeed, one of the few methods to produce AM at high frequencies is with the *PIN modulator* (*Figure 4-18*). The PIN diodes are forward biased by the -Vcc. This bias is subtracted from or added to by the negative and positive alternations of the modulation signal, which varies the resistance of the PIN diodes and, since these diodes are in series with the master oscillator and the load, modulates the amplitude across the AM OUT terminals.

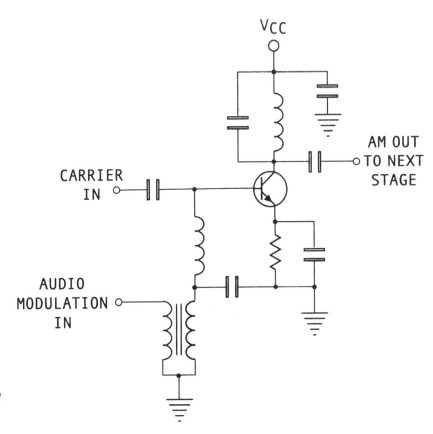

Figure 4-17. Low-level
base modulation.

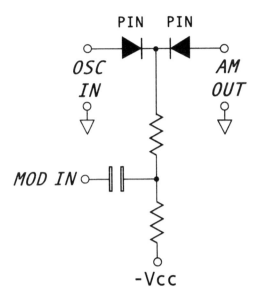

Figure 4-18. *A PIN AM modulator.*

As mentioned above, the second technique employed in AM is *high-level modulation* (*Figure 4-19*), which modulates the RF carrier in the very last stage, generally using extremely efficient Class C RF amplifiers. High-level modulation offers excellent linearity but, unfortunately, the audio modulator stage must be capable of supplying a peak voltage almost equal to V_{CC}, as well as half the power of the final RF power amplifier, for 100% modulation. This means that significant power must be generated in the inefficient linear audio stages.

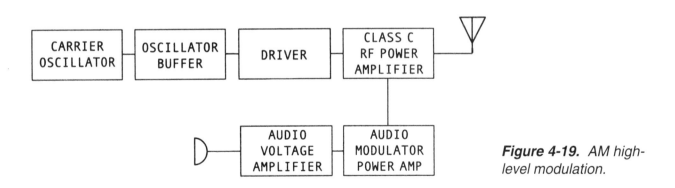

Figure 4-19. *AM high-level modulation.*

Figure 4-20 demonstrates the typical high-level collector modulator circuit. The transistor and its bias components furnish non-linear Class C RF power amplification, while the audio signal, which is in series with the collector supply voltage through the secondary of the audio modulation transformer, either adds to or subtracts from the collector voltage, which will then cause either an increase or decrease in collector output current. Since this amplifier is biased at Class C, only pulses are outputted across the RFC. These pulses are then coupled into the tank network by C_4, where they are changed into a complete AM waveform (See *Pi filter*, *Flywheel effect*, and *Tank circuit*).

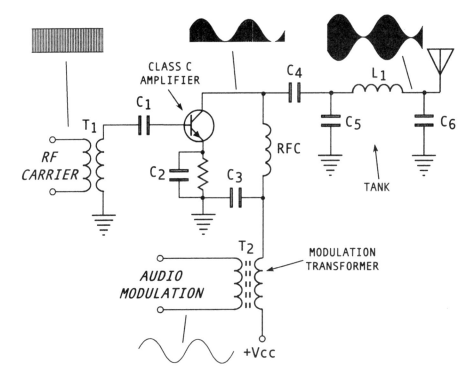

Figure 4-20. *An AM high-level collector modulator.*

Figure 4-21 shows a high-powered tube final AM stage employing plate modulation (comparable to a transistor's collector modulation). Class C is used for this last RF amplifier because the final must be non-linear to produce the necessary modulation sidebands that create AM. The output tank reproduces the complete carrier, with its attendant sidebands, from the pulses of this Class C final amplifier, causing the output tank to oscillate at its natural resonant frequency.

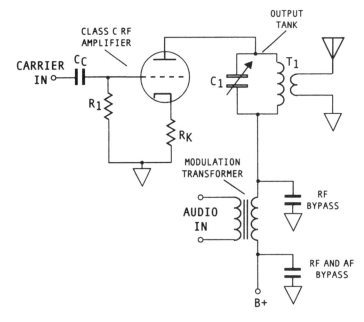

Figure 4-21. *An AM vacuum tube plate modulation stage.*

The audio modulator itself would be a linear single-ended Class A, or a double-ended AB or B push-pull, power audio amplifier. The modulation transformer's secondary is in series with the final's plate supply voltage, with the audio modulation adding to or subtracting from the plate supply, which then varies the amplitude of the output sine wave, creating AM.

To achieve solid and reliable 100% modulation with high-level modulation techniques, both the driver and the final are typically modulated, with both operating Class C. Another common method modulates both the plate and the screen-grid of a pentode vacuum tube to achieve high-efficiency, reliable 100% modulation.

High-level modulation is used in all but the most powerful AM transmitters. With very high-powered transmitters it becomes increasingly difficult to efficiently increase the audio power needed for high-level modulation, necessitating the use of low-level modulation techniques.

4-4. AM Disadvantages

Some disadvantages of AM are as follows. Most of the power is in the carrier, which is not needed to supply the intelligence. The bandwidth is also double what it needs to be, since only one sideband is needed for transmission and reception of the intelligence. The phase relationship between the carrier and the sidebands must be exact; this is hard to maintain under many conditions (if this phase relationship is not sustained, severe fading will result).

As an example of wasted AM power: With 100% modulation and a 100W carrier, the total AM power would be 150W, with 25W in *each* sideband. Even when no speech is occurring, 100W is completely wasted by the carrier, with another 25W wasted when modulation occurs.

Most, if not all, of these problems are corrected with the use of *Single Sideband Suppressed Carrier* (SSSC or SSB). SSB needs only half the bandwidth — the power conserved in not transmitting the carrier and one sideband can be placed into transmitting the single information-carrying sideband signal. Also, since there is less bandwidth with only one carrier, there is less noise received by the SSB receiver. As well, far less fading occurs because no phase relationships must be maintained between the USB, LSB, and carrier. Additionally, an SSB signal is only generated when modulation is present, since the carrier is *always* suppressed at the beginning stages within the transmitter, further reducing wasted power.

CHAPTER 5
Frequency Modulation

FM was created to overcome the limitations of AM, which is essentially that of excessive noise. Noise is generally caused by unwanted amplitude variations, which can be removed by amplitude limiters in the FM receiver.

The method of producing FM can be accomplished by two methods. *Direct FM*, which directly modifies the frequency of the FM carrier in step with the modulation's amplitude variations; and *phase modulation*, which varies the phase of the carrier, creating *indirect FM*. Both methods cause the end effect of modulating the frequency of the carrier, and are both categorized under the term *angle modulation*.

5-1. FM Fundamentals

Modulation is the process of superimposing information on a carrier wave, be it voice, music, etc. Demodulation is the process of removing this information and conveying it to any number of transducers; such as a speaker or headphones for voice and music, or through to digital circuits for processing or storage.

FM performs this process of modulation by varying the frequency of the RF carrier in sync with the changing amplitudes of the modulating (baseband) signal. When the modulated carrier reaches the receiver, these baseband-created frequency variations are stripped off of the RF carrier and converted into audio amplitude variations, then amplified and fed into the appropriate transducer.

5-1A. FM Modulation

In FM the *amplitude* of the modulation signal varies the *frequency*, and not the amplitude as in AM, of the carrier (*Figure 5-1*). As the modulation signal amplitude increases, so does the amount of *frequency deviation* (the amount the carrier deviates from its center frequency in *one direction* during frequency modulation). The frequency of the carrier without modulation, or the *center frequency*, will remain at the transmitter's pre-set carrier frequency until modulated.

When modulation occurs, the carrier increases and decreases in frequency, depending on the amplitude of the baseband signal. As the audio swings positive, the carrier increases in frequency; as the audio modulation swings negative, the frequency of the carrier swings below its rest frequency.

As the *frequency* of the *modulating audio* varies, the rate that the frequency-modulated carrier crosses the rest frequency varies at this same audio rate. For instance, if the modulating audio tone is 3 kHz, then the carrier will swing past its rest frequency three thousand times in one second.

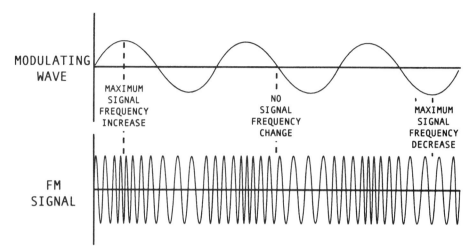

MODULATING WAVE

MAXIMUM SIGNAL FREQUENCY INCREASE

NO SIGNAL FREQUENCY CHANGE

MAXIMUM SIGNAL FREQUENCY DECREASE

FM SIGNAL

Figure 5-1. *The effect of sine-wave modulation on an FM signal.*

The percent of modulation for FM is controlled by government regulation, and not by any natural limitations as it is in AM. 100% modulation in narrowband (voice) communications, for instance, is considered to be around 5 kHz deviation. For wideband (FM broadcast), 75 kHz deviation is the maximum allowed. FM deviation is also linear: If a certain FM deviation is 75 kHz, and is considered to be 100% modulation, then half that, or 37.5 kHz, is 50% modulation.

If the amplitude of the baseband signal causes the FM deviation to go above 100%, more sidebands are produced, which increases bandwidth, causing possible interference with adjacent channels.

When frequency modulation is viewed on an oscilloscope the carrier can be seen not to vary in amplitude when modulated, but merely in frequency (*Figure 5-2*). This is evidenced by a shortening and lengthening of the signal's wavelength, causing a blur on the display due to the rapid frequency changes. (Since wavelength equals the speed of light divided by the frequency, or $\lambda = C/f_r$, then any change in wavelength directly corresponds to a change in frequency.)

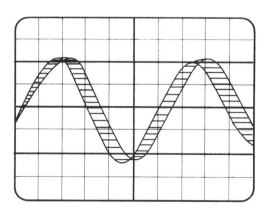

Figure 5-2. *Frequency modulation as seen in the time-domain.*

5-1B. Output Power and Significant Sidebands

The combined power or voltage in a frequency-modulated signal does not change when modulated or unmodulated. FM transmitter power always remains constant during modulation. Any sidebands produced must obtain all of their power from the carrier, which then must obviously lose power in the creation of these sidebands. For example, when an FM transmitter is unmodulated, a certain carrier is 100W. Therefore, when this RF carrier is modulated, the carrier must give up some (or all, depending on sideband amplitude and number) of its power to these sidebands; and the carrier and its *significant sidebands* must all add up to a total of 100W.

During this modulation process, an infinite number of sidebands are generated. This is because the carrier is shifted through an infinite number of sundry frequency (or phase) values by the constantly varying modulating frequency, consequently producing an infinite number of sideband frequencies (the amplitude of even a single test-tone frequency will, of course, vary in a sinusoidal manner with an *infinite* amount of amplitude variations throughout a single cycle).

Considering *infinite* is an impossible concept both mathematically and in practice, the idea of *significant sidebands* must be used. These significant sidebands are simply any sidebands that have an amplitude that is 1% or more of the unmodulated carrier amplitude. Any sideband that is less in amplitude is not considered to be of consequence.

At certain *modulation indexes*, which is the ratio between the carrier's instantaneous frequency deviation divided by the instantaneous frequency of the modulation (or

$$\frac{FREQUENCY\ DEVIATION}{AUDIO\ MOD\ FREQUENCY}$$

the sidebands will contain all of the energy, and the carrier will disappear completely (*carrier null, Figure 5-3*). The greater the amplitude of the modulation the larger the number of significant sideband frequencies generated, while the sidebands are spaced on each side of the carrier, and from themselves, by an amount equal to the frequency of the single-tone modulation signal, as shown in *Figure 5-4*. But, unlike AM, more than one pair of sidebands are produced for each single-tone modulation (*Figure 5-5*).

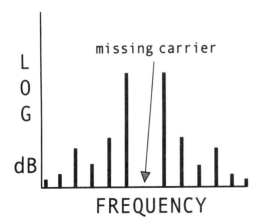

Figure 5-3. *An FM carrier null occurs at certain modulation indexes.*

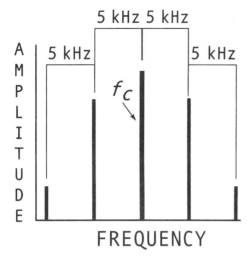

Figure 5-4. *Single-tone modulation sideband spacing from the FM carrier.*

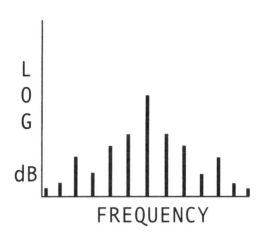

Figure 5-5. *Multiple sidebands produced by a single-tone baseband signal in FM.*

By using the chart of *Table 5-1*, the amplitude and number of significant sidebands can be found. Since the modulation index equals the frequency deviation divided by the frequency of the modulation, by simply calculating this index, the number and amplitudes of all significant sidebands, as well as the carrier amplitude, can be easily found. Using the calculated modulation index, look under the table's *Modulation Index* column and then across. The relative amplitudes of the carrier and each sideband, as well as their number, are displayed (any numbers with the negative sign means that a phase reversal has occurred).

To find the *bandwidth* of an FM signal, multiply by *two* the number of significant sidebands by the maximum modulating frequency (BW = 2N · $f_{m(max)}$).

FM frequency swing and bandwidth are quite different. FM *bandwidth* and sideband generation is a *Bessel function*, which is a complex mathematical process, while frequency *swing* is strictly a matter of the amount of frequency movement caused by the modulation.

MODULATION INDEX	CARRIER	PAIRS OF SIDEBANDS TO THE LAST SIGNIFICANT SIDEBAND							
		1ST	2ND	3RD	4TH	5TH	6TH	7TH	8TH
0.00	1.00	--	--	--	--	--	--	--	--
0.25	0.98	0.12	--	--	--	--	--	--	--
0.5	0.94	0.24	0.03	--	--	--	--	--	--
1.0	0.77	0.44	0.11	0.02	--	--	--	--	--
1.5	0.51	0.56	0.23	0.06	0.01	--	--	--	--
2.0	0.22	0.58	0.35	0.13	0.03	--	--	--	--
2.5	-0.05	0.50	0.45	0.22	0.07	0.02	--	--	--
3.0	-0.26	0.34	0.49	0.31	0.13	0.04	0.01	--	--
4.0	-0.40	-0.07	0.36	0.43	0.28	0.13	0.05	0.02	--
5.0	-0.18	-0.33	0.05	0.36	0.39	0.26	0.13	0.05	0.02

Table 5-1. *Carrier and sideband amplitudes at different modulation indexes.*

A modulation index of 0 is a unmodulated CW carrier, and produces no sidebands (*Figure 5-6*). As the modulation index increases, the sidebands begin to take up more bandwidth, as shown in *Figure 5-7* for a modulation index of 1.5. This is why the frequency of modulation and its amplitude must be controlled in FM communications to lower bandwidth requirements and adjacent channel interference. The two-way (narrowband FM) radio modulation index is typically kept to a maximum of 2 (or 5/2.5) or under, with a maximum allowed frequency deviation of around 5 kHz and a maximum audio modulation frequency of approximately 2.5 kHz — a bandwidth of between 12 and 20 kHz is normally considered adequate to pass sufficient sideband power for full voice intelligibility. In FM broadcast, or wideband communications, maximum deviation is 75 kHz, the maximum audio frequency is 15 kHz, with a deviation ratio of 5 (or 75/15). Maximum bandwidth is 200 kHz, with two 25 kHz guard bands on either side of its full 150 KHz frequency swing to minimize adjacent channel interference.

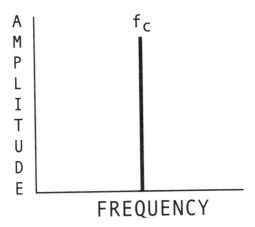

Figure 5-6. *An unmodulated FM (CW) carrier (modulation index of 0).*

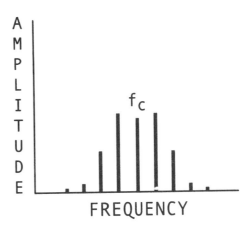

Figure 5-7. *An FM signal with a modulation index of 1.5.*

5-1C. FM and AM Comparisons

There are many advantages that FM has over AM. The most important are: Much better noise immunity due to limiting employed to remove any AM noise on the received FM signal, as well as the use of pre-emphasis (boosting the higher frequencies at the transmitter) and de-emphasis (attenuating these overemphasized frequencies at the receiver) to lower high-frequency noise components. Furthermore, because of the *capture effect*, most interfering signals near or at the same frequency as the desired signal are rejected. Because there is no delicate modulation envelope as in AM, FM does not need linear amplifiers, except in the audio stages, but can use Class C high-efficiency amplifiers. Also, FM can be modulated by low-level modulation techniques, needing very little modulation power (only a few fractions of a watt of audio power is needed to modulate an FM transmitter) — this makes for much better transmitter efficiency.

However, FM does have its disadvantages. Wider bandwidth is needed due to increased sideband production over that of AM. More elaborate and expensive transmitters and receivers are required, mainly on account of the typically higher-frequency operation and increased stability requirements over that of AM (use is ordinarily restricted to frequencies above 30 MHz because of FM's wide bandwidth requirements). Also, unless operated through repeaters, FM is a short range, line-of-sight communications medium, since skywave propagation (bouncing the signal off of the ionosphere) causes distortion of the FM wave.

Some important terms to remember:

Center frequency: The frequency of the RF carrier with zero percent modulation. Also called the *rest frequency*.

Frequency deviation: The amount the carrier deviates from its center frequency in *one direction* during frequency modulation.

Frequency swing: The total deviation of the modulated carrier on *both* sides of the center frequency during frequency modulation.

Modulation index: The ratio between the carrier's instantaneous frequency deviation divided by the instantaneous frequency of the modulation. Used when a single tone at a constant deviation is transmitted.

Deviation ratio: The ratio between the maximum frequency deviation (100% modulation), divided by the maximum audio modulation frequency.

Capture effect: An FM receiver seizes and amplifies the stronger FM signal and rejects another weaker signal at the same or nearby frequencies.

5-2. FM Receivers

Modern FM receivers are usually based on the superheterodyne principle (*Figure 5-8*), and contrast little from AM receivers — except in their method of demodulating and limiting (removing any AM variations of) the incoming FM signal.

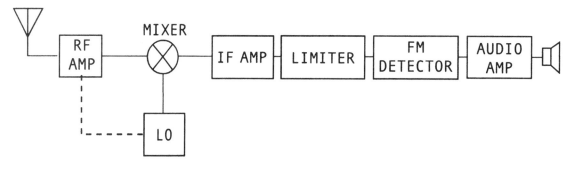

Figure 5-8. *A basic FM superheterodyne receiver.*

Furthermore, the use of de-emphasis at the receiver, and the initial pre-emphasis at the transmitter, is also typically found in most FM systems and not encountered in any other form of communications equipment. It is employed to increase the higher-frequency but low-in-power audio signals. This allows the audio signals to overcome any interfering noise voltages.

Entire FM receivers are now commonly available on a single chip (*Figure 5-9*). This IC lacks only certain support components and an antenna, an audio power amplification stage, and a transducer (speaker or other device).

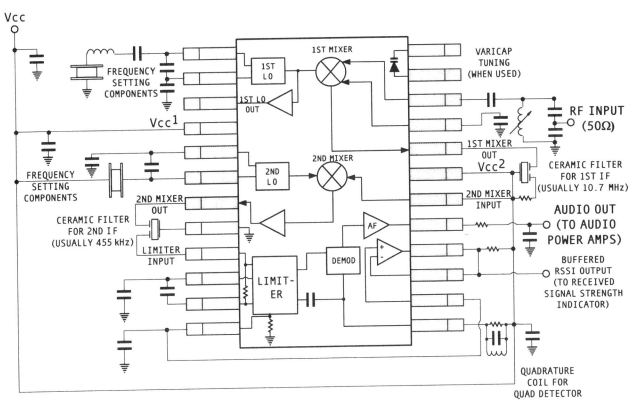

Figure 5-9. *Nearly an entire FM receiver on a single chip.*

5-2A. *Detectors*

Unlike AM receivers, which detect and amplify the change in amplitude of an incoming signal, a technique of detecting a change in frequency and converting this into a change in amplitude, for reproduction by the speaker or headphones, must be used in FM receivers. The older methods used for this purpose, such as the Foster-Seeley discriminator and the slope, double-tuned, and ratio detectors, are rarely utilized today — more modern techniques have replaced them.

An FM detector works as such: When a varying (modulated) IF is inserted into the demodulator, the output is a varying audio voltage, which is then amplified to drive the speakers. However, when the IF is unmodulated, and thus at its center, or rest, frequency, no varying output will occur, consequently creating no amplitude output into the receiver's speaker.

Most FM detectors are now usually in IC form. The most common IC detector designs used today are the *phase-locked loop demodulator*, the *pulse-averaging discriminator*, and the *quadrature detector*.

The *phase-locked loop demodulator* (*Figure 5-10*) has become much more common due to the availability of low-cost PLL integrated circuits. The PLL receives the FM intermediate frequency into its phase detector, which then compares this incoming signal with the output of the VCO. Since the IF is constantly varying in frequency, if frequency modulation is being received, an "error" voltage, equivalent to this frequency deviation, is output from the detector's low-pass filter and DC amplifier as a demodulated audio signal. The higher the frequency deviation, the higher the error voltage output, and the louder the subsequent output audio from the receiver's speaker. A phase-locked loop demodulator needs little aligning, no exterior inductors and few bias components, and has very low distortion and a great S/N ratio.

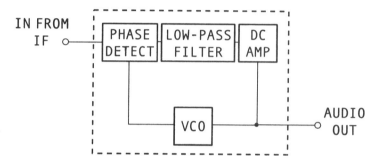

Figure 5-10. *An FM phase-locked loop demodulator.*

The *pulse-averaging discriminator* is a top-of-the-line IC demodulator (*Figure 5-11*). The frequency-modulated IF enters the first stage of the discriminator, which is an overdriven amplifier used as a limiter, outputting rectangular and constant amplitude pulses that are equal in frequency to the incoming FM IF. These pulses are then fed into a *monostable multivibrator* (a type of non-sinusoidal oscillator that outputs a single preset duration and amplitude output pulse for each pulse at its input), which outputs its pulse into a low-pass filter. The higher the FM IF frequency deviation (or, to put it another way, the higher in amplitude the original modulating signal), the more pulses that are output per second, and the higher in voltage C_1 charges up to. Because the load is in parallel with C_1, the load will also have this higher voltage across its terminals. The lower the FM IF frequency deviation (the lower in amplitude the original modulation), then the more time C_1 has to discharge across the load, and the lower the voltage will be across that load.

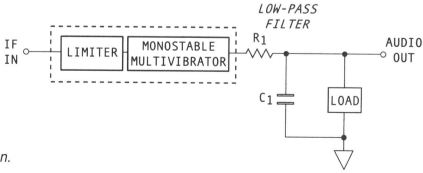

Figure 5-11. *A pulse-averaging discriminator for FM demodulation.*

In a *quadrature detector* (*Figure 5-12*), while one of the most difficult FM detectors to understand, is one of the most common. The first stage of a quadrature detector is usually a limiter (which removes AM from the FM signal, lowering noise output), followed by a phase detector (some quadrature detector ICs do not have a separate limiter, but use a phase detector that has been overdriven, thus easily driving it into saturation and cutoff), and out to a low-pass filter.

The "quad" detector functions as follows: The limited FM IF is fed into the input of the phase detector, which acts as an AND gate. When both inputs are high, there is an audio output, if one or more is low, then there is no audio output. In other words, both inputs must be of the same phase (same frequency) for any audio output to occur. If there is no modulation present in the IF, then both inputs will be 90° out-of-phase with each other; thus there will be zero output. The phase shift between the inputs occur due to the reactance of the tuned circuit, consisting of the tunable quad coil and C_2, and its phase relationship with C_1. The tuned circuit is of a very high resistance with a non-reactive nature at resonance (which is at the FM rest, or unmodulated, frequency), and the low value C_1 is, being almost purely capacitive and of a very high reactance to the IF, at a 90° phase shift when compared to input "1", which has a 0° phase shift. As frequency modulation occurs in the IF the phase relationships between the inputs change. This is because the tuned circuit goes off resonance due to this change in frequency, which alters the reactance of the tank, changing the phase relationship between the tank and C_1 and, consequently, input 2 as compared with input 1.

The output is a series of pulses into a low-pass filter, which vary in width depending on the frequency swing of the frequency-modulated IF. The pulses are averaged and the resultant audio is output across to the next stage.

Tuning for rest frequency is accomplished by tweaking the external adjustable quadrature coil.

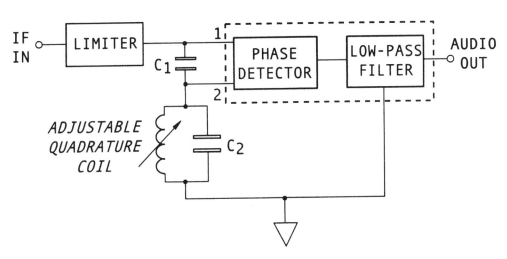

Figure 5-12. An FM quadrature detector.

5-2B. Limiters

A separate limiter can be used in front of any detector that does not have built-in limiting. The limiter is utilized to remove most of the noise-generating AM components (*Figure 5-13*), or *incidental AM.* Two limiter stages are normally found in higher-quality receivers.

A limiter is basically an overdriven Class A or C single-ended amplifier with a tuned output circuit, used after the last IF stage of an FM receiver, to clip off any amplitude variations in the received signal. The limiter cuts off the amplitude peaks of the signal, in some cases creating a ragged square wave. The signal is turned back into a sine wave by the tank circuit, which filters out the harmonics caused by the clipping and recreates the original sine wave.

Figure 5-13. *Types of noise: (a) Noise spikes and; (b) noise riding on FM waveforms.*

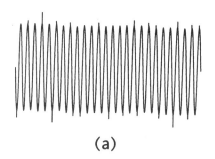

(a)

(b)

A simple tank circuit itself in Class C amplifiers can also act as a limiter by removing many RF or IF amplitude variations — if the Q of the tank is high. A high-Q tank will be narrow enough to pass the desired frequency variations and reject most amplitude variations, which are riding on the FM waveform, as being outside of its bandwidth, thus removing most of the incidental AM (*Figure 5-14*). Also, another way to think of the effect a high Q-tank has on amplitude variations is that the tank will not lose enough energy during flywheeling in time for the next pulse, which may possibly decrease in amplitude. This amplitude variation will not be noticed because the high-Q tank continues to "ring", so any lowering of signal amplitude will not be output to the next stage.

Figure 5-14. *Effect of a high-Q amplifier on AM component.*

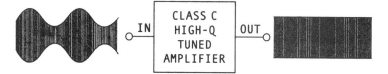

5-3. FM Transmitters

FM transmitters (*Figure 5-15*) can be significantly different in design than the average AM transmitter: Low-level modulation is always used with, commonly, Class B or C frequency multipliers, employed even after modulation — and most FM transmitters are designed to function only in the VHF and higher area of the radio spectrum.

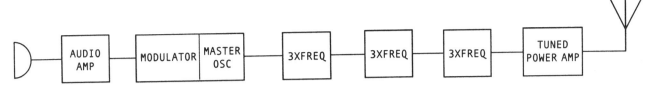

Figure 5-15. *A basic fixed-frequency FM transmitter.*

5-3A. Modulation Circuits

An FM transmitter must use some technique to vary the output frequency in response to amplitude variations in the baseband signal, while allowing little or no amplitude variations in the carrier. There are two basic ways to perform this function: *direct*, or frequency modulation, which directly changes the frequency of the carrier oscillator in proportion to the amplitude of the baseband signal; and *indirect*, or phase modulation, which indirectly creates FM by varying the phase, and thus the frequency, of the carrier oscillator's output frequency.

Direct modulation can be created by the use of a *varactor modulator*. Varactor modulators use varactor diodes to vary the frequency of an oscillator to create FM. Varactors, also referred to as varicaps, work on the same principle as the junction diode. When a reverse bias is placed across a semiconductor diode, the depletion region widens. Think of this region as the dielectric, since it acts as an insulator. The thicker the dielectric, the less capacitance. So, increasing reverse bias on any semiconductor diode decreases its capacitance, and decreasing the reverse bias decreases the depletion region (equivalent to the plates of a capacitor being placed closer together) and increases capacitance. A varicap is a diode that is optimized to perform this very important function.

A common FM modulator is the *Pierce varactor oscillator* as shown in *Figure 5-16*; but the same idea can be applied to any *Hartley* or *Colpitts*-based oscillator. Resistors R_1 and R_2 set the Pierce varactor's reverse bias, and thus the oscillator's center, or rest, frequency. When an audio signal is applied at the oscillator's input port, it adds to or subtracts from the varactor's reverse bias, which changes the capacitance of the circuit, altering the modulator's center frequency (FM).

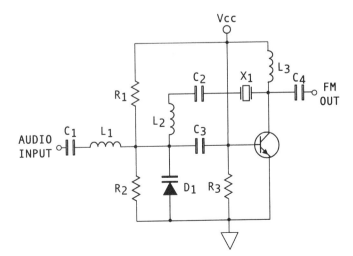

Figure 5-16. *A Pierce varactor (oscillator) modulator for FM production.*

A varicap that is in series or parallel with an oscillator's crystal can only deviate the center frequency by a few hundred hertz, and is therefore used mainly in narrow-band FM applications. Still, to increase both the frequency and the value of deviation, one or more stages of frequency multiplication may be utilized.

Reactance modulators (*Figure 5-17*) can also be employed to generate direct FM. They are used to vary the frequency of the master oscillator (MO), but cannot typically be used in *crystal controlled* MO's.

Reactance modulators are placed across the LC frequency-determining tank of the master oscillator. An audio signal is then fed into the reactance modulator, which is biased so that it appears to be *either* a purely capacitive or, in certain circuits, as a purely inductive reactance to the MO's tank (a reactance modulator must be designed to be *purely* reactive, or it will introduce AM components that, when transmitted, cause noise at the receiving end). This audio signal varies the current through either a transistor, FET, or tube of the reactance modulator. As the current of the active element increases and decreases, the reactance of the active element changes, varying the MO's tuned circuit reactance in equal proportion, thus altering the output frequency of the MO around its resting frequency, creating direct FM.

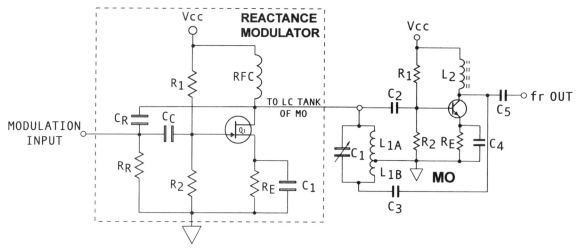

Figure 5-17. A reactance modulator for direct FM production.

The reactance modulator is capable of very wide frequency deviations when used in LC oscillators, as well as having inherently very little distortion. However, since it cannot control a crystal MO, it must use other frequency-stabilizing circuits, such as *AFC* (*Automatic Frequency Control, Figure 5-18*).

FM transmitter AFC inspects the output from the frequency multipliers (or the output directly from the MO), and sends this sample into a mixer, comparing it with a crystal oscillator reference frequency. If the carrier frequency is found to be off of its design frequency, the mixer will output the difference between the two frequencies into the discriminator. The discriminator will then output a correction voltage (since this new frequency will be different from its preset rest frequency) into the reactance modulator, changing its reactance, which forces the MO back on frequency.

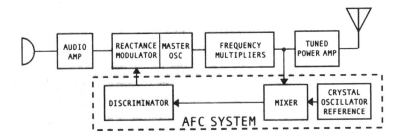

Figure 5-18. *A type of automatic frequency control (AFC) employed in some non-crystal FM transmitters.*

A phase-locked loop can also be used to generate direct FM (*Figure 5-19*), and is employed in many of the newer frequency-synthesized transceivers. The PLL produces and maintains a rock steady rest frequency, while any modulation injected at the input to the voltage-controlled oscillator (VCO) varies the PLL modulator's output frequency in step with the modulation voltage, thus creating FM.

As stated, phase modulation (PM), or indirect FM, can be used to produce frequency modulation. In many FM applications it is employed more than direct FM because of PM's ability to utilize a crystal oscillator or frequency synthesizer as the MO, thus giving the FM transmitter the ability to maintain a very accurate and steady output frequency — especially in narrowband use. PM has limited frequency deviation abilities, however, when compared to a reactance modulator, so it is ordinarily used with large amounts of frequency multiplication.

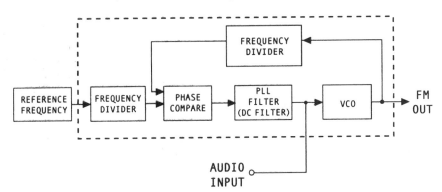

Figure 5-19. *An FM phase-locked loop modulator.*

With PM, the *output* of the transmitter's MO (*Figure 5-20*), or its buffer, is modulated, and not the input, as in direct FM. This allows the MO to be crystal or PLL controlled, as well as protecting the MO from any deleterious effects of the modulation — resulting in an extremely stable oscillator. The phase of the RF carrier from the MO or its buffer is changed, creating a leading and lagging time difference between the resting carrier frequency and the modulated carrier frequency; which is the same as varying the frequency of the carrier.

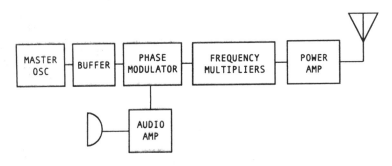

Figure 5-20. *A phase modulated (indirect FM) transmitter.*

CHAPTER 5

In *Figures 5-21 and 5-22* two simple PM circuits are shown. Recall that an RC network can alter the output phase angle of any input signal, inasmuch as a pure resistance has a phase angle of 0°, while a pure capacitance has a phase angle of -90°; and combining the two in a circuit will create a phase angle anywhere between 0° and -90°, depending on the values of the resistance and the capacitive reactance.

In *Figure 5-21* the output will lead the input by any value between 0° and -90° by varying the ratio between X_C and R. If, by the application of a modulating voltage, either the value of the resistance or the capacitive reactance could be changed, then phase modulation would take place. This can be accomplished by replacing the resistor with an FET or a bipolar transistor (*Figure 5-22*), which acts as a voltage or current-controlled variable resistance. As the modulating signal amplitude increases and decreases at the input, the transistor's resistance decreases and increases respectively, varying the ratio between the resistive element (the transistor) and the reactive element (the capacitor), thus changing the phase angle of the output.

Since the output is also amplitude-modulated due to the varying audio bias on the gate of the active component (the transistor), the stages after the modulator are biased saturated Class C, which removes any generated AM components (See *Limiters*).

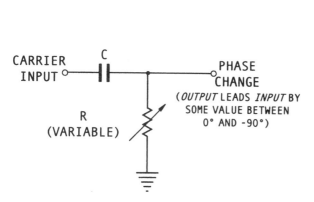

Figure 5-21. *Principle of phase modulation.*

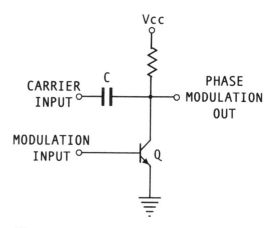

Figure 5-22. *A basic transistor phase modulator.*

A practical FET based phase modulator is shown in *Figure 5-23*, with the audio input injected into the low-pass filter consisting of R_1 and C_2. This filter is used to remove PM's inclination to increase the carrier's frequency deviation as the frequency of the modulating signal increases. The MO rest frequency is placed at the gate of the FET, and through C_3 and across the tank, which consists of L and C_5. Through the interaction of C_3, the tank, and the FET, PM, as well as AM, is created. The amplitude modulation is removed by Class C amplifiers or frequency multipliers placed after the modulator.

A varactor diode and a resonant tank can also be used to modify the phase of the output of the MO (*Figure 5-24*). When the audio amplitude across a varactor diode changes, the capacitance of the varactor, and thus the resonant frequency of the tank, is varied, creating phase modulation. The tank consists of L_1, C_3, and the varactor, D_1. The varactor's bias is then added to or subtracted from by the modulating audio signal while the carrier is sent through C_4 and into the tank. The tank, due to this changing bias on the varactor, alters its capacitance, and thus the resonant frequency of the tank circuit. When the tank is on frequency, then the circuit is purely resistive with no reactive components, since both X_C and X_L cancel each other at resonance — with a resultant phase angle of zero. When the audio signal across the diode is changed the entire reactance of the tank is altered, and will become either capacitive (if the frequency decreases), or inductive (if the frequency increases), producing lagging and leading phase shifts.

The bias dropped across R_2 sets the diode's quiescent (rest) reverse bias, which fixes the tank's resonant frequency to that of the unmodulated carrier. R_1 and C_1 form a low-pass network, which removes PM's proclivity to increase the output frequency swing for the higher audio modulation frequencies, while the RFC is used to minimize the loading of the tank, which would decrease its Q. This phase shifter is also followed by either Class C amplifiers or frequency multipliers to remove any amplitude variations produced by the phase-shifting.

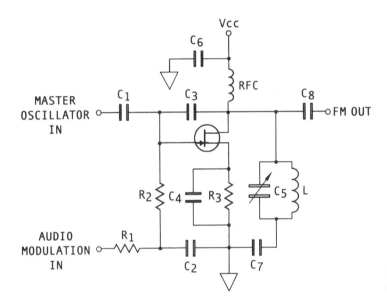

Figure 5-23. A practical phase modulator using an FET.

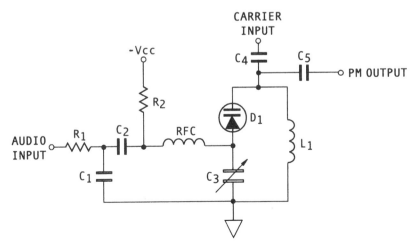

Figure 5-24. A varactor phase modulator.

CHAPTER 5

5-3B. Frequency Multipliers

Frequency doublers and triplers, used extensively in FM transmitters, are typically basic tuned output non-linear (Class B or C) amplifiers. They are utilized to multiply the frequency-modulated carrier obtained from a stable crystal oscillator (the master oscillator) that would have difficulty working dependably at such high frequencies.

As well, multipliers are able to increase the deviation of any frequency modulation present. A carrier oscillator's frequency and its deviation might be multiplied by thirty or more times, as needed. A carrier frequency that started out at 6 MHz with an FM deviation of 150 Hz would be changed, if multiplied by a final 30X chain of multipliers, into an output frequency of 180 MHz with a deviation of 4,500 Hz.

Figure 5-25 shows a common Class C amplifier, run into its saturation curve, functioning as a frequency multiplier. Distortion of the fundamental is created by the non-linear operation of the Class C amplifier, thus generating a signal that can be rich in harmonics. The input tank is tuned to the fundamental, while the output tank is tuned to an exact harmonic of this frequency. The output tank not only passes the chosen harmonic on to the following stage, but also attenuates the now unnecessary fundamental and any amplitude variations (AM).

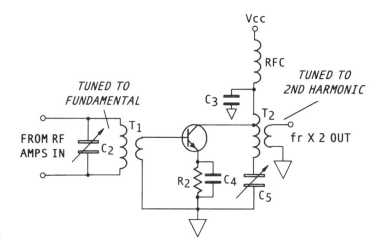

Figure 5-25. *A Class C frequency multiplier.*

Another type of frequency multiplier is the *push-push doubler* (*Figure 5-26*). One FET receives a 180° out-of-phase signal input as compared to the other FET — yet their outputs are tied together (in parallel), so twice as many pulses occur into the output tank circuit as would occur in a normal push-pull circuit. Because the tank is tuned to the second harmonic, a frequency multiplication then takes place.

Yet another common method of frequency multiplication is the *tripler varactor frequency multiplier* (*Figure 5-27*). As the input signal is coupled into the input tank, which is tuned to the fundamental frequency, the varactor is successively switched on and off, which distorts the input signal, creating harmonics. The 2nd harmonic is sent to ground through the series bandtrap filter, while the 3rd harmonic is passed on to the secondary of the 3X fundamental tuned tank circuit.

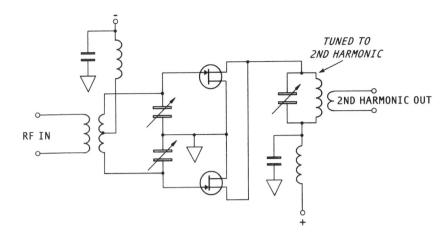

Figure 5-26. *An FET push-push frequency doubler.*

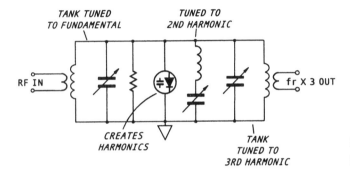

Figure 5-27. *A varactor frequency multiplier.*

Any component that can function in a non-linear area of its operation, even a simple signal diode (*Figure 5-28*), may be used in a multiplier string.

Active frequency multiplier amplifiers, despite their high frequencies, do not typically need neutralization, since their input and output circuits are tuned to different frequencies.

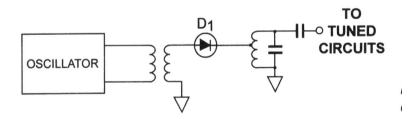

Figure 5-28. *A basic small signal diode frequency multiplier.*

5-4. Selective Calling

Selective calling is a method that can be employed in FM to keep all receivers in a network "off" until it is desired to contact all or some of these receivers. This function can be performed by the use of *CTCSS*, or *Continuous-Tone Coded Squelch System*.

Many FM transceivers use *Channel Guard* and *Private Line*, which are proprietary selective calling systems used to unsquelch the receiver's audio section by the caller's transceiver transmitting a continuous subaudible tone on the RF carrier below 300 Hz, which is under the passband of the receiver's audio amplifiers.

Figures 5-29 and 5-30 show the general idea of CTCSS. At the transmitter, the subaudible tone, at a very low frequency (a few hertz, depending on frequency multiplication), is injected, along with the audio modulation, into the modulator, where it is sent to the frequency multipliers and then to the finals and out to the antenna, where it is radiated. The receiver picks up the original audio as well as the subaudible tone. The audio is sent on to the audio amplifiers and speaker, while the subaudible CTCSS tone is tapped off and sent to a bandpass filter to separate its frequency from any other tones present, then to a rectifier and low-pass filter, where it is changed into a DC control voltage to switch off the squelch, which activates the audio amplifiers. The receiver operator is now able to listen to the open channel for as long as the CTCSS is being transmitted, which will be for as long as the transmitter operator is pressing the PTT switch.

Another type of selective calling uses bursts, a pulsed tone, or a combination of tones, usually sent only at the beginning of a transmission, at frequencies anywhere between 600 to 3,000 Hz (depending on the system), to open the squelch and allow reception of the FM signal.

Still another selective calling method, referred to as DPL (*Digital Private Line*) and DCS (*Digitally Coded Squelch*), are a non-CTCSS mode of Private Line. With these systems a digital word is subaudibly sent by the transmitter every time transmission is chosen. The digital word is then recognized by the onboard ROM in the DPL or DCS receiver. The receiver compares this word with the word stored on-board, and unsquelches if it is found to be correct.

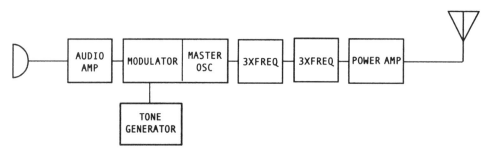

Figure 5-29. *CTCSS tone generation in a transmitter.*

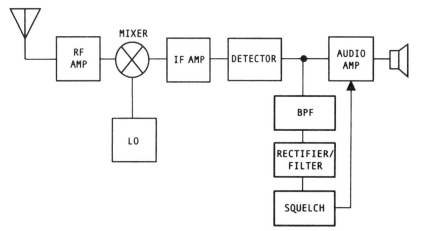

Figure 5-30. *Detection of the CTCSS tone in a receiver.*

Before transmitting with any type of selective calling equipped transmitter, the mic is taken off-hook to listen for channel traffic (the radio automatically sets itself to normal squelch mode). If using a hand-held, its squelch can be turned to *Carrier Squelch*, which will allow the CTCSS receiver section to open squelch if any strong signal is received. In this way, the operator will not interfere with any ongoing radio traffic, and transmission can then proceed as with any transceiver.

A few things must be kept in mind when working on a CTCSS-equipped transceiver. When aligning or testing an FM transmitter section, the subaudible tone will be modulating the carrier by about 20%, whether or not there is any speech modulation present, confusing any desired readings if it is not expected. Some CTCSS produce an FM deviation of approximately 750 Hz, thus allowing a maximum narrow-band deviation of the speech component of only ±4.45 kHz. Also, when testing a CTCSS receiver, a CTCSS tone generator (or *CTCSS encoder, or toner*) must be used to open the squelch (unless there is a switch to turn CTCSS on and off), or the receiver might appear dead.

5-5. Pre-Emphasis and De-Emphasis

Pre-emphasis and *de-emphasis* is a technique employed in most FM systems to lower noise as heard in the receiver.

Pre-emphasis (*Figure 5-31*) overamplifies the higher audio frequencies in a stage just before modulation in the transmitter. Generally a simple high-pass filter is used. Since higher-frequency components have less energy in speech and music than lower-frequency components, the higher frequency will now not be masked by either internally or externally generated noise.

At the receiver, de-emphasis (*Figure 5-32*) is performed on the audio, right after the demodulator, which attenuates the high frequencies created by the pre-emphasis back to natural levels. This also, of course, significantly lowers the amplitude of any high-frequency noise present in the signal.

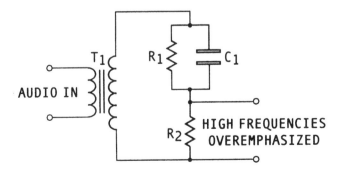

Figure 5-31. *A pre-emphasis circuit in an FM transmitter.*

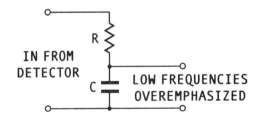

Figure 5-32. *A de-emphasis circuit in an FM receiver.*

5-6. FM Transceivers

An entire no-frills FM handheld transceiver is shown in *Figure 5-33*. Most of the circuits of this chapter, as well as many of the support circuits explained in Chapter 11, are in this block diagram.

For transmission, the *PTT* (*Push-To-Talk*) button is depressed, supplying power to the transmitter section and removing power from the receiver. As the operator speaks into the microphone, the transmitter's 1st audio voltage amplifier increases the very low audio voltage, which is then further boosted by the 2nd audio amplifier while also furnishing amplitude limiting (which can consist of two back-to-back diodes shunting any high amplitude audio to ground). The low-pass filter (LPF) stage supplies frequency limiting and removes any harmonics created by the prior amplitude limiting. Maximum frequency deviation and significant sideband generation and spacing, and thus bandwidth, are consequently restricted.

This limited audio is fed into the varactor modulator, where the bias on the varactor, and thus the frequency of its oscillator, is varied by the audio voltage. The oscillator's output tank can be tuned to some multiple of the varactor's crystal oscillator (See *Pierce varactor oscillator*) to increase its output frequency (frequency multiplication). Its output frequency can be varied by the channel select switch, which allows the operator to switch in different frequency-determining crystals — or a PLL can be used. This frequency is output into a frequency tripler amplifier to increase the carrier frequency, as well as any deviation, and then sent into a buffer amplifier. The buffer amplifies, filters, and isolates the varactor modulator's oscillator and tripler amplifier from the next stage, the transmit mixer. This frequency, as well as the offset oscillator's frequency, is mixed in the transmit mixer, creating sum and difference frequencies.

The offset oscillator is able to choose from three crystals, switched in by the operator, which allows the transmitter to be either at the same frequency as the received signal (referred to as *simplex operation*, when communication is directly from transceiver to transceiver), or to offset the transmit frequency from the receive frequency by ±600 kHz for *repeater* utilization (a repeater receives on one frequency and retransmits your signal on another).

The sum of the mixing products is then filtered by the BPF (bandpass filter) and fed into the predriver, the driver, and the power amplifier.

The output of the power amplifier is sent into the diode RX/TX selector, which consist of diodes that are forward-biased by a DC voltage, allowing the transmitter's signal to pass, while other diodes are reverse-biased into the receiver section, disconnecting the receiver on transmit.

The LPF and *wavetrap* further attenuates all harmonics, as well as assuring that the most powerful harmonic, the second, has added attenuation.

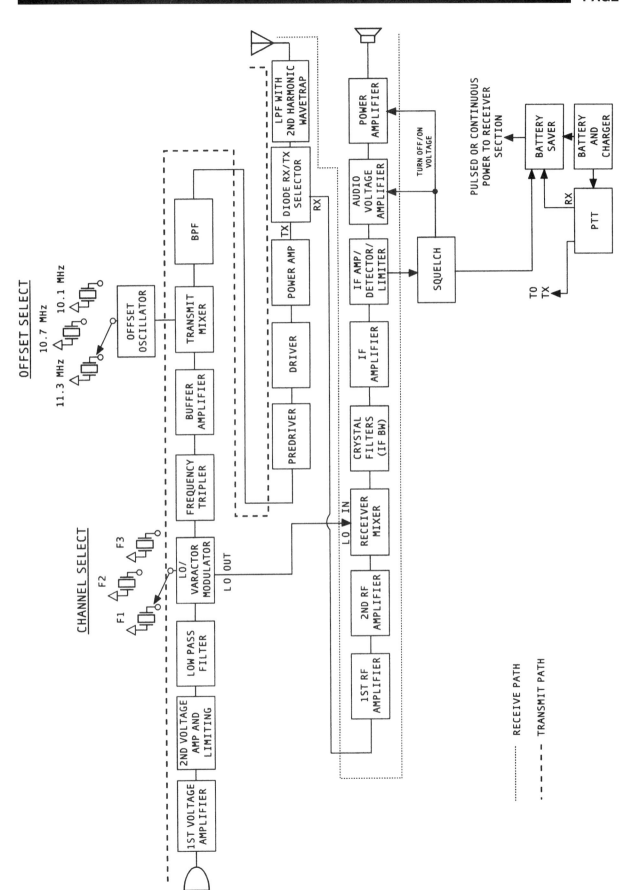

Figure 5-33. *A complete FM hand-held transceiver.*

When the PTT button is released, the receiver section is supplied with power, while power to the transmitter section is removed. The received signal passes through the LPF and wavetrap filter, which protects the receiver from overload due to strong adjacent UHF signals. The receiver's power supply forward biases other diodes in the diode RX/TX selector, allowing the received signal to pass through to the 1st RF amplifier, while blocking any signal from entering the transmit section.

The high-frequency 1st and 2nd RF amplifiers provide a limited amount of gain and selectivity. The amplified RF signal is sent into the receiver mixer, where it mixes with the output of the combined local oscillator/varactor modulator. The difference frequency of 10.7 MHz is filtered and narrowed by the IF crystal filters and sent into the IF amplifiers (which, in a double-conversion VHF receiver, might have a first IF of 45.15 or 43.1 MHz, and a second IF of 450 kHz), then into the combination IF amp/detector/limiter, where the IF is further amplified as well as limited to remove AM noise, and demodulated.

There are two outputs from the amp/detector/limiter — one to the audio amplifiers and one to the squelch section. The squelch section compares the incoming signal level with a preset reference voltage to discern if the signal is of an adequate amplitude to turn on the receiver's audio amplifiers. If the signal is of sufficient amplitude the audio amplifiers are turned on — and the detected audio is amplified and fed to the speaker for reception by the listener.

An additional circuit, the *battery saver* stage, rations battery power during periods when no or only low-level signals are being received. It pulses the receiver on and off five to ten times a second until the squelch circuit detects that a sufficient signal has been received, at which time the power becomes continuous.

CHAPTER 6
Single-Sideband Suppressed-Carrier

As lower cost transmission and receiving equipment for the generation and reception of single-sideband suppressed-carrier signals have become available it has turned into the preferred method for long-range communications in specialized and amateur applications, and is used for voice signals all the way from low frequency up to UHF.

6-1. SSB Fundamentals

Single-sideband suppressed-carrier is a type of amplitude modulation that transmits only a single sideband, while the carrier and the other sideband are heavily attenuated (*Figure 6-1*).

This is a huge improvement over standard AM, in which the carrier, which contains no information and is thus useless, and the two sidebands, each containing duplicate information, are transmitted.

SSB also needs only half the frequency spectrum as standard AM for transmission, since one sideband is suppressed. The power conserved in not transmitting the carrier and one sideband can be better utilized in transmitting the single-sideband signal. This also means less noise is received due to this decrease in bandwidth (*Figure 6-2*). Fading is less of a problem because complex phase relationships between all of the transmitted elements (both sidebands and the carrier) need not be maintained, considering only one sideband is being transmitted. As well, a transmitted signal is only generated when modulation is present, as the carrier is suppressed within the transmitter, saving even more power.

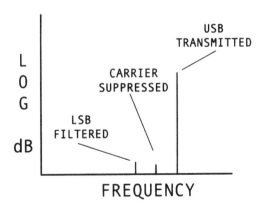

Figure 6-1. *The frequency spectrum of an SSB signal.*

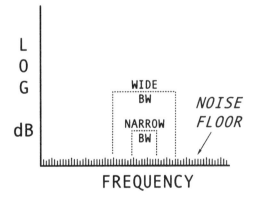

Figure 6-2. *A narrower SSB bandwidth over standard AM equals less received noise.*

6-1A. SSB Modulation

Single-sideband is produced at the SSB transmitter (*Figure 6-3*) by injecting a carrier and a modulating audio signal into a *balanced modulator*. The balanced modulator mixes these inputs, which results in lower and upper sidebands, while suppressing the carrier through phase cancellation or common-mode rejection techniques. The balanced modulator then outputs the resultant double-sideband suppressed-carrier (DSB) signal to the next stage.

Keep in mind that the original carrier must mix with a modulating signal to produce any side-bands. Without modulation present there will be no output — a sideband must be created for SSB transmission, since the carrier itself is severely attenuated within the balanced modulator.

The next stage of the SSB transmitter can consist of highly selective filters to pass either the upper or lower sideband or, through phase-cancellation methods, attenuate the undesired sideband using twin balanced modulators and phase-shifter circuits. The SSB signal can then be amplified and sent out through the antenna.

As only low-level modulation methods are put to use in SSB, linear amplifiers must be utilized throughout to maintain the delicate modulation envelope.

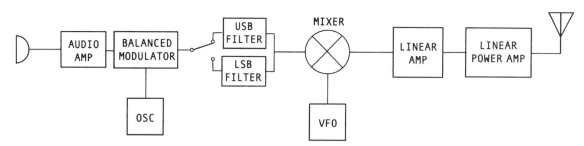

Figure 6-3. *A basic filter-type SSB transmitter.*

At the SSB receiver (*Figure 6-4*), the signal is amplified and inserted into a non-linear mixer, called a *product detector*, where it is mixed with the output of a *carrier oscillator* (or *BFO*) to supply the missing carrier. The output is an audio signal, which is then amplified and sent to the speakers.

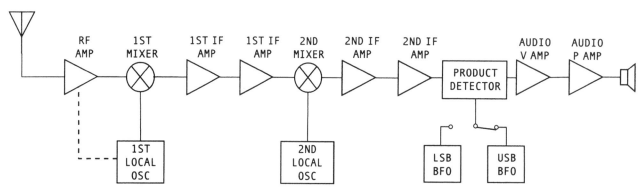

Figure 6-4. *A dual-conversion SSB receiver.*

When single-tone modulation is applied in SSB, a constant amplitude and frequency (CW) signal is produced, as in *Figure 6-5*. A two-tone baseband signal produces the classic SSB modulation envelope oscilloscope display of *Figure 6-6*, with the amplitude of the RF modulation envelope contingent on the audio modulation level. When overmodulation occurs, the two-tone signal begins to flat-top (*Figure 6-7*), creating severe distortion and spurious outputs.

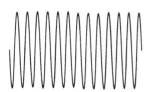

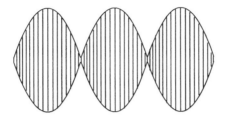

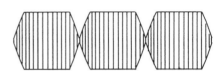

Figure 6-5. *A single-tone SSB signal.*

Figure 6-6. *A two-tone SSB signal showing the modulation envelope.*

Figure 6-7. *Overmodulation of an SSB two-tone signal.*

6-1B. SSB Output Power

In SSB, as in AM, the peak envelope power (PEP) is the measurement of the peak (maximum) instantaneous power, with 100% modulation applied, of the transmitted signal, and is calculated by

$$\frac{V^2_{RMS}}{R}, \text{ or } V_{RMS} \times I_{RMS}, \text{ or } I^2_{RMS} \times R$$

of the peak values at 100% modulation.

Unlike AM, there can be no power measurements without modulation, since there is no output possible with SSB unless the transmitter is modulated and the sidebands are produced.

6-1C. Standard AM and SSB Comparisons

SSB transceivers have become the preferred method of communicating at extreme distances, and have now virtually replaced all AM transmitters for long-range two-way voice communication. SSB generation and reception, however, demands more expensive and complex equipment than AM. The SSB transmitter is low-level amplitude-modulated, but circuits must be included that suppress the carrier and filters out the extra sideband, while the receivers must contain added circuitry that reinjects the carrier, at a very steady frequency, before the baseband signal is demodulated.

6-2. SSB Receivers

An SSB receiver (See *Figure 6-4*) is basically the same as any superhet receiver. The main difference being in the detection of the SSB signal as well as a few other design features.

Some SSB receivers use a *wideband* first IF with a fixed-frequency crystal LO that passes many frequencies, and only tunes for the desired signal through a VFO (*Variable Frequency Oscillator*) during the second IF conversion (See *Dual-conversion*). Older receiver units also use some type of *varactor*-controlled *AFC* to control drift in either the first or second LO, whichever section uses the VFO.

The SSB receiver of *Figure 6-4* is a double, or dual-conversion receiver. The low-level RF signal is picked up by the antenna, boosted by the RF amplifier, and injected into the first mixer where it is heterodyned with the first LO to form the high *image-frequency* rejecting first intermediate frequency. This is outputted, filtered, and sent into the IF amplifiers to increase gain and improve selectivity.

This first mixer stage typically has a LO that uses a VFO or frequency synthesizer, since it must normally be tunable. The first IF is then injected into the second mixer, which converts and filters it into a lower second IF. The second IF amplifiers supply most of the receiver's selectivity and sensitivity. The second LO is normally a fixed-frequency crystal oscillator. This second intermediate frequency is then sent into one input port of a balanced modulator (BM). The balanced modulator is used to detect the SSB signal after reinsertion of the carrier by either the LSB or USB BFO (Beat Frequency Oscillator), depending on which sideband is the desired one for reception, which is then injected into the second BM input port. The BM's output port sends the resulting difference frequency, which is now an audio signal, into the audio amplifiers and on to the speaker.

6-2A. SSB Detection

When a balanced modulator is utilized for the purpose of SSB detection, it is frequently referred to as a *product detector* and, since a balanced modulator is nothing more than a non-linear mixer, many other types of non-linear mixers can be employed for this purpose.

Considering the carrier is missing from the received signal, a new carrier must be inserted at the same frequency as the missing carrier would have been after the second conversion in a dual-conversion receiver, or after the first and only conversion of a single-conversion receiver. This new inserted carrier frequency will be at a much lower value than the one that would have been transmitted, since it would have gone through one or two down-frequency conversions. An oscillator called a BFO generates this missing carrier.

The IF, with the audio information being carried on its sidebands, is injected at one input of the product detector, while the BFO frequency is injected into the other. The output of the product detector is the low difference frequency of the beating of these two signals, which is a reproduction of the original audio.

The other non-audio signals are eliminated by different means, depending on the type of product detector used. The BFO frequency can be suppressed by the canceling action of the "balancing-out" of the in-phase reinserted carrier when a balanced modulator is employed, or through low-pass filtering if standard mixers are utilized — while the IF and the sum frequency can be passed to ground by the LPF. Any frequency higher than the desired audio is out of the passband of the audio amplifiers, further helping to attenuate any high-frequencies components. The audio itself is sent to the audio amplifiers and on to the speaker or headphones.

The BFO is normally a crystal oscillator, since extreme stability is vital: If the BFO is off frequency by more than 100 Hz, or its output amplitude is not substantially larger than the received sideband, the resulting signal becomes difficult to read. Most receivers use twin switchable BFOs so that the desired sideband, either upper or lower, can be chosen for amplification to the speaker. Furthermore, a small variable tuning capacitor or varactor circuit can be utilized to slightly tune the BFO up or down to vary the tone of the received audio for increased clarity. This is called an *RIT* (Receiver Incremental Tuning) or *clarifier control*.

6-2B. SSB AGC

Most AGC in an SSB receiver is similar to other RF receivers. The detected audio is tapped after the product detector (audio-derived AGC), filtered to DC and amplified, and used to control the gain of the RF and IF amplifiers (*Figure 6-8*).

AGC in SSB is switchable for a long or short attack time, depending on whether speech (long pauses between words) or Morse code (short pauses between "words") is being received. This is normally accomplished by using different RC time constants and switching in different value resistors. A low value resistor is adopted for CW reception to remove the AGC control voltage quickly; and a high value resistor for SSB reception to hold its AGC control voltage for a longer time period. SSB voice AGC, however, must have fast attack times with slow decay times. Without fast attack an annoying AGC *thump* will occur at the start of each word.

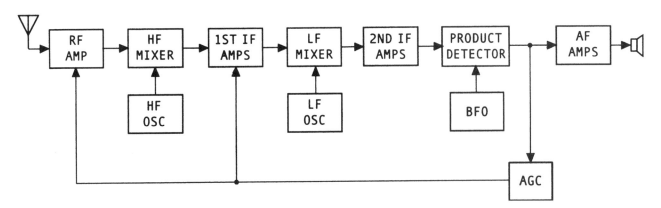

Figure 6-8. *AGC in an SSB receiver.*

Delayed AGC is common in SSB. This type of AGC does not reduce the gain of any stage unless a sufficiently strong signal is received, avoiding the problem of lowered gain even with weak input signals. A Zener can be employed in this role, and will hold the AGC voltage to some constant value until a signal of adequate strength is received to activate the AGC circuits.

Other SSB receivers may utilize an AGC that is tapped off of the second IF, amplified, rectified, filtered to DC, further amplified, and applied to the appropriate RF and IF stages.

6-3. SSB Transmitters

SSB is almost always generated at a much lower frequency than the transmitted RF sideband frequency that is finally injected into the antenna. This assures easier filtering of the unwanted sideband, while the desired sideband is increased in frequency by heterodyning in a later stage.

The SSB transmitter of *Figure 6-9* uses two crystal oscillators in the LO to supply the stable low-frequency carrier in a frequency appropriate for USB or LSB transmission.

As shown, the carrier from the LO and the audio frequencies are fed into a balanced modulator (BM), which is nothing more than a doubly-balanced mixer. Doubly-balanced mixers offer high port-to-port isolation, which allows for almost no coupling of the separate input/output frequency's (LO and IF and AF) energy between the input/output ports. The BM suppresses the carrier and the audio frequencies, while outputting the resultant upper and lower sidebands created by the non-linear mixing action.

The carrier's input from the LO into the balanced modulator is usually about ten times the amplitude of the audio signal for distortion reduction.

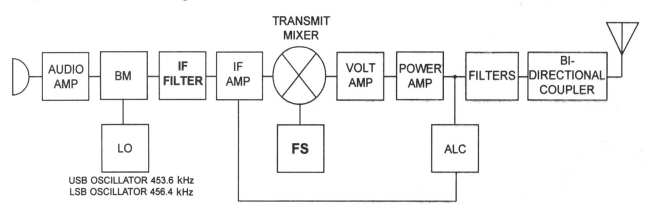

Figure 6-9. *A single-filter SSB transmitter.*

Then, at the output of the balanced modulator, either the LSB or the USB is filtered by the *IF filter*, which is a method of sideband suppression called the *single-filter method* (described later, *Figure 6-18*). Or the USB or LSB can be chosen by switching in a very sharp filter (*Figure 6-3*), with the RF carrier generated from a single-frequency LO. Either way, the selected sideband is sent to an IF amplifier and on to the transmit mixer for frequency translation into a higher frequency.

The VFO for the transmit mixer is commonly a PLL or DDS frequency synthesizer (FS) to supply a wide tuning range for the preferred output operating frequency.

A tuned circuit at the output of the transmit mixer can be set to attenuate either the resulting sum or difference frequency. Ordinarily, another tuned circuit in the tuned linear voltage amplifier stage supplies further attenuation of the unwanted sum or difference frequency, while providing amplification for the desired frequency. These amplifiers are biased Class AB or B linear or, more rarely, Class A. This sideband signal is then further amplified by the final linear wideband push-pull power amplifier (some SSB RF output amplifiers may each be *stagger tuned* to a different frequency, so as to pass the total bandwidth of the SSB signal).

The signal is placed into a low-pass filter to attenuate any harmonics present, placed in the *bidirectional coupler* (which maintains an impedance match between the final amplifier and the antenna), and inserted into the antenna, where it is radiated.

6-3A. Balanced Modulators

The most common balanced modulator is the *diode-ring* type of *Figure 6-10*. Between two center-tapped input and output transformers are bridge-connected diodes. The carrier is placed at the primary of T_1, while the audio is injected at the center tap of T_2. The carrier and audio are suppressed within the modulator, with the double-sideband (DSB) output taken from T_2's secondary. The carrier is suppressed by the action of T_2: The diodes force the carrier current to flow equally between the upper and lower sections of the primary of T_2 and out of or in to the grounded center tap, creating equal but opposite magnetic fields in each section of the winding between T_2's grounded center tap. This causes the carrier to be canceled, or balanced out — thus no carrier energy is coupled to the secondary of the transformer.

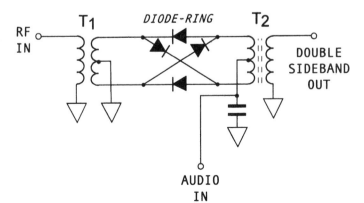

Figure 6-10. A diode-ring balanced modulator for SSB.

Obviously, for proper carrier suppression, the components of a balanced modulator must be perfectly matched. Balanced modulators are available in module form, with the diodes and transformers already matched internally; or as an IC using *differential amplifiers* (*Figure 6-11*). The IC differential amplifier rejects the common-mode RF carrier signal because there is no difference in potential between the outputs of transistors Q_1 and Q_2 for the in-phase carrier. The modulation injected into the base of the constant-current amplifier Q_3 disturbs the balance of the circuit, creating non-linear mixing between the carrier and the modulating audio, producing both upper and lower sidebands and suppressing the carrier (See *Differential amplifier*).

Carrier suppression with the average balanced modulator is 40 dB, while suppression capabilities of up to 60 dB is not uncommon.

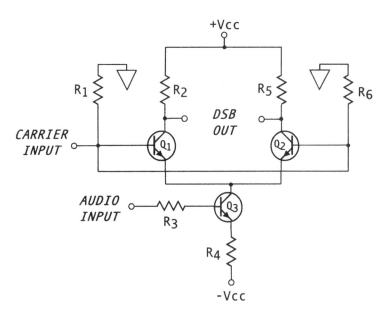

Figure 6-11. *The differential amplifier balanced modulator.*

6-3B. Sideband Attenuation

The undesired sideband out of the balanced modulator can be attenuated in one of two ways: by the *filter* or the *phasing method*.

The desired single sideband can be selected and sent on for amplification by the application of a *sideband filter*, while the undesired sideband is suppressed because it is out of this filter's passband. There are two types of crystal filters commonly used for this purpose: The *crystal-lattice* type (*Figure 6-12* and *Figure 6-13*) and the *crystal ladder* type (*Figure 6-14*; see under *Filters: Crystal and ceramic*, for a further explanation of these two configurations). Also, expensive and low-frequency (under 600 kHz), but high in quality (very narrow and steep bandpass characteristics), *mechanical resonant* (*Figure 6-15*) type filters are found in some new SSB transmitters and many older units.

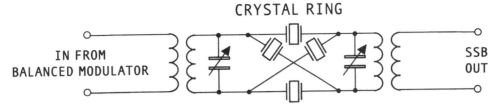

Figure 6-12. *A type of crystal-lattice sideband filter.*

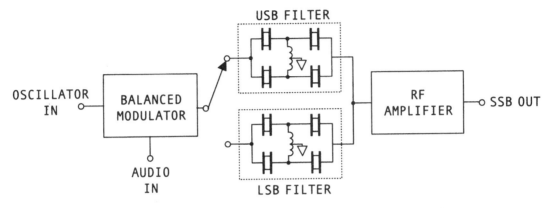

Figure 6-13. *Another common crystal-lattice sideband filter.*

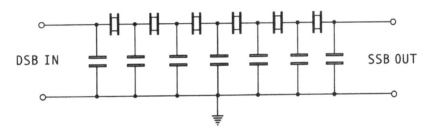

Figure 6-14. *A crystal ladder sideband filter.*

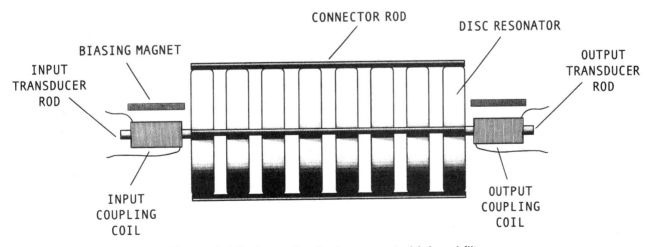

Figure 6-15. *A mechanical resonant sideband filter.*

The mechanical filters consist of metal disks and rods in a single sealed assembly, with the DSB signal fed into an *input coupling coil*, which has been placed in the permanent magnetic field of a *biasing magnet*. An induced magnetic field produced by the DSB input signal either adds to or opposes the permanent magnetic field, producing movement of the coupling coil and its attached *transducer rod*. The *resonant disks* are coupled together by *connector rods*, with the first and last resonant disks attached to the transducer and connector rods. The resonant disks, which are cut to exact dimensions, will only vibrate at their natural frequency, which is dependent on their diameter and thickness. If the frequency at the filter's input is at the natural resonant frequency of the discs, the resulting vibrations will vibrate the attached output transducer rod, which will move the output coil in synchronicity. This coil is also in the magnetic field of a permanent magnet, so its movement produces an induced output signal voltage that is a duplicate of the input resonant frequency. The sideband that is not within the passband of this filter is severely attenuated. A mechanical filter is shown in an IF strip in *Figure 6-16*.

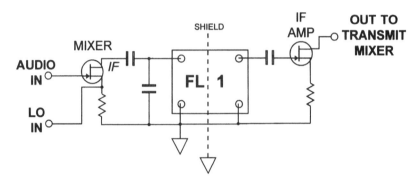

Figure 6-16. *Mechanical resonant SSB filter in-circuit.*

An additional type of mechanical filter is called the *flexure-mode* mechanical filter (*Figure 6-17*). An input signal is injected into a ceramic crystal transducer, which converts this varying signal voltage into a mechanical vibration. The vibrations are conveyed to an iron-nickel alloy bar on which the crystal is attached. This alloy bar twists, transferring this flexing to another crystal and alloy bar combination through coupling rods, creating a vibration in the output piezoelectric crystal, which causes a corresponding output signal voltage at this filter's resonant frequency.

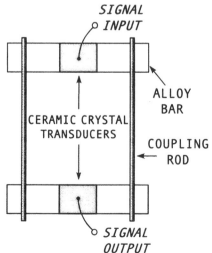

Figure 6-17. *A flexure-mode mechanical sideband filter.*

Another filter technique of producing SSB uses two carrier frequencies and one SSB filter (*single-filter SSB, Figure 6-18*), with the carrier frequency shifted to send one sideband through the sideband filter. The other sideband is now out of the filter's passband and is consequently suppressed.

The second, but less common, manner of suppressing both the carrier and one of the sidebands is the *phasing method (Figure 6-19)* for SSB generation. This arrangement uses a quadrature phase shifter to break the audio into two constituents that are out of phase by 90°. The carrier is also split and phase shifted by 90°, with one -45° audio and one -45° carrier constituent inserted into one of the balanced modulators, while the +45° audio and +45° carrier constituent is inserted into the other balanced modulator. The balanced modulators attenuate the carrier and outputs a DSB signal to the RF combiner, which adds the two signals. Since there is a phasing difference between these input signals, one sideband is intensified, while the other is canceled. The output is SSSC.

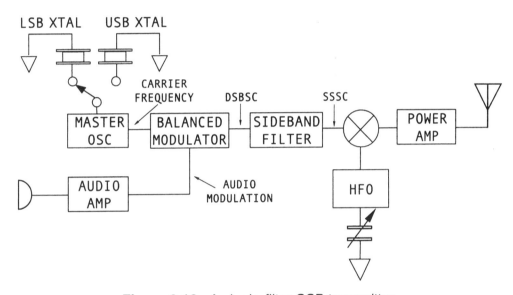

Figure 6-18. *A single-filter SSB transmitter.*

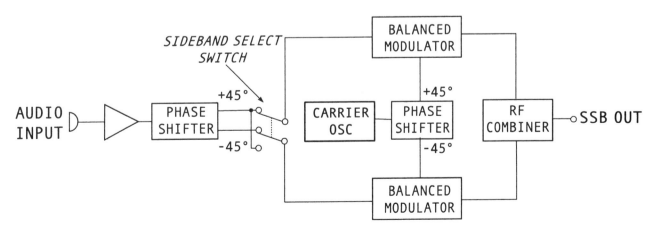

Figure 6-19. *A filterless phasing method for SSB production.*

6-4. NBVM

There are some SSB transmitters that transmit at a bandwidth of only 1.3 kHz, and are called NBVM (*Narrow Band Voice Modulation*). These transmitters are designed to send only the frequencies of the human voice that benefit clarity (considered to be between 300 to 600 Hz, and 1600 to 2600 Hz), while rejecting all other frequencies.

A special receiver, outfitted with the proper compensation circuits, must be utilized to receive the special transmitted signals, since only the frequencies of 300-600 Hz and 1600-2600 Hz have been transmitted, with the 1600-2600 Hz frequencies being converted to 700-1600 Hz within the transmitter. In order for the transmission to be readable, the receiver needs to convert these back to the proper frequencies.

CHAPTER 7
Other Communication Applications

There have been various processes adopted that employ the basic AM, SSB, FM or PM communication techniques that enable us to increase the reliable range and/or information transfer rate of an RF signal over that of common simplex single-channel radio communication methods. Additionally, small, low-cost, and simple-to-operate equipment has opened up long and short-range RF communications to the untrained operator.

Many of these communication methods are relatively new, while some have been available for years. *Multiplexing* is important for conserving spectrum space and permits economical multi-channel communications to take place over a single transmission path. *Data communication* techniques are the perfect medium for efficient digital and analog high-speed and high-quality information transmission. *Satellite* use has revolutionized almost all forms of voice, data, and video long-distance communications. *Spread spectrum* is a developing and expanding form of RF communications that can be reliably employed in a very crowded spectrum with little signal degradation. *Cellular radio* has made instant mobile telephonic wireless communications a low-cost and viable communications medium. *CW* is an old and dependable long-range communications means that is still quite popular. *Repeaters* have made low-powered VHF, and above, terrestrial line-of-site communications possible over long distances, especially for FM.

7-1. Multiplexing

Multiplexing refers to the ability to send two or more channels of information over a single channel. There are two methods of communication multiplexing: *Frequency-division multiplexing* and *time-division multiplexing*.

7-1A. *Frequency-Division Multiplexing (FDM)*

FDM has the ability to transmit multiple signals over one wide channel by converting each desired signal into a different narrow frequency span, and transmitting them all simultaneously within that single wide channel (*Figure 7-1*). Bandpass filters are used at the receiving end, set to the proper frequencies, to remove the desired intelligence from the appropriate narrow frequency channels.

FDMA (*Frequency-Division Multiple Access*) is a closely related technology, and is used in most standard analog cell phone systems. It differs from simple FDM in that FDMA is dynamic, and can change its channel allocations as needed by the communication system. For instance, when a mobile cell phone ends its call it no longer needs the frequency it was just transmitting on, so that frequency is now released to the common pool to be re-used by another transmitter. In simple FDM, all transmit and receive frequencies are fixed.

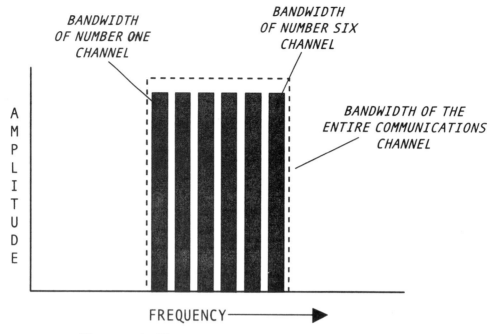

Figure 7-1. The frequency spectrum of an FDM signal.

In a simple FDM transmitter (*Figure 7-2*) each baseband voice signal modulates a different carrier. This modulated carrier is then fed into a linear mixer, which does not create sum and difference frequencies, but allows the original signal frequencies to be maintained and separated without generating any new signals. This linear mixing action combines all the channels into a single complex signal, which is then fed into a wideband transmitter and sent out over the air on one broad channel. AM, FM, PM and SSB can all be used in FDM.

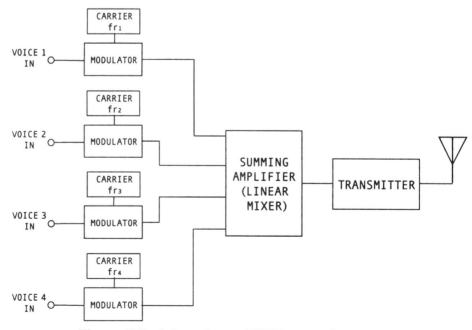

Figure 7-2. A four-channel FDM transmitter system.

The FDM wideband receiver (*Figure 7-3*) amplifies and demodulates this signal, while the bandpass filters are each tuned to a specific carrier. The filters pass these modulated carriers on to the detectors, which demodulates and outputs the original baseband signals to the appropriate transducers. FDM is typically used for analog information transfers.

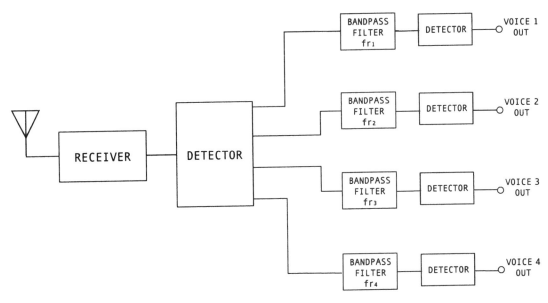

Figure 7-3. *A four-channel FDM receiver system.*

7-1B. *Time-Division Multiplexing (TDM)*

TDM transmits one signal in one time period, and another signal in another time frame, over a single channel. In a TDM transmitter a repetitive time period is reserved for each specific voice or data channel. It is basically a time-sharing approach. One channel of information is sent, then the next, then the next, until it returns to the first channel and the process repeats — all on a single radio channel.

A very common modulation technique for TDM is the digital method called *PCM* (See *Pulse-code modulation*): To transmit voice (*Figure 7-4*), the analog waveform is sampled at set time periods by an analog-to-digital converter (ADC), which outputs a parallel digital word for each discrete analog amplitude level. This is then sent to a parallel-to-serial data converter and on to a *multiplexer*, which relays the PCM, at repetitive time slices, through to the transmitter.

The multiplexer can be compared to an electronic rotary switch which is programmed by the timing and control circuits to place a digital word, waiting at the input of the multiplexer, to its single output. The timing and control circuits then instruct the multiplexer to send the next word, waiting at the next input, to its single output. This process continues until all of the inputs have been serviced; and the cycle then repeats.

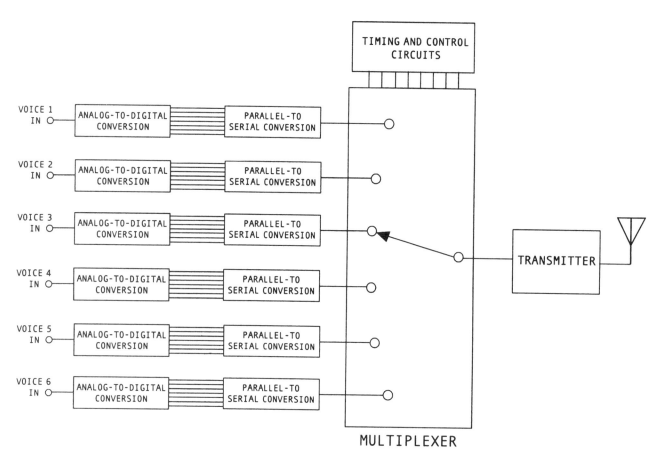

Figure 7-4. *A TDM voice transmitter showing multiplexer action.*

The PCM transmitter then converts this information into AM or FM PCM (or other more sophisticated modulation schemes), and supplies synchronizing pulses to help the receiver properly decode the PCM. The transmitter thereupon sends the signal out over the air in AM constant amplitude pulses (no pulse = 0, pulse = 1), or as an FSK signal (f_1=0, f_2=1). No ADC would be needed if digital data is being sent.

The TDM receiver (*Figure 7-5*) picks up the transmitter's signal, as well as the synchronizing pulses, and demodulates them. The receiver's output is sent through a Schmitt trigger, which reshapes the now ragged pulses, and into a demultiplexer. The demultiplexer places the binary data, located at its single input, into the appropriate output port for serial-to-parallel conversion — and on to digital-to-analog conversion (DAC) and filtering for output into a suitable transducer. Or, if the original baseband signal was digital data, D-to-A conversion is bypassed and the data is moved into either storage or on to a computer for immediate processing.

TDMA (Time-Division Multiple Access) is closely related to simple TDM, and is used in digital cellular phone systems. TDMA gives up to three different cellular calls special time slots (called *frames*) on the same frequency to communicate. Each slot may only be a fraction of a second long.

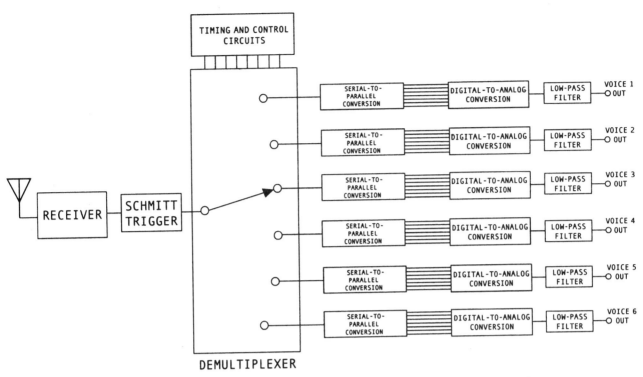

Figure 7-5. A TDM voice receiver capable of receiving six channels.

7-2. Data Communications

Data communications by the use of *pulse modulation* entails the transmission of analog or digital modulation signals by varying either the width, amplitude, frequency, position, or phase of a pulse — or even whether a pulse is present or not. The pulses can then be decoded at the receiver. There are two basic methods to produce pulse modulation. The older *analog pulse modulation* (*APM*), and the more modern and more commonly used *digital pulse modulation* (*DPM*).

A method for transmitting binary information, as well as voice and video that has been digitized, at higher speeds than would normally be possible, is accomplished by employing *quadrature phase-shift keying* (QPSK), as well as phase-shifting combined with amplitude modulation (*quadrature amplitude modulation*, or QAM). This allows for the transfer of large quantities of information over space satellites or terrestrial wireless microwave links.

7-2A. Pulse Modulation

Analog pulse modulation permits the production of high values of peak power for short time periods, granting smaller transmitters higher output powers. A tiny sample of the modulating signal is taken by any of the methods below, which is then allowed to continuously modify a certain characteristic of a pulse. Since the carrier can be analog modulated by changing its width (PWM), amplitude (PAM), position (PPM), duration (PDM), or timing (PTM), then any of these methods can be adopted to impress pulsed information onto the RF carrier. Three of the most common of these techniques are shown in *Figure 7-6*.

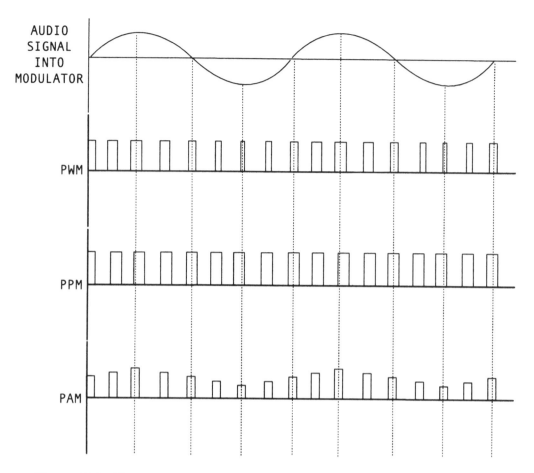

Figure 7-6. *Three of the most popular analog pulse modulation methods.*

Taking the example of PWM: The analog audio signal is sampled and converted to varying width pulses, depending on its amplitude (the higher the amplitude of the audio, the wider the pulse), and sent on to a modulator where these pulses are converted to RF constant-amplitude (but varying width) pulses by the switching on and off of the transmitter for different time periods. A suitable receiver receives this signal, converts the RF constant-amplitude PWM pulses into an analog audio signal for reception by the listener.

Fast RF pulses with rapid rise and fall times have the disadvantage of consuming a lot of bandwidth. This is due to the harmonic generation caused by the rectangular-like wave-shape of the pulses. The bandwidth needed is usually available only in the microwave region.

An added use for PWM is in special transmitters. With the help of pulse-width modulation methods, transistors or FETs can be adapted to the finals of some high-powered amplifiers where, in the past, only vacuum tubes could have been used. By employing pulse-width modulation, transistors will be supplying current for something less than 100% of the time — but at the expense of a much more complex semiconductor circuit than that furnished by high-powered vacuum tube finals.

This Class "D" amplification method transforms the audio signal into a pulse-modulated signal train by using a pulse-width modulator. The output of this modulator is a rectangular signal that varies in pulse width and frequency to that of the changing modulating baseband signal. The transmitter's power amplifier is turned on and off by these pulses, allowing for a higher RF output due to the lowered duty cycle necessary. The PWM signal is then passed through tuned circuits and harmonic filters, which restore the original shape of the audio modulated carrier wave for transmission through the antenna and into space.

Pulse-code modulation (PCM), a digital method for obtaining pulse modulation from an analog signal, has essentially replaced most of the analog techniques for the transmission of voice and video.

Unlike the analog methods, PCM is virtually noise-free and is able to punch through when atmospheric conditions are very poor, with little information loss. PCM is simply a way of converting analog data to digital data and then using other modulation schemes, such as FM, AM, or PM, or any in combination, to output a binary stream of data.

PCM is obtained by assigning a voltage level, on the modulating audio waveform, a specific digital value (*Figure 7-7*). For example, if the waveform of *Figure 7-7* were sampled at point four, then a PCM modulator would code that amplitude as binary data 100. Or, if a signal has an amplitude of 3V and 5V at the sampling points, then a pulse train representing 011 and 101 would be output from the PCM modulator. In practice, many more sampling points must be used. For example, to send an 8-bit word (a common length), 256 sampling levels must be employed.

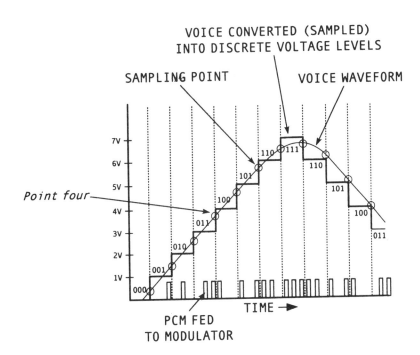

Figure 7-7. Conversion of an analog waveform into discrete voltage levels and PCM.

A block diagram of a simple single-channel PCM RF communication system is shown in *Figure 7-8* and *Figure 7-9*. Either AM (CW or a single-tone modulation) or FM (FSK) can be adopted as the medium for the PCM transmission and reception.

Figure 7-8 is the transmitter section of a non-multiplexed PCM/AM or PCM/FM system: The audio is placed into the ADC for digital conversion (for digital signals the ADC is bypassed), are transferred to a parallel-to-serial converter, then on to an AM or FM modulator to change this binary data into amplitude pulses or to frequency-shifted "pulses". This signal is placed into the transmitter, which amplifies and filters the output for transmission from the antenna and into space.

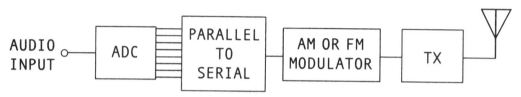

Figure 7-8. *A basic single-channel PCM transmitter.*

The signal is then received by the antenna of the PCM receiver of *Figure 7-9*. The PCM is amplified by the receiver, detected by the demodulator, sent for pulse shaping by the Schmitt trigger — which squares the now ragged pulses. These serial pulses are placed into the serial-to-parallel converter, which outputs parallel data into the digital-to-analog converter (DAC) for conversion back into an analog signal. The DAC output is filtered by a low-pass filter to create an almost exact reproduction of the original audio signal.

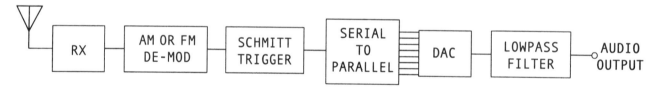

Figure 7-9. *A basic single-channel PCM receiver.*

7-2B. PSK and QAM Modulation

Techniques that have much in common with the above AM and FM modulation methods, but can be much higher in transmission speed, are the multi-level encoding schemes, such as QPSK and QAM.

It is important to keep in mind that the wider the bandwidth of a channel, the more information that can be passed within that channel over a given time. A narrower communications channel means less information can be passed within a certain time frame — high amounts of data can still be sent through a narrow communications link, but the time it takes to pass this information will be longer.

A way of helping to overcome some of the above limitations is to utilize multi-level encoding schemes. In other words, each change of state of a digital signal can indicate two or more different binary values instead of the usual two-level binary (high=1 and low=0) arrangement. This standard two-level binary system only allows a change of state that indicates either a 1 or a 0, while with multi-level encoding each change of state may indicate a 00, a 01, a 10, or a 11, permitting for twice the information to be transferred within the same bandwidth and time period.

Most of these digital modulation methods have been developed for use in microwave digital radio high-capacity information transfers. A basic digital radio that employs some of these high speed processes is shown in *Figure 7-10*. The *coder* chip (*See* PCM) digitizes the incoming voice signal, which is sent to the modulator. The modulator can employ any type of QPSK or QAM method (*explained below*), which affects the phase of the carrier oscillator output, or both its phase and amplitude. The output of the modulator is sent through a bandpass filter and into a mixer — with its accompanying transmit oscillator — for up-conversion. The resulting frequencies are then placed into the power amplifier and bandpass filter and out to the antenna for transmission through space.

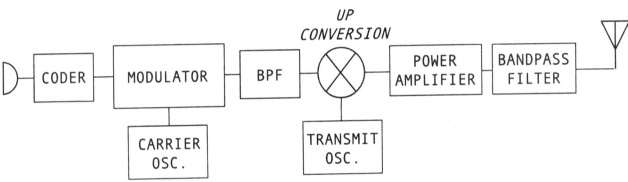

Figure 7-10. *A digital radio transmitter.*

The receiver of *Figure 7-11* receives this signal, filters it through a BPF, then further amplifies it with the RF amplifier, which passes it into the down-conversion circuits. Down-conversion transforms the very high RF into a low IF of either 70 or 140 MHz, which is then filtered by the IF BPF and amplified by the IF amplifiers. The output of the IF amplifiers is sent for demodulation. The recovered signal is then fed into the decoder, which converts this digital input into a reconstructed analog output.

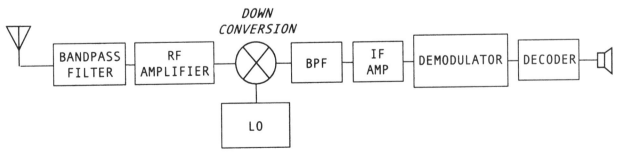

Figure 7-11. *A digital radio receiver.*

The above modulator can consist of any one of many different two-level encoding methods, such as PSK (*Phase-Shift Keying*), which changes the phase of the carrier to denote a digital "0" or "1". Either a 90° or 180° change of phase of the carrier signal marks the beginning of a digital "0" or "1" (*Figure 7-12*) or, in modern *BPSK* (*Binary Phase-Shift Keying*), the constant amplitude carrier's phase is shifted between 0° and 180° for a "0" (0°) and a "1" (180°) in response to the coder chip's digital output (the spectrum of a BPSK RF digital signal as displayed on a spectrum analyzer is shown in *Figure 7-13*). There are other methods, such as *DPSK* (*Differential Phase-Shift Keying*). DPSK is similar to BPSK but can employ simpler circuitry since DPSK does not need to reinsert an exact on-frequency carrier, which is furnished by the BPSK's elaborate carrier recovery circuitry, to supply the phase comparisons for proper data recovery. DPSK instead compares phase shifts from a received bit to an earlier received bit. Or *BFSK*, which is *Binary Frequency-Shift Keying*, and is simply a digital modulation scheme in which a $F_1=1$ and a $F_2=0$ (*Figure 7-14*).

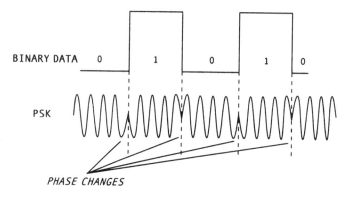

Figure 7-12. *A phase-shift keying waveform in response to the modulating binary data.*

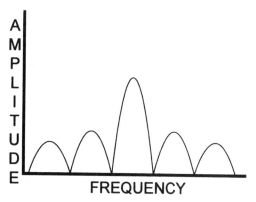

Figure 7-13. *A BPSK spectra.*

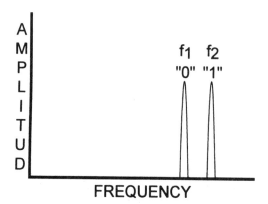

Figure 7-14. *Binary frequency-shift keying.*

The common multi-level *QPSK* (*Quadrature Phase-Shift Keying*), uses four different phase values, such as 0°, 90°, 180° and 270°, on a constant amplitude carrier to furnish two bits for each change of phase (00, 01, 10, 11), instead of 1 bit (1, 0) as in the above PSK techniques. This method would obviously provide twice the information within the same bandwidth and time period. *QFSK* is *Quaternary Frequency-Shift Keying*, where each frequency equals a different symbol: $f_1=00$; $f_2=01$; $f_3=10$; $f_4=11$ (*Figure 7-15*), is also used.

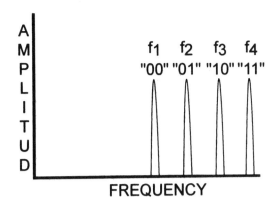

Figure 7-15. Quaternary frequency-shift keying.

But perhaps the most popular modern method for transmitting data at very high rates utilizes a combination of amplitude and phase modulation, and is referred to as *QAM* (*Quadrature Amplitude Modulation*). QAM is the major long-distance digital communications technique employed for transmitting data over the air across terrestrial microwave links.

QAM applies different phase shifts to the carrier, with each of these phase shifts having the capability to have two or more amplitudes. Each amplitude-phase combination can represent a different binary value. For instance, if a signal had a phase shift of 180° and an amplitude of +2, then it may represent a digital value of 111; if the phase is shifted to 90° with a amplitude of -1, then a 010 may be represented. This is referred to as 8-QAM, which uses four phase shifts and two carrier amplitudes for a total of eight possible states, or 3 bits (000, 001, 010, 011, 100, 101, 110, 111), that can be transmitted. 8-QAM is quite common, but is being supplanted by higher QAM values.

Figure 7-16 clarifies the basic idea of this type of modulation on what is referred to as a *constellation diagram*, in this case for a 16-QAM scheme (which supports 4 bits per AM/PM change). Constellation diagrams can display both amplitude and phase changes with their given binary coded values.

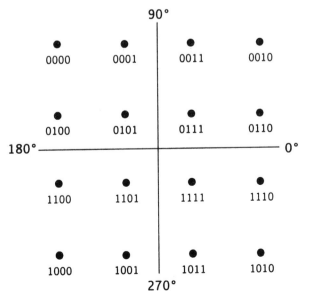

Figure 7-16. A constellation diagram for 16-QAM.

As the number of possible AM/PM states are increased, more data can be sent within an allotted bandwidth or time frame because more bits per change can now be encoded. However, as AM/PM states increase, noise starts to become more of a consideration, since each state becomes very close together. High QAM values can be subjected to crippling interference under relatively low noise conditions. This is why high noise environments, such as that encountered in satellite communications, usually adopt the simpler schemes, such as BPSK and QPSK.

Low-noise terrestrial *radiolinks* using 16, 32, 64, 128, and even 256 QAM are now in operation. These newer methods can transfer data at far higher rates than older and simpler digital systems within the limited bandwidth constraints of most information channels.

7-3. Satellite Communications

Communication satellites are simply a repeater and antenna combination that amplify an incoming signal and retransmits this signal at an extremely high altitude (*Figure 7-17*), allowing for a very large area of coverage.

Most are in geosynchronous orbit around the earth at an altitude of 22,300 miles, which allows the satellite to revolve once around the planet in a single twenty-four hour period. This permits the satellite to remain stationary over a chosen point above the earth, day in and day out. An earth-bound antenna does not have to track a geosynchronous communications satellite as the earth revolves — it merely needs to be aimed once during installation and then locked in place.

The extreme distance of the satellite from the receiver and transmitter, however, has a negative effect: The earth station antennas, transmitters, and receivers must be of high gain to effectively access these satellites to obtain a usable signal-to-noise ratio.

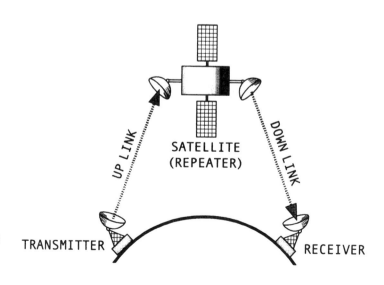

Figure 7-17. A communications satellite with its earth stations.

As shown in *Figure 7-17*, the earth-bound transmitter sends a signal to the satellite (the *uplink*), where the satellite amplifies and translates (changes the frequency of) the signal, generally to a lower frequency, and retransmits this frequency (the *downlink*) to an earth station receiver. Without this *frequency translation* the very low-powered signal received at the satellite would be inundated by the far more powerful satellite transmitter's frequency. The transceiver of a satellite that performs this function of amplification and translation is referred to as a *transponder*.

All modern satellites utilize some form of multiplexing (either TDM or FDM) as well as multiple transponders set to different channels, allowing for the handling of huge amounts of information. Satellites also exploit very wideband transponders, which permit any earth transmitter's signal within the transponder's bandwidth to be received and retransmitted. In view of this wide bandwidth, and the need for small, highly directional antennas, microwave frequencies are the only practical section of the RF spectrum exploited for satellites. These frequencies are normally in channels between about 4 and 14 GHz.

7-3A. Frequency Reuse

Frequency reuse can double the amount of channels that a single satellite can retransmit within its band of allotted frequencies. A separate bank of transponders are employed, with one bank using a vertically polarized antenna, while the other bank uses a horizontally polarized antenna. Each bank utilizes the same frequencies as the other bank, but because a transmitting antenna of the opposite *polarization* as the receiving antenna cannot induce a voltage into that antenna, very little interference occurs.

When *circularly* polarized antennas are operated for frequency reuse, they are wound in the opposite sense for each bank of transponders, inducing little or no voltage in the receiving antenna, thus also isolating each set of transponders.

Another method, called *spot beaming*, uses a highly selective antenna to service two different stations, at separate locations on the earth, with the same set of frequencies.

7-3B. The Transponder

A simple transponder is shown in *Figure 7-18*. The uplink antenna receives the signal. This extremely feeble signal is then amplified by a low noise amplifier (LNA) to obtain a decent S/N ratio, mixed with the LO frequency to acquire a sum and difference frequency, fed to another amplifier, then out to a bandpass filter (BPF) to achieve the proper frequency translation for the downlink channel. The downlink frequency's strength is increased by the power amplifier, which is typically a *traveling wave tube* (TWT).

TWTs are not completely linear and produce *intermodulation distortion products*: The input signal mixes with any nearby signal or its harmonics, creating undesired sum and difference frequencies — which must be removed by another bandpass filter. This output is then fed into the downlink antenna for retransmission to an earth station.

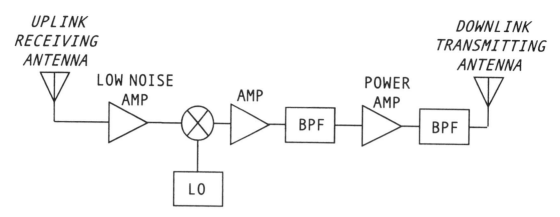

Figure 7-18. *A single-conversion satellite transponder.*

Other satellites use transponders with *double-conversion* circuits, or employ *regenerative repeater* transponders (*Figure 7-19*). Regenerative repeater transponders are transponders that are more like separate receiver-transmitters than transceivers. A signal is picked up by the receiver, amplified, converted, and demodulated. The output of the demodulator is a reproduction of the original digital, voice, or video signal transmitted by the earth station. This modulating signal is then further amplified and fed into the modulator of the transmitter section, which impresses the carrier with this baseband information. The modulated carrier is sent to the final power amplifier and BPF and out to the downlink antenna.

The regenerative repeater scheme offers better signal-to-noise ratios, with increased and cheaper amplification of the uplink signal possible (since the low-frequency baseband signal is far easier to amplify than a higher RF or IF of the single and double-conversion types).

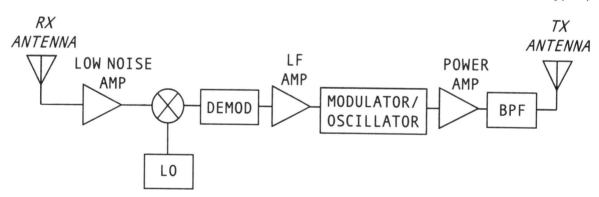

Figure 7-19. *A regenerative repeater transponder.*

7-4. Spread Spectrum

Spread spectrum (SS) is a communications system that spreads its desired signal randomly over a wide bandwidth of the frequency spectrum. Any receiver must be synchronized with the transmitter in order to receive the transmitted SS signal.

SS is most commonly used in data communications over wireless *LAN*s (Local Area Networks), as well as in industrial and military applications; and recently in consumer 900 MHz cordless phones and wireless biomedical monitors.

SS is low-noise, resistant to jamming, eavesdropping, man-made interference (QRM), and certain types of fading. Band utilization is also maximized because the transmitter does not "own" any single frequency.

Digital data as well as voice transmission are both equally easy to accomplish in SS.

There are different ways to implement spread spectrum — with the two most prevalent being *frequency hopping* (FHSS) and *direct-sequence spread spectrum* (DSSS).

Spread spectrum's exploitation for commercial operation is in three specific frequency bands between 902 and 5,850 MHz, and for the amateur services above 420 MHz.

7-4A. Frequency Hopping

A common SS technique is referred to as frequency hopping spread spectrum, with the transmitter shown in *Figure 7-20*. The digital data to be transmitted feeds its parallel bits into a parallel-to-serial-converter, which outputs the necessary serial binary pulse trains into a frequency-shift keyer (FSK) modulator, which converts this square wave binary data into two varying frequencies (*Figure 7-21*).

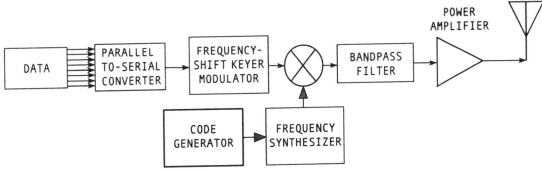

Figure 7-20. *A frequency hopping spread spectrum transmitter.*

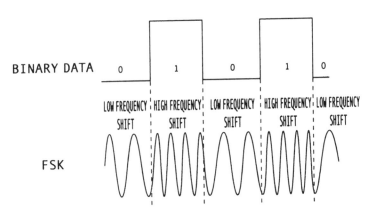

Figure 7-21. *An FSK waveform in response to modulating binary data.*

One frequency denotes a binary "0", while the other frequency represents the binary "1". FSK is an FM system with two frequency states — any frequency can be used for these two states, but are typically in the low audio range.

The FSK modulator then outputs its stream of two tones, representing the serial binary data, into a mixer. A frequency synthesizer, usually a PLL, is driven by a code generator (or computer with the proper hardware and software) called a *pseudo-noise generator* (*PN*). The combination of the frequency synthesizer and the code generator performs the function of a *transmit oscillator* that is constantly changing its frequency output at preset times. The code generator supplies the PLL with divide-by-N counter control words. The output of the doubly-balanced mixer is the sum and difference frequencies of the signal and the transmit oscillator frequencies. The sum *or* difference frequencies can be filtered out and applied to a power amplifier and sent to the antenna for transmission.

The time that the code generator allows the frequency synthesizer to remain on any single frequency is called the *dwell time*, and is typically between 10 and 400 milliseconds (*Figure 7-22*), differing by design. Depending on the programming in the code generator, the carrier may jump to fifty or more frequencies before the original sequence is again repeated.

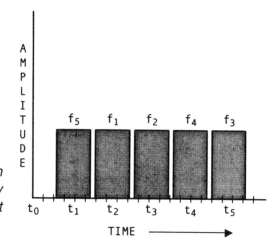

Figure 7-22. Frequency shifting sequence in spread spectrum communications: Frequency f_5 dwells at time t_1, then shifts to frequency f_1 at time t_2, etc.

A receiver must, of course, have the same programming in its code generator, as well as be in synchronization with the transmitter, to receive the transmitted signal. The receiver of *Figure 7-23* is capable of receiving this SS signal. A synchronizing code is received by the antenna, setting up the receiver to be locked in time-step with the transmitter. The receiver's LO, consisting of the frequency synthesizer, is now ordered to change frequency over time by the code generator, which keeps it in step with the frequency changes occurring within the transmitter. The output of the mixer is always at the fixed tuned IF. The IF section then amplifies this intermediate frequency, inserts it into the frequency-shift keyer demodulator, which outputs serial binary data, in square wave form. This output is usually sent to a serial-to-parallel converter for insertion into a type of processing equipment, such as a computer.

Figure 7-24 presents a spectrum analyzer display of a frequency hopping spread spectrum signal.

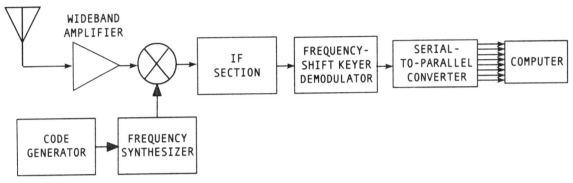

Figure 7-23. *An SS frequency hopping receiver.*

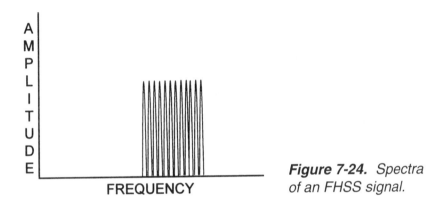

Figure 7-24. *Spectra of an FHSS signal.*

7-4B. Direct-Sequence

The direct-sequence spread spectrum, also referred to as *code division multiple access*, or *CDMA*, is a very popular spread spectrum technique.

CDMA itself is very new and used in some cellular digital phone and PCS systems. CDMA works by assigning each user a specific code: Only the receiver with the same code will be able to decode and listen to the RF signal, while the other CDMA transmitted signals will not be able to interfere with the desired received channel since all signals in CDMA are spread and are of low power density. In this way a single communication channel may be employed by many users without interference.

The DSSS method of producing spread spectrum for non-cellular systems is shown for one type of DSSS transmitter in *Figure 7-25*. The modulating signal is placed into a logic gate (an X-OR) along with the high-frequency digital code generator output. The output signal from the logic gate is then fed into a modulator along with a transmit oscillator output. The modulator can produce any of the PSK or QAM modulation outputs, but commonly uses BPSK. Since the modulator is a doubly-balanced mixer, the RF carrier is attenuated and the output now consists of various random wide-ranging sidebands, or as a "spreading" of the resulting signal (*Figure 7-26*). The output signal is *DSBSC* (*Double-Sideband Suppressed-Carrier*), which produces a very wide signal that appears to other standard receivers as noise.

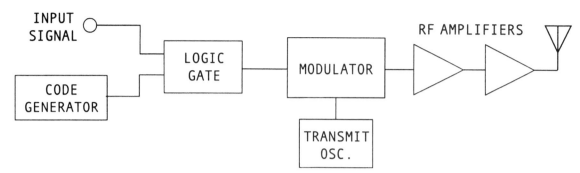

Figure 7-25. A direct-sequence spread spectrum transmitter.

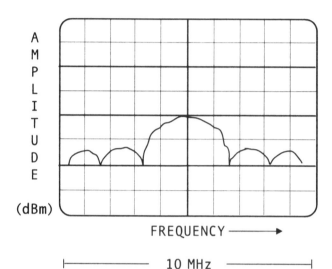

Figure 7-26. A DSSS signal
in the frequency-domain.

10 MHz

The DSSS radio receiver (*Figure 7-27*) picks up this wideband "noise", sends it to a doubly-balanced mixer, where it is combined with a comparable digital code produced by the code generator. The code generator supplies the same code as that generated by the transmitter. Synchronization of this code generator is maintained by sync circuits, which can strip off special synchronizing clock pulses from the transmitter's signal (or by other complex methods). The resultant output of the mixer is now a "despread" signal, which is transferred to a regular receiver where it is demodulated and fed to a speaker — or to some digital device for further processing or storage.

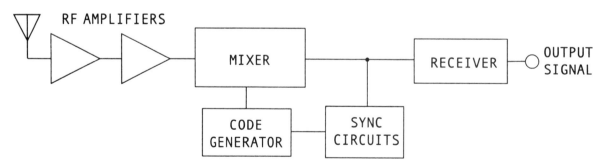

Figure 7-27. A basic DSSS receiver.

7-5. Cellular Communications

Instead of using one large repeater to receive and retransmit radio signals, as in older mobile phone communication systems, *cellular radio* employs many multiple repeaters, or *cell sites* (*Figure 7-28*).

Each cell site contains a low-powered transceiver designed to have limited range. This range overlaps with the coverage area of the next cell site, which operates at a different frequency. Frequencies are only repeated several cell sites down so that interference is minimized; but frequency availability and use is maximized.

All of these cell sites are connected and controlled by the *MTSO* (*Mobile Telephone Switching Office*) through regular telephone lines or microwave links, which also passes the cellular phone signal to the wired or wireless number called by the cell phone user.

As a mobile user passes from one cell site into the coverage area of another, *hand-off* occurs. The cell site and the MTSO monitor the signal power of the user's cell phone. As the phone's signal begins to fade, the cell site sends a signal to the MTSO, which then orders the neighboring cell sites to monitor the signal strength of the user's phone. The cell site that has the strongest signal indication is ordered by the MTSO to pick up the user's phone signal, but on a different frequency. This instantly frees up the prior cell site for another call, and is generally not noticed by the cell phone user.

Carrier frequencies are in the 800 MHz band, with over eight hundred fully duplexed channels obtainable. When more cell sites are needed as the subscriber base increases, each cell site can lower its output power as more sites are installed, which increases the cell sites available for usage, while nullifying any possible interference between these sites.

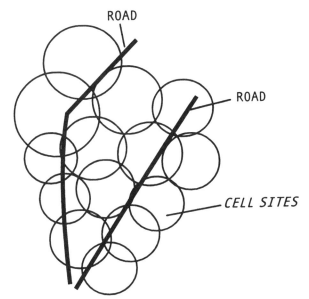

Figure 7-28. *Overlapping cell site coverage.*

7-5A. The Cellular Transceiver

A complete working analog cellular radio is shown in *Figure 7-29*.

After the caller inputs a telephone number through the keypad and picks up the unit, the phone automatically attempts to institute communications with a cell site. After communication is established, the MTSO is automatically sent ID numbers (such as the ESN, or the *Electronic Serial Number*, and the MIN, or the *Mobile Identification Number*) for that particular cell phone. These ID numbers are then compared and validated by a data base for authentication.

The user presses the *send* key and the system finds a free channel. Information on what frequency and signal strength the handset should use is sent to the cell phone from the MTSO through the cell site. The MTSO then makes contact with the desired number as dialed by the cellular phone user.

The user speaks into the microphone, where the voice level is raised by the audio amplifier. The frequency synthesizer has already been commanded by the MTSO through the logic control unit (LCU) to feed a certain carrier frequency into the transmitter section's mixer. Thus a channel has been effectively selected by the MTSO. This carrier frequency is placed into the phase modulator, where it is modulated by the audio and further amplified by the voltage and power amplifiers and inserted into the microstrip *directional coupler*. A directional coupler is simply a device that passes most of the microwave power that is placed at its input, while tapping off a small sampling of this input power to another of its ports. This tiny specimen of input power is fed to a power output sensor, where it is used by the APC (Automatic Power Control). As well, the level of this power is sent to the MTSO, which can then order the APC circuit to decrease output power if it feels this energy is excessive, or to increase it as needed — up to a maximum of three watts. The directional coupler feeds most of the signal into a low-loss *duplexer*, which can consist of two notch filters or two bandpass filters.

The duplexer allows the cell phone to use one antenna for both transmitting and receiving by isolating the receiver section from any of the transmitter section's RF power — but permits the transmitter to output its RF energy and the receiver to obtain an incoming signal from the same antenna. Since the transmit frequency is separated by 45 MHz from the receive frequency, the filters effectively notches or filters out the unwanted transmit frequency, preventing receiver overload and damage.

When a signal is received by the cell phone, it proceeds from the antenna into the duplexer, which sends it into the RF amplifier and into the mixer. The first LO frequency has already been set by the MTSO through the logic control unit to the proper receive channel, as output from the frequency synthesizer at port LO_1. This frequency mixes with the signal frequency, producing the first IF (either 82.2 or 45 MHz), which gets amplified and filtered. This IF is then input into the second mixer to obtain the second IF (either 10.7 or 0.455 MHz).

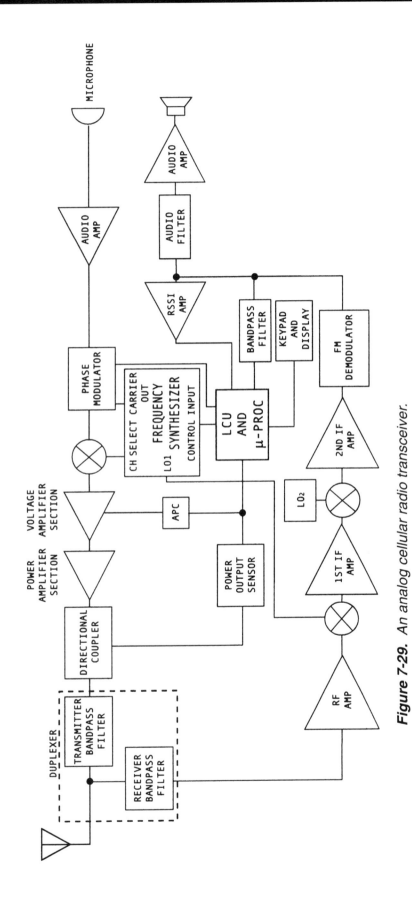

Figure 7-29. An analog cellular radio transceiver.

The second IF is detected by the FM demodulator to recover the voice modulation, then amplified and sent to the speaker, while the bandpass circuit filters this demodulated output to recover the *control signals* for the logic control unit. These control signals were generated by the MTSO.

The RSSI (*Receive Signal Strength Indicator*) amplifier feeds the strength of the received signal, as a DC level, into the logic control unit, which outputs the proper indicator signals to the MTSO through the transmitter section to keep the MTSO updated. The MTSO can then, as needed, switch the cell phone to another cell site if signal strength is lagging.

The microprocessor section controls the 12-key DTMF tones (Dual-Tone Multiple Frequency) for dialing out, drives the LCD or LED display, stores the last number dialed and frequently-called numbers, and other common functions.

7-6. CW Communications

CW (continuous wave, or "A1") communications equipment is still extensively used in amateur radio, and a basic knowledge of its operation is certainly required of any RF technician.

There are two basic methods used to transmit Morse code. One entails employing amplitude modulation and impressing the carrier with a 1000 Hz tone for reception by any receiver (*Modulated Continuous Wave*, or *MCW*). This technique is rarely used today.

The dominant method is *interrupted CW*, which involves switching on and off the carrier of the transmitter to form the dots and dashes necessary to transmit the desired information (*Figure 7-30*).

All amplifiers can be biased for highly efficient Class C operation in a CW transmitter. Thus, little power is consumed unless an actual dash or dot is being sent. However, linear stages are often used so that SSB can be transmitted, as well as the higher-quality code output inherent with linear stages. Also, since there is little concern about maintaining the various frequency's phase relationships, as in AM and FM transmissions, there is less fading of the received signal, while there is significantly lower noise received due to the very narrow bandwidth of a CW signal. Moreover, a single tone is far easier to read than the complex waveforms of speech under poor receiving conditions.

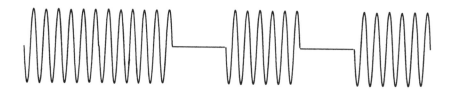

Figure 7-30. *An RF carrier turned on and off to transmit information in Morse code.*

7-6A. CW Receivers

The standard CW receiver (*Figure 7-31*) is of the superheterodyne type, and varies little from AM, SSB, or FM receivers. Nevertheless, in order to properly receive CW, a receiver must incorporate a BFO (Beat-Frequency Oscillator) — in fact, because all SSB receivers have a built-in beat-frequency oscillator, they are automatically able to receive high-quality code. Code can even be heard through a simple AM diode detector receiver, but the signal is a very mushy, hard-to-read humming sound.

To obtain its signal a CW receiver's BFO mixes with the IF in a *product detector*, which is basically a non-linear mixer. The BFO is set to almost the same frequency as the incoming IF, but is offset by a prefixed frequency, usually about 800 Hz. This offset means that the output of the mixer will be a 800 Hz sine wave tone — or the difference between the BFO's frequency and the intermediate frequency. The IF, the BFO frequency, and the sum frequency are too high to pass through or to be amplified by the audio amplifiers.

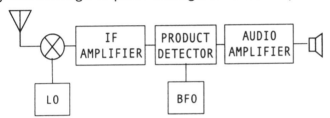

Figure 7-31. *A basic superheterodyne CW receiver.*

7-6B. CW Transmitters

A CW transmitter is basically an amplified carrier oscillator. A method is needed to switch on and off this carrier, usually by removing or supplying a path for the DC power to a stage or stages — either directly or through a relay or semiconductor switch (a *keying transistor*).

Many keying methods can be utilized. Some allow the operator, when employing a straight telegraph key, semi-automatic or fully automatic "bug", or even a computer, to interrupt the emitter or cathode return of the final amplifier (*Figure 7-32*). As soon as the operator depresses the sending key, the DC return is re-established, allowing the stage to become operational and transmitting the carrier into space. All of the emitters or cathodes can be tied together to switch the entire oscillator and amplifier chain on and off (*Figure 7-33*) — or just the oscillator itself can be turned on and off. Additionally, the Vcc to the collector or plate of the amplifiers can be controlled (*Figure 7-34*) by either removing or supplying the DC power.

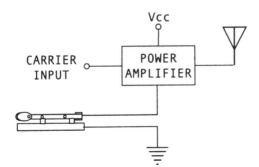

Figure 7-32. *Keying a CW transmitter by breaking and re-establishing the emitter or cathode return of an amplifier.*

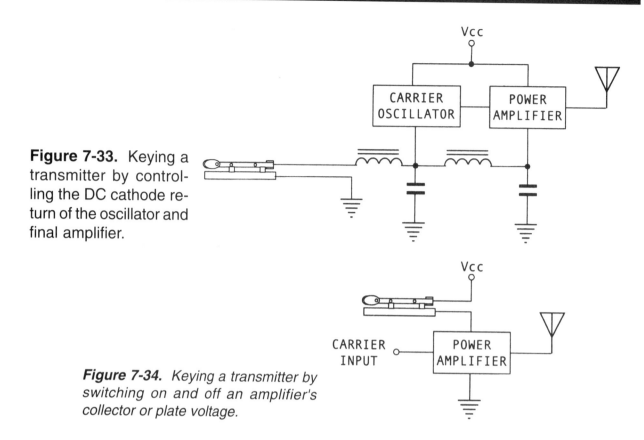

Figure 7-33. Keying a transmitter by controlling the DC cathode return of the oscillator and final amplifier.

Figure 7-34. *Keying a transmitter by switching on and off an amplifier's collector or plate voltage.*

A very common keying method is called *blocked-grid keying* (*Figure 7-35*). This procedure supplies a bias that cuts off the transistor's or tube's base or grid, allowing no collector or plate current to flow until the Morse key is depressed, shorting out the blocking voltage to ground and allowing the stage's normal bias to permit amplification of the transmitter's CW signal.

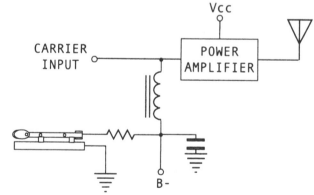

Figure 7-35. *Blocked-grid keying for a CW transmitter.*

7-6C. CW Spurious Responses

If the carrier is too rapidly switched on or off (*Figure 7-36*), the fast rise and fall times created will cause side frequencies to be generated, producing what is known as *key clicks*, which can be heard by the receiver of the code. This will also cause unacceptable adjacent channel interference. If the rise and fall times at make and break are made softer, or extended, as in

Figure 7-37, then the production of spurious radiation will be eliminated. This is usually accomplished by placing a capacitor across and a small choke in series with the keying device.

Chirp is a condition in a CW transmitter caused by the transmitter's oscillator being pulled off frequency when keyed (FMing), creating a bird-like chirping sound and adjacent channel interference. This is generally produced by poor regulation, a buffer or oscillator component failure, poor design, or other power supply problems pulling down the oscillator when keyed.

Figure 7-36. *CW waveform with a rapid rise and fall. Will cause key clicks.*

Figure 7-37. *A filtered CW transmission. No key clicks will be evident.*

7-6D. A Solid-State CW Transceiver

Figure 7-38 shows the basic parts, in block diagram form, of a simple CW transceiver; although nowadays these circuits, or similar circuits, are ordinarily part of a complete SSB transceiver and share many of their common circuits (See *Circuit sharing*). Furthermore, almost exclusive use of frequency synthesis for variable frequency tuning applications is predominating in the modern CW and SSB transceiver.

Looking at the block diagram of *Figure 7-38*, when the T/R switch is set to *transmit*, a relay driving transistor energizes an electromagnetic relay. This relay switches the antenna into the transmitter circuit and out of the receiver circuit while also supplying power to the transmit section — this too places an offset capacitor in the VFO circuit to provide a transmitter offset of approximately 800 Hz by slightly varying the capacitance of the VFO's frequency-setting components.

The input to the audio pre-amp is also grounded through a muting transistor during this period to provide the necessary receiver muting. When the telegraph key is then closed for transmission of a dot or a dash, the driver's collector is supplied with its Vcc through a switching transistor, turning it on. An 800 Hz tone is also produced by the sidetone generator for use by the operator, and amplified by the audio power amplifier to drive the speaker. The sidetone indicates to the operator when the telegraph key has been depressed and aids in the sending of code.

The VFO's output is fed to a buffer amplifier, to isolate the oscillator from any possible pulling effect of subsequent stages, and sent into the balanced mixer. The buffered HFO (*Heterodyne Frequency Oscillator*) frequency is injected into the other input port of the balanced

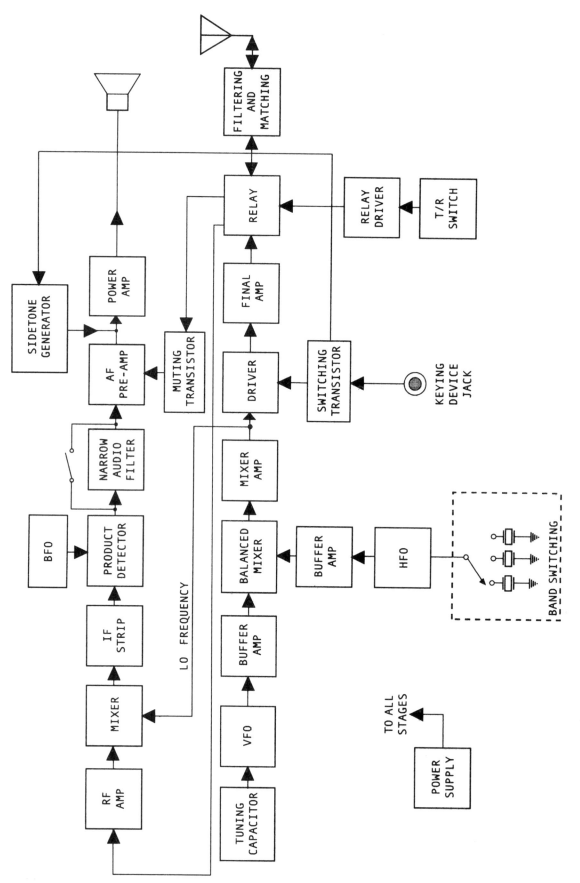

Figure 7-38. A complete multi-band CW transceiver.

mixer. The HFO selects the desired band through the operator manually switching in frequency-determining crystals. The VFO is mixed with this crystal-controlled HFO frequency. The output is amplified and filtered by the mixer amplifier to obtain the VFO minus the HFO — or the difference frequency. With this type of *pre-mixing* scheme, any frequency within a certain range can be tuned by the air-dielectric *tuning capacitor* for an almost drift-free output from the LC VFO.

The signal is sent through the driver, filtered, and transferred into the final amplifier and out through an adjustable LC filter and matching network and into the antenna.

When the T/R switch is set to *receive*, the receiver section is un-muted, the antenna is switched into the receiver and out of the transmitter circuit. Any incoming signal is now filtered by the filtering and matching network and amplified and further filtered by the tuned FET RF amplifier. It is then sent to the mixer to heterodyne with the local oscillator frequency from the mixer amplifier output; which produces a sum and difference frequency. The difference frequency is able to pass through the tuned output circuit of the mixer where it is amplified by the IF strip, which supplies most of the receiver's selectivity and amplification.

The IF is then coupled to an IC-balanced product detector. A crystal-steady BFO mixes with the IF, creating an audio tone difference, which is sent to the active narrow audio filter to restrict the received bandwidth if interference and noise is a problem. If not, the narrow filter can be switched out.

The tone is transferred to the audio pre-amp for voltage amplification, and then into the final amplifier to power the speaker.

7-7. Repeaters

Repeaters are used to extend line-of-sight communications over long ranges (*Figure 7-39*).

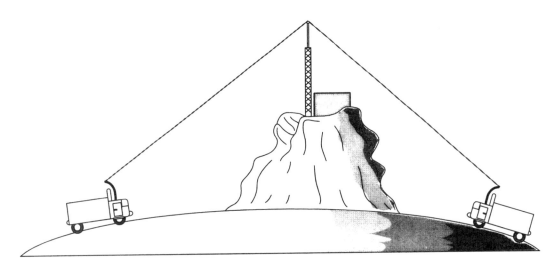

Figure 7-39. *Repeaters are typically placed on high ground to extend communication ranges.*

Repeaters have been operated by police, fire, the amateur services, and local trucking companies for years, typically in FM. Long-haul, high capacity telephone transmission by complex microwave link repeaters utilizing multiplexing and exotic modulation schemes is also commonly practiced.

The repeater concept takes advantage of increased antenna height and re-amplification to extend a small low-power and low-lying transmitter's communications range beyond the curvature of the earth or over any obstruction, such as mountains.

One type of repeater is shown in block diagram form in *Figure 7-40* and displays the antenna, duplexer, receiver, controller, and transmitter sections common to most repeaters. The repeater user will transmit on the repeater's *input frequency*, which is the frequency of the repeater's receiver section, and, if the repeater so requires, a *CTCSS* or a *DTMF* tone, riding on the user's transmitted carrier, relays to the repeater's *controller* an access code to obtain repeater access and switch on the repeater's transmitter section.

If no access code is required by the repeater, then the controller is replaced by a simple *carrier-operated relay* (COR), which will allow access to the repeater by any transmitter sending on the proper frequency with a sufficiently powerful signal. The COR is activated when the receiver's squelch is overridden, and will then open the repeater's transmitter section for retransmission of the received signal.

The repeater's dedicated receiving antenna, or a single common antenna for both receiving and transmitting, can be employed to receive a radio operator's signal. The single common antenna must operate with a *duplexer*, which permits the receiver section to receive signals while allowing the transmitter section to send signals, with both the receive and transmit frequencies separated, or *split* (common split-frequency spacings between the input and output frequencies are 600 kHz and 5 MHz, depending on carrier frequency). This eliminates the massive interference that would result by using a single common receive and transmit frequency. Retransmission of the received signal is at the *output frequency* into a transmitting antenna or into the duplexer for transmission out of the common antenna.

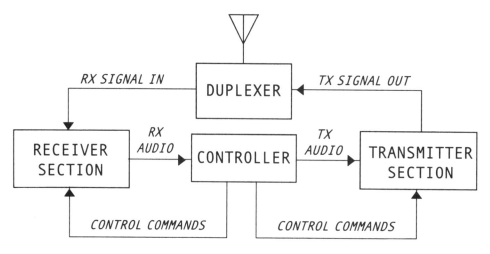

Figure 7-40. *A single-antenna repeater.*

A typical twin-antenna repeater must employ high-Q tuned cavities (See *Cavity resonators*) in the separate transmitter and receiver feedlines to decrease any receiver desensitization created by the simultaneous operation of the two units in such close proximity.

7-7A. Radiolink

As stated above, transmission of telephone conversations from coast-to-coast can be implemented with digital microwave multiplexed relay (repeater) stations, referred to as *radiolink*, and the newer *ISDN* (Integrated Services Digital Networks) systems, with each link able to carry thousands of simultaneous communication channels at up to 155 Mbits per second using special modulation techniques (however, not all radiolinks have been completely digitized as yet).

Radiolink microwave repeaters must be placed every 20 to 40 miles on tall towers using high-gain dish antennas (*Figure 7-41*). Radiolink is adopted instead of high-capacity optical fiber cable if there are no continuously available installation routes for the cable; or if bad terrain would make installation too expensive or difficult; or when quick installation of the communication's link is a priority.

To transmit huge amounts of data, even with time-division multiplexing techniques, will consume significant bandwidth. This demands the use of the microwave bands, commonly of 4, 6, 7, 8, 11, 14, or 18 GHz (short-range inter-city links are being implemented in bands between 15 and 50 GHz that provide high-capacity transmission capabilities). Above 18 GHz rain attenuation of a long-haul microwave signal becomes a significant problem.

Different radiolink systems apply different modulation methods, such as AM, FM, PM — or the most popular modern technique of all, AM/PM, referred to as *QAM* (See *Digital pulse modulation*).

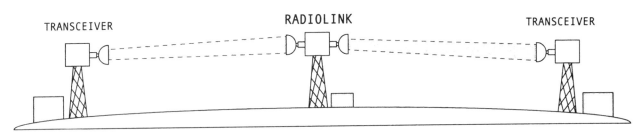

Figure 7-41. A radiolink system.

CHAPTER 8
Microwaves

Microwave communications are utilized mainly by telephone companies employing multiplexed microwave terrestrial and space-located repeater stations (See *Radiolink* and *Satellite communications*), instead of copper or fiber-optic cable. As well, certain electronic television news gathering (*ENG*) and amateur services applications employ microwaves for their communications needs, while modern satellite cable TV delivery need not be made to a single distribution point as in the past, but also directly to the cable subscriber's personal receiving dish (satellite TV).

Microwaves are in the high end of the radio frequency spectrum between 3 and 300 GHz. Normal RF amplifiers cannot be used in this microwave region because of an increase in degenerative feedback created by the interelectrode capacitance and *transit time* problems of the transistors and vacuum tubes. This causes the gain of the amplifier, as the frequency increases, to become poor (See *Amplifier frequency limitations*). Also, because the values of any reactive components must become smaller and smaller as the frequency is raised, their physical dimensions in any amplifier circuit reach a point where they are no longer viable. In addition, stray reactances of even a single piece of wire or the reactance of a common resistor can adversely affect circuit operation. Special components, certain wiring considerations, and different designs for amplifiers must therefore be implemented.

8-1. Microwave Semiconductors

Microwave semiconductor amplifiers and oscillators are reaching higher and higher frequencies and are displacing the once dominant vacuum tube in low-power microwave applications. With the utilization of exotic materials and designs, microwave transistors are ever increasing in power output, frequency range, and importance in a field that was, until recently, dominated by the traveling-wave and klystron tubes.

Furthermore, for some years now, microwave diodes have been able to operate at high microwave frequencies as oscillators, mixers, switches, and detectors.

8-1A. Microwave Diodes

Schottky barrier diodes (also called *hot-carrier,* or *HCDs*) are used as microwave detectors, rectifiers, harmonic generators, doubly-balanced modulators, and mixers. The Schottky diodes are mechanically strong and have a low forward barrier voltage, enabling reception of low signal levels at up to 100 GHz.

The *point-contact* diode is able to perform as a microwave detector and mixer. A very fine wire (a *cat's whisker*) is utilized as the cathode, which makes contact with the silicon anode chip (*Figure 8-1*). This point of contact has extremely low capacitance due to the small surface area involved. The diode needs only a low forward voltage, which makes it perfect for low signal applications. It is, however, very fragile physically, and cannot survive large signal currents or mechanical impacts.

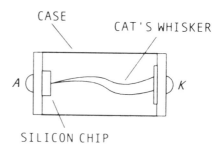

Figure 8-1. *A microwave point-contact diode showing its internal structure.*

Gallium arsenide *varactor diodes* are utilized in the microwave region, commonly in frequency multipliers, as are *step-recovery (snap) diodes.* In the frequency multiplier circuit of *Figure 8-2*, multiplication is obtained as the signal at the input turns on and off the varactor, or a step-recovery diode, creating numerous harmonics due to the diode's non-linear response to this input frequency.

This circuit does not contain discrete components as a lower-frequency circuit might, but is instead etched on a circuit board in the form of *microstrip* or *stripline*, with the bandpass filter for the output frequency and the bandstop filter for the second harmonic commonly being cavity resonators. Unlike in an active frequency multiplier, there is an insertion loss — however, this circuit requires no external power source to operate.

PIN diodes are operated in the microwave region as current-controlled resistors used to attenuate an incoming signal, as AM modulators (See *PIN modulator*), or to switch signal paths between circuits. Attenuation of an incoming signal is varied depending on the amount of forward DC bias placed across the diode. More forward bias decreases the resistance, while a decrease in forward bias increases the diode's resistance. Signal switching works by supplying a forward bias to the diode, which turns on the device, permitting a signal to cross along with the DC bias. A reverse DC bias turns off the device, blocking the signal.

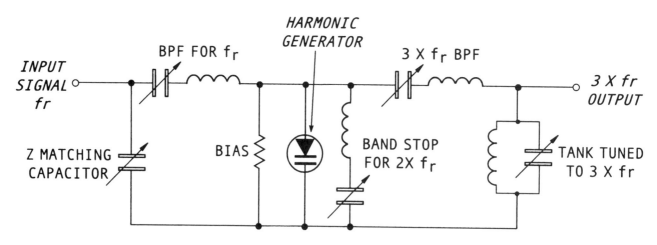

Figure 8-2. *A harmonic generator for frequency multiplication.*

Gunn diodes are used as microwave oscillators, with their frequency of oscillation dependent on the transit time of the electron through the device and, when placed in the proper resonant cavity, the diode can be made to oscillate at up to approximately 50 GHz. Its output power is contingent on the diode's design frequency — the thinner the semiconductor, the higher the frequency, but the lower the output power, while the lower this operating frequency, the higher the output power possible (up to 3 to 4 watts).

The Gunn device has a very limited frequency adjustment range when placed in a cavity with a *tuning-screw* (*Figure 8-3*), which varies the cavity's reactance. For much wider Gunn diode frequency variations a *YIG crystal sphere* (Yttrium-Iron-Garnet, *Figure 8-4*) is utilized, which tunes the frequency of the cavity in proportion to YIG coil current through an electromagnet. The bias across a *varactor tuning diode* (*Figure 8-5*) can also be changed to slightly modify the output frequency of the Gunn oscillator.

Other diodes are available for use in oscillators, such as *impatt* diodes, capable of high output powers (25 watts), but at the deficit of high noise generation; *trapatt* diodes, employed usually in high-power pulse-mode only; and *barritt* diodes, which are low-noise, but at low power levels of a few milliwatts.

Tunnel diodes can be adopted as a high-speed switch or made to oscillate at microwave frequencies. However, they have a low power output and become unstable with temperature changes. Tunnel diode use has declined over the years since their introduction, with the Gunn diode assuming most of their functions.

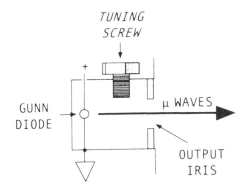

Figure 8-3. *A Gunn oscillator with tuning-screw frequency adjustment.*

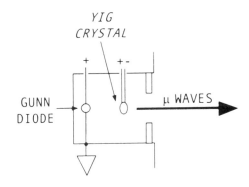

Figure 8-4. *A Gunn oscillator with YIG frequency adjustment.*

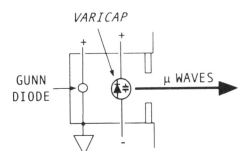

Figure 8-5. *A Gunn oscillator with varactor frequency adjustment.*

8-1B. Microwave Transistors

Gallium Arsenide FETs (*GASFETs*, sometimes referred to as *MESFETS*) are depletion-mode FETs, and are more popular than microwave bipolar transistors. They must be used instead of the transistor if amplification above 10 GHz is needed — and are capable of amplifying up to 20 GHz. Also, FETs have higher gain, increased input resistance, better thermal stability, and lower IMD and noise generation than bipolar microwave transistors.

The *power MESFET* is a high-power GASFET with an output up to 50 watts at 1 GHz and several watts at 15 GHz.

Microwave bipolar transistors are operated in amplifiers and oscillators, are made only as the silicon NPN type, and are usually placed on 50 ohm microstrip. Maximum frequency is around 12 GHz, but are generally used at 8 GHz and below. Most high-frequency transistors have four leads, such as the common *SOE* (*stripline-opposed emitter*) package that uses two emitters (*Figure 8-6*).

Another such type of high-frequency transistor is the *junction-tetrode transistor* of *Figure 8-7*. This transistor uses two base leads. The extra base terminal is supplied with 0.4V, which decreases capacitance and base resistance and thus increases high frequency response.

Many VHF-and-above solid-state devices are vulnerable to burnout by accidental application of high signal levels. As well, high-frequency transistors are easier to damage with electro-static discharge due to their smaller and/or thinner junctions. This is why most inputs to the RF stage of a VHF receiver are protected by diodes (*Figure 8-8*).

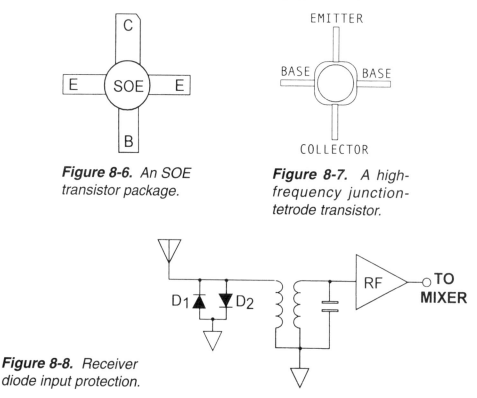

Figure 8-6. *An SOE transistor package.*

Figure 8-7. *A high-frequency junction-tetrode transistor.*

Figure 8-8. *Receiver diode input protection.*

Older high-frequency transistors, such as the *point-contact transistor*, being basically constructed with a base of silicon and a collector and emitter made from two fine wires, are now rarely seen.

Transistors made for UHF or VHF may cause parasitics if used at HF, so it is a good idea to replace all RF transistors with an equivalent model in all RF circuits.

8-2. Microwave Tubes

For higher power and frequencies than can be obtained from semiconductors, vacuum tubes must be employed. Microwave tubes adopt the principle of *velocity modulation* to amplify a signal: Alternately increasing and decreasing the speed of a stream of electrons causes the electrons, when their velocity is decreased, to give up energy to an output probe.

8-2A. Klystrons

Twin-cavity and multicavity *klystrons* are used as microwave amplifiers or oscillators. *Figure 8-9* shows a two-cavity klystron. A cathode inside the electron gun is heated, releasing a stream of electrons toward the positively charged beam collector plate. This beam of electrons must pass through two cavity resonators, which set the resonant frequency of the klystron. The first cavity (the *buncher*) is the input, the second (the *catcher*) is where the output is tapped off. As the electron beam passes through the first cavity the input signal needing amplification velocity modulates this stream, which then excites the second cavity into oscillation at the cavity's resonant frequency. This power is tapped off by a conductive loop. Thus the original input's signal frequency causes the cathode's direct current electron beam to be transformed into an alternating stream of electrons, causing amplification of this low-power input signal.

Adding more cavities (*multicavity klystrons*) increases the amplifications of the input signal and can be used to increase its relatively narrow bandwidth.

Klystrons are capable of output powers of thousands of watts with tubes that can be up to five feet in length. Frequencies of operation can commonly reach between 3 to 100 GHz.

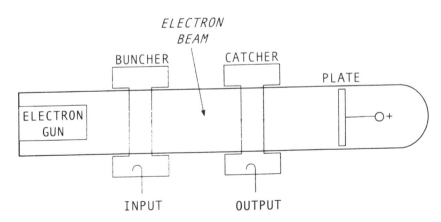

Figure 8-9. A two-cavity klystron microwave tube.

For the klystron to function as an oscillator, some of its output signal is injected in-phase into its input port, creating positive, or regenerative, feedback — and thus oscillations.

The *reflex klystron* is a single-cavity klystron used as a low-powered microwave oscillator. It is normally found only in older equipment, with *Gunn diodes* being the replacement device in modern circuits.

8-2B. Magnetrons

The *magnetron* is a high-powered, highly efficient narrow-bandwidth microwave vacuum tube oscillator, usually found in radar transmitters and in some microwave transmitters.

The magnetron consists of a uni-directional device (basically a diode) in which the magnetic field of a very powerful magnet passes (*Figure 8-10*). This magnetic field forces electrons to escape from the heated cathode (See *Thermionic emission*), causing them to travel in a spiral path around the rod-shaped cathode, which is surrounded by the positively charged anode. Due to the varying magnetic and electric fields effecting the electrons, they are sped up and slowed down, releasing energy into the resonant cavities, causing oscillations. Some of this power is removed by a wire loop probe placed in one of the cavities.

The magnetron is able to deliver up to 1 kW at 1 GHz, and approximately 20W CW at 10 GHz.

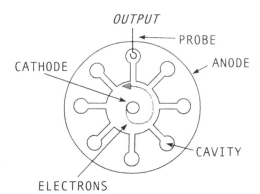

Figure 8-10. *Operation and internal structure of a magnetron.*

8-2C. Traveling-Wave Tubes

Traveling-wave tubes (TWT) are used as amplifiers or oscillators up to a few hundred GHz. They are radiation hardened (for satellite use), very reliable, have a wide bandwidth and high gain, and can generate up to 20,000 watts of power. It is the premier microwave communication's tube in use today.

Electrons are emitted from the heated cathode (*Figure 8-11*), into the acceleration anode, which accelerates and focuses the electron stream towards the positively charged collector plate. The input signal to be amplified is now injected into the helix. The helix slows down this signal to the speed of the electron beam, causing their two fields to interact, producing velocity modulation. More electrons are slowed down than are sped up, which induces in-phase voltages into the helix, causing amplification of the original signal. The output is taken from the opposite end of the helix.

For use as an oscillator, some of the TWT's output is injected in-phase into its input, creating positive, or regenerative, feedback.

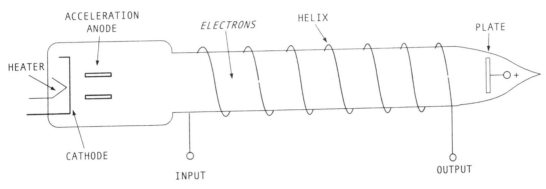

Figure 8-11. *Operation and internal structure of a TWT.*

8-3. Microstrip, Stripline and Cavity Resonators

Microstrip and *stripline* are both operated as short transmission lines and as tuned circuits and high-Q filters, while high-frequency components (BJTs, FETs, diodes) can be mounted directly on the microstrip and stripline's PCB (Printed Circuit Board).

Cavity resonators are found, instead of conventional LC tuned circuits, at microwave frequencies, and are employed as bandpass or bandstop filters, or to set the frequency of an oscillating active device.

8-3A. Microstrip and Stripline

Microstrip (*Figure 8-12*) is usually one-quarter or one-half of a wavelength long, and is unbalanced. Due to its unshielded construction it can radiate RF. Balanced stripline (*Figure 8-13*), because of its twin ground planes, does not. Both commonly have a dielectric made of a printed circuit board of fiberglass, polystyrene, or Teflon, while the conductor and ground plane are commonly made of PCB copper foil strips. Microstrip is cheaper and easier to fabricate since it can use standard PCB manufacturing techniques. Their characteristic impedances are ordinarily between 10 to 100 ohms.

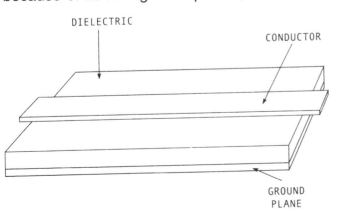

Figure 8-12. *Structure of microstrip.*

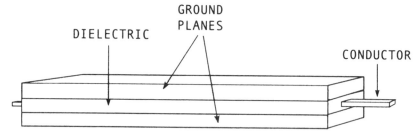

Figure 8-13. *Structure of stripline.*

As stated above, microwave components can be soldered to the printed circuit board along with the PCB's integral printed microstrip or stripline. The *stripline high-Q filter* (*Figure 8-14*), a common microwave bandpass filter, is also easily created by using a small tuning capacitor and copper traces on a PCB and enclosing the entire assembly in a shielded case. It has low insertion loss (around 1 dB), a narrow bandwidth, and steep skirts. Extra selectivity can be obtained by coupling together two or more such stripline filters.

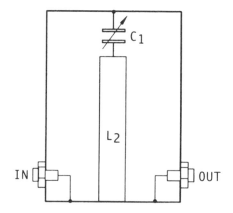

Figure 8-14. *A popular stripline high-Q bandpass filter used at microwave frequencies.*

8-3B. Cavity Resonators

A device called a *cavity resonator* can be used as a tuned circuit at microwave frequencies. The resonator consists of a hollow metal cavity that has the ability to select its own natural resonant frequency and reject all others. Volume of the cavity dictates this natural resonant frequency. Due to a very high Q, it has a very sharp filter response.

A resonant cavity can be operated as a tunable bandpass filter driven by a waveguide through coupling apertures (*Figure 8-15*), or coax driven by 1/4 wavelength probes (*Figure 8-16*); or as the frequency-controlling element in some microwave oscillators.

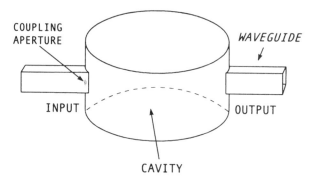

Figure 8-15. *A waveguide-driven cavity resonator.*

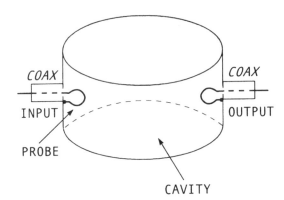

Figure 8-16. *A coaxial-driven cavity resonator.*

The cavity resonator is commonly employed in repeaters to lower *desensitization* (a close unwanted frequency can cause an overloading in the first RF amplifier or the first mixer of a receiver, producing a loss of sensitivity to the desired signal), and can be placed in both the repeater's transmitter and receiver feedlines.

A *duplexer*, a device that allows one antenna to be used for transmitting and receiving, can also utilize resonant cavities to pass the incoming signal from the antenna to the receiver, while allowing the transmitter to place its signal on to the antenna for transmission. Duplexers employ two cavities with each resonant to a different frequency — one frequency for the received signal, the other for the transmitted signal.

Cavity resonators can also be made tunable by mechanically changing their volume, as in *Figure 8-17*. As the tuning knob is rotated in one direction the flat disk will lower, decreasing the volume of the cavity, thus increasing the operating frequency of the resonator. As it is rotated in the opposite direction the volume is increased, thus lowering its natural resonant frequency.

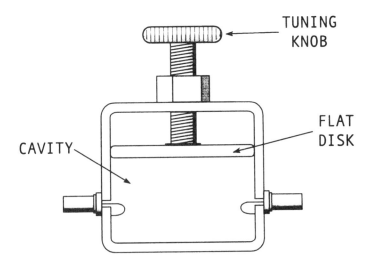

Figure 8-17. *A tunable cavity resonator.*

CHAPTER 9
Antennas and Transmission Lines

Antennas convert alternating current into electromagnetic waves and inject these waves into free space; or receive electromagnetic signals from free space and convert them into an alternating current. There are many types of antennas available, and their design depends on the transmitting or receiving frequency, the directivity required, cost considerations, and space limitations.

Transmission lines are used to transfer this alternating current to or from a receiver or transmitter with minimal loss and maximum rejection of any external interference.

9-1. Antennas

A resonant antenna acts as a series resonant circuit. When an antenna is cut to a 1/4 or 1/2 wavelength, or typically multiples of a 1/2 wavelength, then maximum current flows through the antenna elements, thus giving maximum signal strength (*Figure 9-1*).

Figure 9-1. *Equivalent dipole antenna circuit.*

Radiation patterns, however, vary with the transmitter's harmonic frequency when the antenna length remains the same. In *Figure 9-2a* a dipole is being run on its fundamental frequency as a normal half-wave antenna; *Figure 9-2b* shows the same antenna operated on its second harmonic as a full-wave antenna; *Figure 9-2c* demonstrates the dipole running on its third harmonic as a one and a half-wave antenna.

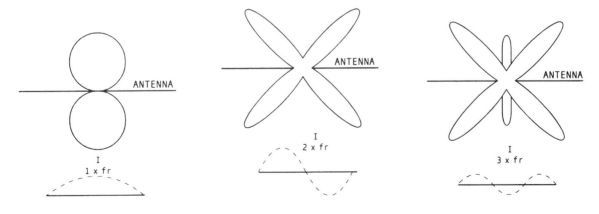

Figure 9-2. *Radiation patterns of dipole antenna run at its: (a) Fundamental half-wave; (b) as a second harmonic full-wave antenna; (c) as a third harmonic 1-1/2 wave antenna.*

These changes in the radiation pattern of the antenna will obviously affect its directional characteristics, sometimes drastically. Still, an antenna operated at its fundamental as a half-wave, and at its third harmonic as a one and a half-wave antenna, will both give the same low input impedance at their center-fed feedpoint, making harmonic operation possible with a single antenna.

A proper impedance match between the transmitter, the feedline, and the antenna, will prevent the RF energy generated in the transmitter from being reflected back into its finals, possibly causing severe power output loss, transmitter damage, and transmission line insulation breakdown.

The antenna feedpoint impedance will change if the frequency changes; if the frequency stays the same, but the antenna length varies, then the same occurs; as well, as an antenna gets physically closer to the ground its feedpoint impedance lowers (at ground it nears 0 Ω) which, of course, further changes the antenna's input impedance and subsequent reflections.

A perfect match between the transmitter, its feedline and the antenna will occur when the inductive reactances equals the capacitive reactances and cancel (or $X_L = X_C$), and the pure resistances equal, typically, between 50 and 75 Ω (*Figure 9-3*). When this point of matched impedances occur, maximum power will be radiated from the antenna. To put it another way, the antenna must have an equal impedance match to its transmission line's *surge impedance*, or standing waves will result on the mismatched transmission line, and much of the subsequent reflected power will be given up as heat in the line. Surge impedance is the transmission line's natural input and output impedance, which is a function of its physical construction.

When the antenna is resonant and the impedances between the transmitter, feedline, and antenna match, maximum alternating current is produced when the RF signal is injected into the elements of the antenna.

Figure 9-3. *A perfect impedance match between the transmitter, feedline, and antenna, resulting in maximum power output.*

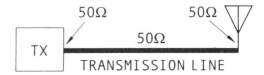

On the antenna, maximum standing waves produce maximum voltage at the ends of the antenna, and maximum current in the middle, as demonstrated using the dipole antenna of *Figure 9-4*. This creates maximum antenna output power. Radiation from the antenna is generated when the frequency reaches a point where the field lines produced by the high-frequency alternating current cannot collapse back into the antenna before the alternating current changes polarity, which usually occurs at around 30 kHz. Some of this energy is radiated into space as an electromagnetic (RF) wave at the speed of light. The receiving antenna's elements are then cut by this wave, inducing a small voltage, which must be amplified and filtered by the receiver's circuitry.

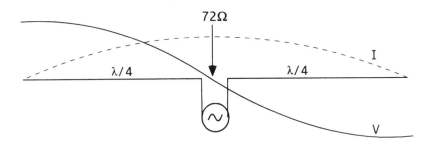

Figure 9-4. *Current and voltage distribution on a half-wave resonant dipole.*

Further, the reason that high-frequency antennas are short, and low-frequency antennas are long, is that the antenna must be long enough to prevent the RF electrons from reaching the end of the antenna and attempting to return before the RF driving source (the transmitter) potential reverses.

The strength of a radio wave that is radiated from an antenna is dependent on the RF current in the antenna. The more current, the more RF power. When maximum current is reached in an antenna, then antenna resonance has occurred. If the frequency from the transmitter changes, the transmitter's antenna eventually becomes non-resonant at some frequency.

9-1A. Polarization

Both sending and receiving antennas should be properly *polarized*. Polarization refers to the orientation of the electric field of the electromagnetic wave through space.

A *horizontally* polarized antenna (*Figure 9-5*, the antenna's elements are parallel with the ground) will receive an electromagnetic wave from a *vertically* polarized antenna (*Figure 9-6*, the antenna's elements at right angles to the ground) only due to a small shift in polarization that occurs over distance. If it were not for this wave change, a vertical antenna could not induce a voltage into a horizontal antenna, and visa-versa.

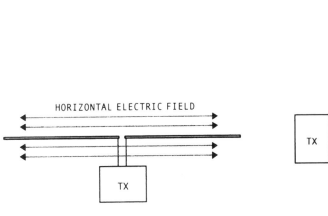

Figure 9-5. *A horizontally polarized antenna showing the horizontal electric fields.*

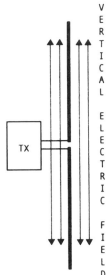

Figure 9-6. *A vertically polarized antenna showing the vertical electric fields.*

A way to overcome this problem is, of course, to make sure both antennas are appropriately polarized or, in the higher-frequency range, the use of a *circular* polarized antenna, such as the *helical* type of *Figure 9-7*, will induce a voltage into either horizontally or vertically polarized antennas with very little loss (about 3 dB). However, when communicating with other helicals, they must be of the same *sense* (direction of winding), or extreme signal attenuation will take place between the transmitting and receiving antennas.

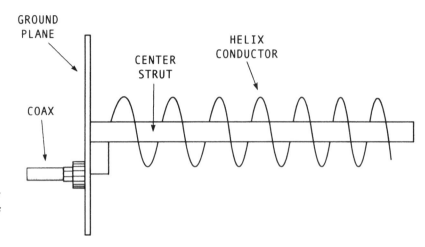

Figure 9-7. A high-frequency circularly polarized antenna of the helical type.

9-1B. Common Antenna Types

In general, a vertical antenna radiates in all directions (*Figure 9-8, omnidirectional*), and is commonly a quarter of a wavelength long, with earth providing the other necessary quarter wavelength through ground reflection (*Figure 9-9*). This is the type regularly adopted for AM, FM, and TV commercial broadcasts, with the vertical tower itself acting as the 1/4 wave antenna, supported by non-resonant *guy wires*.

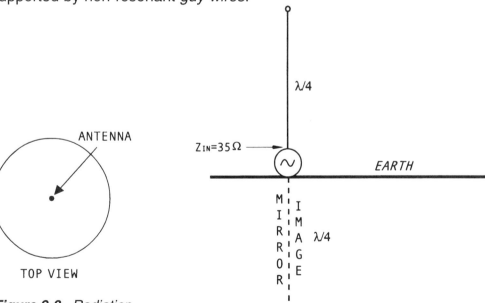

Figure 9-8. Radiation pattern from a vertical antenna.

Figure 9-9. A quarter-wave vertical with earth providing the other quarter wavelength.

Resonant guy wires would act as antennas, radiating the signal or its harmonics and increasing RF losses. The guys are broken up by *strain insulators* placed at the appropriate non-resonant sections of these wires.

Since the 1/4 wave vertical depends on a proper ground for completion of the other missing 1/4 wave, a high-quality ground is essential. This can be a problem in some areas. The use of four or more 1/4 wavelength wire *radials* placed at the base of the antenna and parallel to the earth can be adopted as an artificial ground. This group of radials is called a *counterpoise* or *groundplane*.

Horizontal antenna types have more directional characteristics — all the way from the bi-directional half-wave dipole, with the horizontally polarized radiation pattern of *Figur*e 9-2a, to the highly directional parasitic multi-element Yagi radiation pattern of *Figure 9-10*.

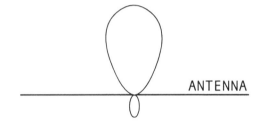

ANTENNA

Figure 9-10. *The radiation pattern of the average Yagi antenna.*

A common parasitic Yagi antenna is shown in *Figure 9-11*. Parasitic refers to an element, made of either wire or tubing, that is not directly driven by the feed line, but has a voltage induced into it by the *driven* element. A single driven element, a dipole, is fed by the transmission line with the radio frequency energy from the transmitter, while one or more progressively longer *reflectors* reflect back to the driven element an in-phase signal that adds to the driven element's radiation. With the addition of one or more progressively shorter *directors* in front of the driven element, the radiation pattern is further enforced, resulting in an extremely directional, high-gain antenna.

Both the dipole and the Yagi can be vertically polarized if the frequency is high enough to allow the antennas to be small enough to be so physically oriented.

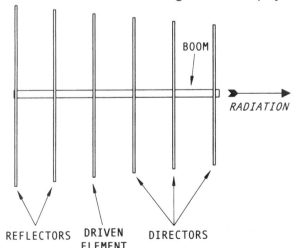

BOOM

RADIATION

REFLECTORS   DRIVEN
                ELEMENT   DIRECTORS

Figure 9-11. *Construction of a popular Yagi antenna.*

Another classification of the directional antenna is the *driven array*, which uses two or more driven elements fed with a certain phase relative to each other, with each element placed at set distances from one another to create either a bi-directional or unidirectional beam pattern. *Phasing harnesses* or *quarter-wave matching stubs* can be used to ensure the proper phasing to these elements. Common driven array examples would be the bi-directional *collinear array*, the bi-directional *end-fire array*, and the omni-directional and wide bandwidth *log-periodic array*.

Special antennas are regularly employed in the microwave region. There are two very popular antennas used in the mid-to-high microwave frequency range: The *dish* and the *helical* antenna.

The dish antenna (*Figure 9-12*) utilizes a spherical or parabolic-shaped focusing bowl with a horn antenna set at the focal point. The dish can be constructed of a solid conductor (usually sheet metal) or wire mesh, with holes 1/8 of a wavelength or less, which act as a solid shield. The horn is attached to a waveguide for transmitting or receiving a signal. The horn itself is merely a flared out portion of the waveguide, similar to a trumpet's bell, which acts as an impedance match between the waveguide and the surrounding space. Gain is dependent on the size of the dish, with the diameter customarily designed to be ten times its wavelength of interest to obtain high gain, low loss, and extreme directivity.

The helical antenna (See *Figure 9-7*) is made of conducting wire or tubing wrapped in a helical manner around a non-conducting center strut, typically fed by a short run of coax placed through a ground plane reflector. The output of the antenna is circularly polarized and, as stated above, the receiving antenna must be wound with the same sense as the sending antenna or little or no reception is possible.

Gain and beam width of the helical does not measure up favorably with the dish antenna, but the helical is inexpensive and easy to construct for non-demanding communications applications.

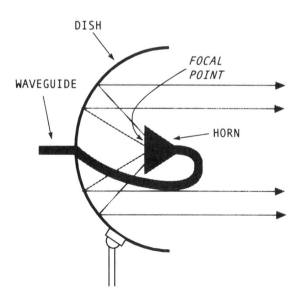

Figure 9-12. A high-gain microwave dish antenna.

9-2. Transmission Lines

Transmission lines, simply, are conductors designed to transfer current from one point to another, without radiating, at a desired impedance. There are two types of RF transmission lines: *balanced*, such as twin-lead; and *unbalanced*, such as coax.

In the microwave region *waveguides* are sometimes employed instead of coaxial cables or twin-lead.

In addition, impedance-matching considerations and matching networks between a transmitter, the transmission line, and its antenna, are vital considerations to lower reflected waves and increase the effective radiated power output.

9-2A. Transmission Line Types

Balanced lines are typically 300 Ω twin-lead (*Figure 9-13*). Unlike unbalanced line (coax) there is no single conductor at ground potential, and each conductor on balanced line has an equal-in-amplitude but opposite-in-phase signal traveling in each line.

Twin-lead is generally used as a feed line to a TV or FM consumer receiver or, rarely, as a balanced feed to a dipole transmitting/receiving antenna. It has very low line losses, is capable of supporting high voltages (high VSWR) on the line before arc-over — but is not commonly available in the proper impedance for most transmitters and receivers without matching networks.

Unbalanced line, such as standard coax (*Figure 9-14*), is shielded to prevent receiving or radiating any signal from the line. The outer shield is at ground potential and the inner conductor carries the RF current.

Coax can be of the flexible or rigid type. Flexible is by far the most prevalent and comes in two basic surge impedances of 50 and 75 Ω, with many other characteristics, such as losses per foot, size, and breakdown voltages of equal importance when choosing coax. These characteristics are shown for each common coax type in *Table 9-1*.

Special coax is available that can be used at frequencies up to 10 GHz.

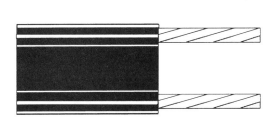

Figure 9-13. *Construction of twin-lead transmission line.*

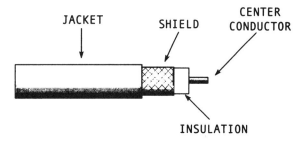

Figure 9-14. *Construction of flexible coaxial cable.*

TYPE	IMPEDANCE (W)	OD (IN)	LOSS AT 100 MHz (dB/100ft)	APPLICATIONS
RG-58	50	0.196	5	CB/HAM — SHORT RUNS, LOW POWER/FREQ
RG-59	75	0.242	3	CB/HAM/TV — MEDIUM POWER/FREQ
RG-8	50	0.405	2	HF/VHF/UHF — LONG RUNS, HIGH POWER
RG-11	75	0.405	2.5	HF/VHF/UHF — LONG RUNS, HIGH POWER
RG-214	50	0.425	1.9	HF/VHF/UHF — LONG RUNS, HIGH POWER

Table 9-1. Common coaxial cables with specifications and applications.

9-2B. Reflected Waves

As stated earlier, with the transmitter's output impedance matched to the transmission line, and the transmission line matched to the input impedance of the antenna, no standing or reflected waves will be present in the transmission line, and no power will be wasted as heat (except in the pure resistance of the line as copper (I^2R) losses). With all impedances matched, the transmission line appears infinitely long, causing no standing waves to be reflected back into the transmitter finals while transferring maximum power to the antenna. This type of line is referred to as *flat line* (*Figure 9-15*). When there are high standing waves present on the transmission line, as in *Figure 9-16*, a very high SWR condition exists. This may damage the line's dielectric and perhaps the finals of the transmitter (especially susceptible are transmitters with unprotected transistor finals).

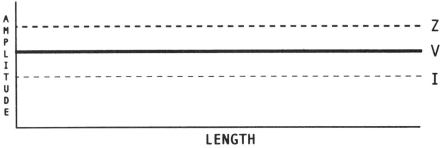

Figure 9-15. Voltage, current, and impedance on a flat line (no standing waves).

9-2C. Waveguides

A *waveguide* (*Figure 9-17*) is a microwave transmission line. It can be a round or a rectangular hollow metal conduit, usually plated internally with silver, made to conduct microwaves from one point to another with a minimal amount of signal loss. The size of the guide deter-

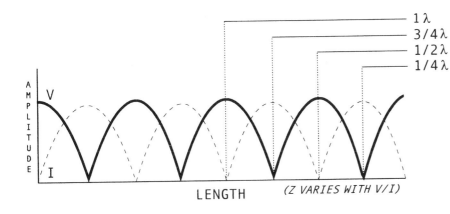

Figure 9-16. *Voltage and current on an open, or non-terminated, transmission line (maximum standing waves).*

mines its working frequency (*Figure 9-18*). Straight 1/4 wavelength probes or loop probes are commonly used to inject or remove the microwave energy from a waveguide.

Waveguides will propagate radiation above their working frequency, but not below their cutoff frequency, basically functioning as a type of high-pass filter.

Wherever feasible, modern microwave designs have eliminated much of the use of waveguides in favor of low-loss semi-rigid coax cable, with 100% solid-copper shielding, to transmit and receive these high-frequency signals.

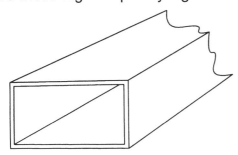

Figure 9-17. *A section of microwave transmission line referred to as waveguide.*

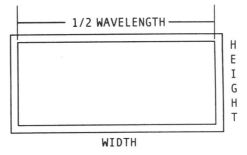

Figure 9-18. *The width of a waveguide sets its working frequency limits.*

9-2D. Antenna Matching Networks

To ensure the proper impedance match (and thus the proper power transfer) between the feedline and the antenna, or between the transmitter and feedline, some type of matching network may be necessary.

The *delta-match* (*Figure 9-19*) technique is applied by tapping, with balanced line, a dipole at a point of matching impedance. Since:

$$Z = \frac{V}{I}$$

there is always a point along an antenna that will match the feedline.

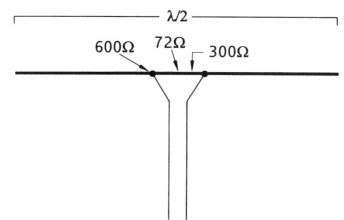

Figure 9-19. *A delta-match for impedance matching between a feedline and its antenna.*

Half-wave matching stubs, which are sections of coax or twin-lead shorted at one end, with the other end connected at the input to the antenna, can be used for this purpose. Both the stub's length, controlled by the *shorting bar*, and the antenna's feed point, controlled by the *slider*, can be made changeable (*Figure 9-20*). By moving both, a combination can be found that will match most impedances.

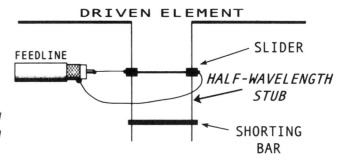

Figure 9-20. *Impedance matching using an adjustable half-wavelength stub.*

Quarter-wave sections of coax or twin-lead, called *Q-sections*, are also commonly utilized to match small impedance mismatches (*Figure 9-21*). By using a section of 50 Ω coax between the 37 Ω antenna input impedance, the 75 Ω coax will be matched to the antenna's feed point. A Q-section value can be calculated by:

$$Z_{OF\ Q\ SECTION} = \sqrt{Z_{LOAD}\ X\ Z_{TRANSMISSION\ LINE}}$$

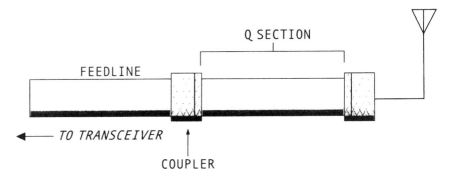

Figure 2-21. *A Q-section used for impedance matching.*

The *pi-network* of *Figure 9-22* is frequently employed to tune out all reactances, leaving the proper pure resistance between the feedline and the antenna. It is primarily used in the finals of tube and tunable transistor transmitters (See *Pi-network*).

Another method for matching quarter-wave verticals is shown in *Figure 9-23*. The antenna slider moves the feedline's insertion point up and down as the variable C_1 is adjusted, creating a proper impedance match.

A *transmatch*, with one type shown in *Figure 9-24*, uses variable capacitors and/or variable inductors to cancel any reactances in the antenna circuit while changing the antenna's resistance, as seen by the transmission line, to an appropriate value (usually 50 or 75 Ω). If inserted between the transmitter and feedline the transmatch will allow the transmitter to see only its proper characteristic impedance. However, any standing waves on the transmission line itself will not be improved.

Baluns (*Figure 9-25*) are also commonly utilized for impedance matching. Most are wideband RF untuned transformers adopted to match balanced to unbalanced line, and visa-versa, by isolating the two types of feedlines while matching any impedance mismatches.

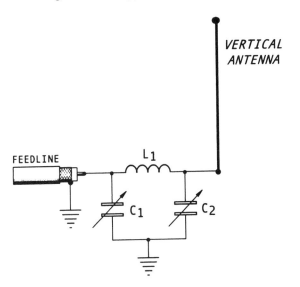

Figure 9-22. *Pi-network impedance matching.*

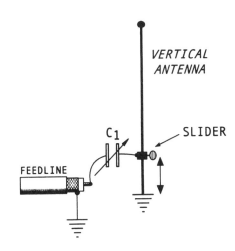

Figure 9-23. *Matching a quarter-wave vertical to its feedline.*

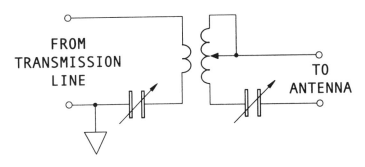

Figure 9-24. *One form of a transmatch. Employed for impedance matching.*

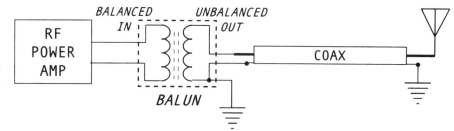

Figure 9-25. *A balun placed between a transmitter's finals and coaxial transmission line.*

9-3. RF Propagation

The method of propagation of radio waves is dependent on the frequency of the carrier and is by three main modes (*Figure 9-26*): *Ground waves*, which travel along and through the surface of the earth; *surface waves*, sometimes referred to as *space* or *direct* waves, travel through the atmosphere in a relatively straight line from transmitting antenna to receiving antenna; and the *sky wave*, produced by a radio signal bouncing (refracting and reflecting) off of the ionosphere.

Only very low frequencies are propagated by ground waves (below 1 MHz) to any significant degree, while surface waves are the primary form of propagation for 30 MHz and above, and extend only to slightly more than line-of-sight (*LOS*) distances. These waves propagate about 1/3 farther than LOS due to the earth's atmosphere creating a bending effect of the radio wave, forming a longer *radio horizon* than an optical horizon.

Sky waves are the principal way low- and high-powered simplex long range RF communications can take place. The frequencies utilized are essentially under 30 MHz, and the *optimum usable frequency* (OUF) varies tremendously with the time of day, sunspot activity and their eleven-year cycle, and the angle-of-radiation from the transmitting antenna. The opti-

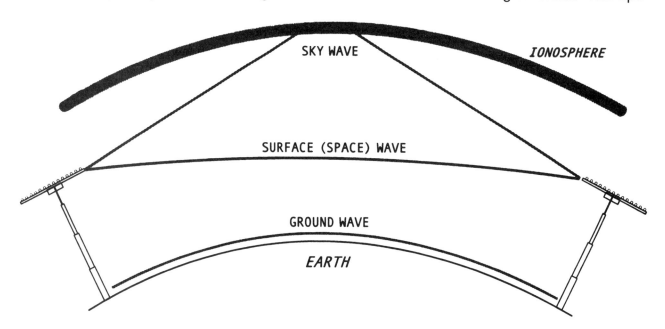

Figure 9-26. *Three modes of RF propagation.*

mum usable frequency is the frequency at which the most reliable communications can take place via the sky wave. It is always between the *lowest usable frequency* (below the LUF the ionosphere absorbs the signal) and the *maximum usable frequency* (above the MUF the signal passes through the ionosphere and out into space).

Multiple-hop transmissions caused by the transmitted RF signal bouncing more than once off of the ionosphere can occur, extending the communications range of a radio signal to many thousands of miles — with relatively little output power.

Troposcatter is a form of dependable long-range communications (about 450 miles) using high-powered UHF transmitters with high-gain antennas. These transmitters send a powerful signal into the troposphere (a section of the atmosphere about six to ten miles above sea level), where the signal scatters. A minute portion of that scattered signal is picked up by a very high gain receiver and amplified.

Due to the high cost of the equipment, the military and large industry are the primary users of this communications method.

CHAPTER 10
Troubleshooting

Troubleshooting is a vital part of the technician and field engineer's job. Unless the equipment and circuits that are being repaired are well understood, troubleshooting down to the component level will be a very time consuming job.

While the all too common under-skilled board-swapper has taken over the lowly tasks of the old vacuum tube-swapper, a technician who can actually troubleshoot down to the component level is a very rare and valued breed. It demands little skill to find a board that displays a red light, indicating a bad board, in a piece of complex equipment. On the other hand, it requires exceptional skill and knowledge to locate the actual *component* that caused that fault. This skill can be obtained by studying, as well as, of course, plenty of practical experience.

10-1. General

Keep in mind that *all* power amplifiers, whether RF or audio, must be terminated by their output impedance (Z_{OUT}), or they can be destroyed when activated. This is accomplished by the use of a resistor, dummy load, or other appropriate resistive terminator. Only then can the amplifier or transmitter be operated and the relevant tests conducted.

Also, many new technicians feel that a sudden fault in a transmitter or receiver, such as low receive or transmit output power, is caused by misalignment of its stages. Seldom is this the case. Lack of alignment is gradual and usually caused by aging components. However, any attempted repair or realignment by an incompetent or inexperienced technician *can* cause a receiver or transmitter to need retuning.

Furthermore, since high power always equals heat, and heat always equals high failure rates, look at the heat producing circuits first; such as a power supply, a transmitter final, or an audio power amplifier, etc.

10-1A. Safety

Safety is always of paramount concern, since one fatal mistake will be your last. Here are some basic safety rules:

1. When working on high current and/or high voltage circuits, place one hand in a pocket when probing the live circuits.

2. Use an isolation transformer when working on transformerless equipment.

3. Discharge all high current/voltage capacitors through a wire and a 100Ω resistor (some capacitors can hold their charge for weeks).

4.   Try not to work alone.

5.   Never place yourself at ground potential when working on live equipment.

6.   Electrolytic capacitors can explode, sometimes for no apparent reason: Wear glasses when working on all circuits with electrolytics under power.

10-1B. Basic Troubleshooting Steps

The basic steps in electronic troubleshooting are: (A) *Define the symptoms* (low power, no power, off-frequency, distortion, etc.); (B) *Restrict the fault to a section* (the IF strip, the frequency converter stage, the audio amplifiers, the power supply, etc.); (C) *Track down the fault to a single bad stage* (the driver, the 1st IF stage, the LO, the final power amplifier, etc.); (D) *Trace the fault down to a bad part* (capacitor, transistor, relay, solder joint, etc.).

First, obtain the service literature and, if you are very lucky, it will contain two or three of the following: A *functional block diagram* (each transistor in the device compromising a separate block); a *servicing block diagram* (each transistor comprising a separate block as well as pictorial waveforms and/or voltages at test points); a *schematic diagram*; and a *parts layout diagram* (which shows the actual physical layout of the parts).

Step A. *Define the symptoms*: First make sure the unit-under-test is plugged in, *that all operator controls are in their proper position,* and that there are no exterior circuit breakers or fuses blown. If fuses are blown, then replace fuses with proper amperage rating as well as the proper voltage rating (which is the maximum rating, when the fuse is in the open condition, before arcing between its contacts occur). Fuses can, and do, fatigue open with time; confirm that fuse does not blow again under full operating conditions. If it does, then open the unit's case for further troubleshooting.

Operate the device and observe incorrect operation or, employing test equipment, search for improper frequency response, distortion, power output, etc.: Is the transmitter area of the transceiver bad? Or is it the receiver area?

Step B. *Restrict the fault to a section*: First, crack the case and *look* for burnt components, broken traces, bad solder joints, smoke, etc.; *smell* for burning resistors, transistors, and overheated transformers; *touch* (carefully), checking for excessively hot components; *listen* for voltage arcing or sizzling of burning parts. This preliminary inspection is extremely important! Also, since modern transceivers usually employ some sort of *circuit sharing*, keep in mind that a fault in a single circuit can sometimes take out both the transmitter and receiver areas.

If the fault is not obvious, then *bracketing* must be used along with *signal injection* or *signal tracing*. Bracketing refers to taking an entire area, such as the transmitter or receiver of a transceiver, that has a good input, but a bad output, and continuing to move your probe from section to section until the problem is confined, within these imaginary brackets, down to the bad section. For example, a receiver could be bracketed until a lack of receiver output is discovered to be located in the IF strip of a receiver, which may contain a chain of three or four amplifiers and numerous filters (*Figure 10-1*).

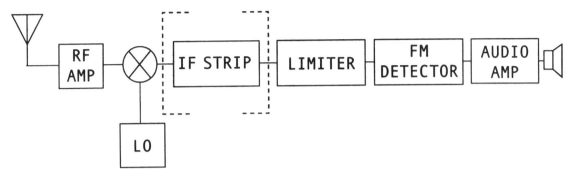

Figure 10-1. *Troubleshooting by bracketing down to a section.*

Step C. *Track down the fault to a bad stage*: Use the functional block diagrams to isolate the problem to an individual circuit by narrowing down the probable bad stage by continued bracketing from stage to stage, instead of from section to section — such as from the speaker to the audio power unit of *Figure 10-2*.

In addition, use all available schematics as needed and check voltage levels and waveforms at the appropriate test points. For example, is the voltage amplifier, driver, or power amplifier in a non-functioning audio amplifier chain bad?

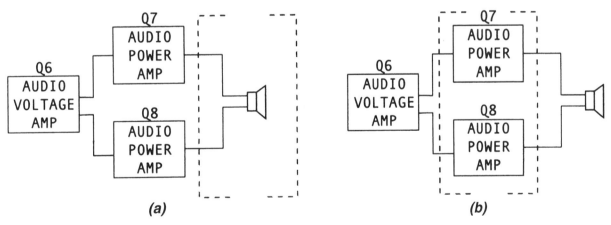

Figure 10-2. *The process of bracketing: (a) First check for a bad speaker; (b) then backup to the power amplifiers.*

Step D. *Trace the fault down to a bad part*: Continue to utilize sight, touch, smell and hearing to check for burnt, overheated, cracked or arcing parts. Employ a scope to observe waveforms and a VOM to check voltages and continuity while referring to the schematic diagram. Check any bias voltages and calculate currents in the bad circuit.

If the schematic is not marked with the proper voltages, use your electronics knowledge to estimate the proper amplitudes while probing. Do these measured amplitudes vary from your own calculations or other similar circuits?

If an individual component is now suspected, make an in-circuit test by checking for the component being possibly shorted or opened. These simple tests assume that the technician has reasonably ruled out other defective components shunting current to ground, or delivering high current levels to the suspected component, or from being open in the suspected component's current path. To find a short in a component take a voltage reading across the component — if the component is shorted, then there will be little or no voltage dropped (if a voltage drop is expected). Check for an open by observing a large voltage drop across the component (if little or no drop is expected). If no parallel paths are present to fool an ohmmeter, then take a simple resistance reading on unpowered equipment across the component. Then, if the component is still suspected, test the part out of circuit. If defective, replace the component with one that is known to be good. Power up unit to check for continued overheating, improper operation, poor waveforms, bad voltages, etc.

The servicing block diagram of *Figure 10-3*, or the schematic of *Figure 10-4* with marked AC waveforms and/or DC bias voltages (also called a *circuit diagram*), if available for the unit-under-test, makes troubleshooting much easier. It is almost impossible to effectively troubleshoot a modern receiver, transmitter or transceiver without at least a simple schematic or block diagram — unless the problem is visually apparent or unbelievably obvious.

Some service manuals may use simple schematics with a separate text list of voltages for each lead or pin of every component (a *voltage table*); a data table on all ICs and semiconductors; circuit descriptions that describe the operation of the major circuits; recommended adjustment and alignment procedures; block diagrams; an exploded diagram of the case and mechanical parts; and a full parts list.

Keep in mind that all indicated AC and DC voltages in such diagrams are usually referenced to ground, and not across a semiconductor junction or component. As well, all indicated voltages are only approximate, since most commercial quality components have rather loose tolerances, not to mention the variable characteristics of semiconductors, even when identical to each other.

Further tests and procedures follow to help in troubleshooting and repairing modern RF equipment.

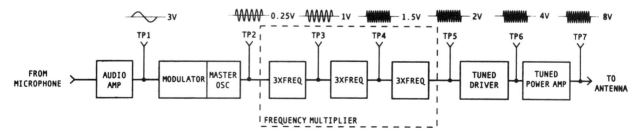

Figure 10-3. A servicing block diagram showing proper waveforms and amplitudes.

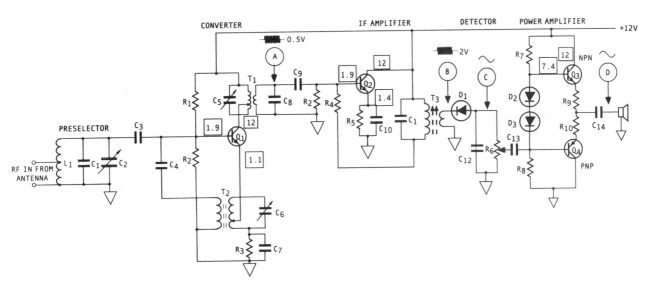

Figure 10-4. *A schematic with DC voltages and AC waveforms.*

10-1C. Signal Tracing

A common technique in troubleshooting receivers or amplifiers is *signal tracing* (*Figure 10-5*).

To perform signal tracing on a receiver, warm up the receiver, the frequency generator, oscilloscope or spectrum analyzer, for twenty minutes. Confirm that the receiver is tuned to the generator's output frequency. Place the antenna or probe of the modulated RF generator near the receiver's antenna or, if de-tuning of the front-end does not result, attach the probe directly to the receiver's antenna input. Start with a generator output of about 100 μV and decrease, if possible, to avoid amplitude distortion. Using an oscilloscope or spectrum analyzer with probe, check each stage's output for the proper level and linearity. No signal, or a distorted signal, at the output means the stage, or its AGC voltage, is defective.

To signal trace a low-powered transmitter, turn on the carrier or, with SSB, modulate with a single or dual tone, and trace from the audio and master oscillator stages to the output, being careful of the final power amplifier stage.

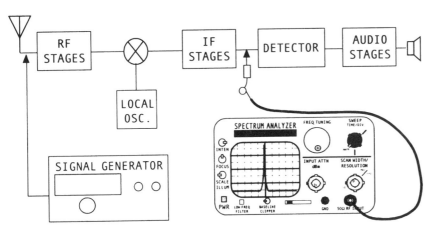

Figure 10-5. *The signal tracing method of troubleshooting.*

10-1D. Signal Injection

A similar technique to the above is called *signal injection* (*Figure 10-6*). To accomplish signal injection, warm up test equipment and receiver under test. Inject the proper frequency, at the proper amplitude (to eliminate amplitude distortion), beginning at the speaker and heading back toward the front end. Listen for a speaker output, or place an oscilloscope across the speaker's leads. The stage that is defective will not send a signal to the speaker; or will have a low output; or will introduce distortion. You *must* change the signal generator's output frequency for each stage being injected (for instance, an AM IF stage would need a signal injected at 455 kHz, while the audio stage would need around 1 kHz, etc.).

When signal injecting into the detector or audio stages of a receiver equipped with a squelch circuit, make sure that the squelch is turned off — or the audio amps may be shut down and appear to be defective when they are not!

When injecting the test signal into an AM or FM receiver, the application of *standard test modulation* values are employed. For AM, a 1 kHz tone of sufficient amplitude to create 100% modulation is used; while for FM, a 1 kHz tone of sufficient amplitude to create a 3 kHz deviation has been adopted. SSB receivers use a two-tone modulation, with each tone being of a matched amplitude. Each tone can be anywhere between 300 Hz and 3 kHz, with a separation of at least 1 kHz between tones and not harmonically related to the other.

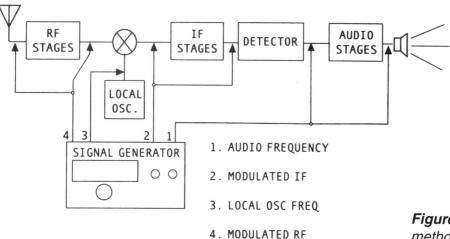

1. AUDIO FREQUENCY

2. MODULATED IF

3. LOCAL OSC FREQ

4. MODULATED RF

Figure 10-6. *The signal injection method of troubleshooting.*

10-1E. Lead Dress

Lead dress (*Figure 10-7*), which is the physical layout of connecting wires, becomes a major consideration in circuits used in high-frequency applications. This is due to the natural capacitance and inductance of the wires.

After repair or alignment, this lead dress must be maintained or unpredictable circuit operation, or actual de-tuning of tuned circuits, can result. As well, lead lengths of components must be kept short for maximum power dissipation ratings.

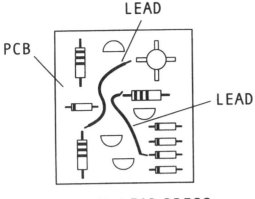

ORIGINAL LEAD DRESS

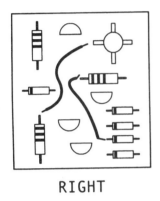

RIGHT

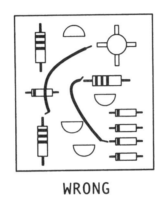

WRONG

Figure 10-7. *Three PCBs showing right and wrong lead dress.*

10-1F. Heat Sinks

Never apply power to circuits under test without the component's proper heat sinks in place — or without the component or IC bolted down to the chassis that is acting as a heatsink. Rapid thermal damage to the component can result. Furthermore, capacitance is added to or subtracted from a circuit if an RF transistor is separated from its heat sink. Also, use only the same thickness insulators between an RF transistor and its heatsink as in the original equipment, or circuit instability or frequency shift might result. This is caused by a change in stray capacitance.

Make sure all heat sink bolts are tight, or proper heat transfer will not occur. Do not over-tighten bolts, as warping or other damage may occur to the heat sink or the semiconductor mounting tab.

Always use the minimum of heat sink compound, since heat sink grease is not that great of a conductor of heat (but air is far worse, so do not *under* apply either!).

Be carefull in certain circuits in that some heat sinks may not be isolated from the collector or drain, and thus may be at a high voltage potential.

10-2. Intermittents

One of the most difficult and frustrating troubleshooting jobs can be the location and identification of an intermittent problem. Intermittents are commonly caused by bad connections: Bad solder joints; loose, dirty, corroded or damaged internal or external connectors; dirty potentiometers; and, of course, bad capacitors or other components.

To check for failing or dirty connectors and pots, shake or tap while the unit under test is turned on and check for any effect. Clean or replace the part as needed.

A simple and very effective technique to check for intermittent operation of a device caused by bad components, cracked PCBs, or bad solder joints, is to thermally stress the suspected circuit areas. While the unit under repair is turned on, chill the general area with freeze spray while checking for proper circuit operation or voltages.

A stubborn intermittent may also have to be further stressed by the application of heat immediately after the circuit is refrigerated. Employ a standard heat gun, switched to low heat, while utilizing a heat concentrator (reducing or constrictor) nozzle for local application. If the unit was already functioning properly, improper operation should now become evident. If the unit was faulty when received, temporary proper operation can result. Either way, localized thermal stresses placed at suspected areas allows troubleshooting of intermittents to be accomplished quickly and simply, down to the component level. (Using cold or heat on *some* SMD PCBs is discouraged due to the possible thermal damage that can be induced by this technique.)

Intermittent problems in circuits are commonly caused by capacitors, especially electrolytics, or broken traces, cold solder joints, or solder joints that have gone bad due to thermal or mechanical stresses (poor solder joints can create a high resistance in the effected joint, producing unusual and unexpected circuit operation).

Also, make sure all ground connections are tight, or odd intermittents can occur as the ground connections go from a low impedance to a high impedance.

10-3. Parts Failure and Their Indications

ICs, since they can be almost fifty times more reliable than separate discrete circuits, are very dependable — and rarely fail. Suspect them, while troubleshooting, as being internally damaged dead last (unless they directly interface with the outside world, in which case these ICs can be damaged by static buildup, lightning, or improper polarity or voltage inputs).

Decoupling and coupling capacitors, as well as trimmer capacitors, on the other hand, fail often. Decoupling capacitors in an amplifier stage (which prevent RF or AF from feeding into the power supply) if open, can cause self-oscillation in that or other stages. If shorted, they can short out the power supply. A shorted coupling capacitor can effect the bias on the next amplifier stage.

As a general rule, a shorted capacitor will affect DC and can affect AC; while an open capacitor will not affect DC voltages, but will affect AC voltages.

Bridging a suspected open coupling or bypass capacitor with another capacitor may find the problem. Also, an excessively warm capacitor can indicate that there is DC current flowing through it, creating power dissipation in the leaky, or partially shorted, dielectric ($P = I^2R$).

When dealing with questionably bad resistors, bear in mind that when a resistor becomes defective it will *usually* open, or partially open, creating a high value of resistance (some resistors are even used as cheap fuses). They will rarely internally short, or become lower in resistance value. However, the most important thing to remember when troubleshooting circuits with a burnt resistor is that it usually means something else failed first (which caused excessive current to flow through the now fried resistor).

If a defective transistor is found, then examine other components that could have contributed to the failure, as transistors seldom go bad unless a defective passive component created the failure (a bad component can force the transistor to conduct excessively high currents, damaging it) — or, possibly, it can be due to poor circuit design, also allowing too much current to flow through the semiconductor. As well, transistor failure can be caused by an overly high drive signal at its input.

Shorts can be one of the most common internal problems with semiconductors. Many transistors, however, due to the internal bond wires connecting their leads to the semiconductor material within their case, may *open* due to high current levels.

When an amplifier is not functioning properly, the problem is usually in the DC bias voltages — check biasing components for any changes in value. Or a preceding stage may become defective, adversely effecting the stage under test. For example, in an RC or LC-coupled two-stage amplifier, the coupling capacitor between stages may short, driving the second amplifier stage into saturation by the direct application of the Vcc of the first stage to the base of the second stage, creating excessive bias. A shorted emitter bypass capacitor or, much more rarely, the emitter resistor itself becoming shorted, can create excessive current (depending on amplifier design) through the decrease in negative bias and the resulting thermal instability, slowly causing temperature-related damage, lowering beta and increasing collector leakage current.

A high Vcc, ordinarily caused by defective power supply regulation, can also destroy a transistor, as can voltage transients (especially during thunderstorms). Some of these component faults are cumulative and slowly destroy the transistor over time, while gradually creating output distortion due to the shift in the amplifier's Q-point bias. Other faults may immediately destroy the semiconductor within microseconds.

10-4. Linear IC Troubleshooting

Since individual components in an IC cannot be examined and, when confronted with many IC leads, troubleshooting of integrated circuits can sometimes be quite daunting to the inexperienced technician. So much so, in fact, that many technicians simply replace the IC before properly checking out other options. ICs have become very reliable, and are extremely dependable — far more so than discrete circuits. This means that most ICs are seldom defective.

First, confirm that the suitable ripple and glitch-free positive and/or negative supply voltage(s) are reaching the Vcc pin or pins, as well as the proper ground to the ground pin (is there a zero resistance path to the grounding point?). Verify that the proper input signal is being input into the proper pin. When testing the chip, and the Vcc is low at its input, then either the supply voltage is low or the chip itself is pulling down Vcc. Also confirm that the output signal is present at its appropriate pin. If the input is good, but there is no output, or an improper output, check the following for defects: Support components, such as open or shorted coupling or bypass capacitors; defective or damaged frequency-determining crystals; bad voltage reference Zeners, etc. All could be pulling down the IC's output, or giving the incorrect output. In addition, a support chip may be feeding improper voltage levels or waveforms to the chip; or the chip's secondary output pins (such as outputs from the chip's internal LO and mixer) may be shorted, affecting the chip's main output.

Push and pull on all suspected ICs during troubleshooting to check for possible bad circuit contacts.

When all support components have been tested and found not to be at fault, then remove the IC and replace with a known good chip. Test for proper function.

10-5. Amplifier Self-Oscillation Troubleshooting

Any amplifier that breaks out into oscillations creates undesirable results, such as lowered output power, overheated parts, and distortion. Two amplifier self-oscillations are encountered: *feedback* and *parasitic oscillations*. The most distinct difference between feedback and parasitic oscillations is that parasitics can occur at almost any frequency, while feedback oscillations are strictly at the amplifier's signal frequency.

10-5A. Feedback Oscillations

An RF amplifier may begin oscillating at its operating frequency due to its internal capacitance feeding an in-phase feedback signal back to its input at a certain frequency. 0° positive feedback in an RF amplifier may occur in a CE amplifier due to the reactive components creating a phase shift in the 180° output frequency. The output frequency thus arrives at the input in-phase, just as in an oscillator.

The lower the amplifier's tuned input and output circuit impedance, or if these circuits are tuned to different frequencies (or if there is only a single tuned circuit), the less likely the RF or IF amplifiers will break out into these feedback oscillations. Undesired external coupling among the input and output components and wiring can cause oscillations, usually demanding shielding between the input and output circuits to stabilize the stage (for instance, if an untuned amplifier's input or output is tightly coupled to a tuned circuit, the amplifier acts as if its input or output were tuned anyway, and may oscillate; or, an amplifier's stability can shift with a change in its source and load impedance, allowing it to also oscillate). An unneutralized RF stage can act as a tuned-collector tuned-base oscillator due to these input and output tuned circuits.

Increasing the frequencies that the amplifier must boost are more likely to cause regenerative feedback, producing oscillations, spurious outputs, distortion, overheating of parts — and an increase in plate or collector current. Self-oscillations can even cause device burn-out due to the added power dissipations created by the oscillations.

If, when tuning an amplifier, the stage does not display smooth voltage or current readings, or if the stage goes into runaway, then oscillations may be occurring. Furthermore, if the RF drive is removed and a signal is still present, then the stage is oscillating. You may be able to find the oscillations by sliding your finger around the (*low voltage and current*) circuit: Your hand may increase or decrease the oscillations.

If a neutralized amplifier stage begins oscillating, check that the neutralizing capacitor or inductor is not open, or that a new feedback path has not been created by conductive dust, fragments or grime (such as soda pop, metal shavings, or a solder blob). If these are not present, then adjust the neutralizing capacitor or inductor as indicated below. Also, a stage will need readjustment of its neutralization if a new tube or transistor has been inserted, since the active device will almost certainly have a slightly different feedback capacitance value.

One type of neutralization procedure employs a tunable inductor, L_n (*Figure 10-8*). The inductor is placed between the collector and the base, with a high-value capacitor (C_b) in series with the inductor to avoid a DC short. The inductor is now in parallel with the collector-to-base capacitance, effectively forming a tank circuit. The inductor is tuned until resonance occurs, causing the tank to be a very high resistance at the undesired feedback frequency, eliminating the in-phase feedback voltage.

Neutralization is also accomplished by adding a capacitor (C_n, *Figure 10-9*) from the transistor's output back to its input. This supplies a bucking out-of-phase feedback voltage to cancel the in-phase feedback voltage creating the offending oscillations or amplifier instability. Tuning the capacitor until the positive feedback is canceled is described below.

To neutralize: At the unneutralized and unpowered amplifier stage, inject the proper amplitude and frequency signal into its input while monitoring the collector or plate tank circuit for this signal (a spectrum analyzer or an oscilloscope, equipped with a loosely coupled *sniffer* probe, can be employed to detect the presence or lack of oscillations). If present, then adjust the neutralizing capacitor or inductor until the stage stops passing the signal from its input to its output, indicating full neutralization. As well, when powered, excess collector or plate current and spurious emissions and distortion will all decrease when neutralization is reached. The neutralizing capacitor is generally adjusted until it is of the same value as the collector-to-base capacitance.

Another common technique to perform neutralization is to inject a drive signal at the amplifier's design frequency into the *output* of the amplifier, while some type of RF test unit, such as a spectrum analyzer (equipped with a special high input impedance probe matched to the low input impedance of the analyzer's front end; or use the common sniffer probe) or an oscilloscope, is attached to the *input* of the unpowered amplifier. C_N or L_n is then adjusted for a null indication on the display, indicating a cancellation of the positive feedback from its output into its input.

High-powered tube amplifiers may use a grid-loading method to neutralize a stage, which effectively lowers stage gain but avoids the need to employ a neutralizing capacitor or inductor. This is commonly accomplished by a grid-to-cathode resistor, or a low tap on the grid tank coil. When proper neutralization of the amplifier has occurred, then maximum drive voltage, minimum plate current, and maximum power output all occur simultaneously.

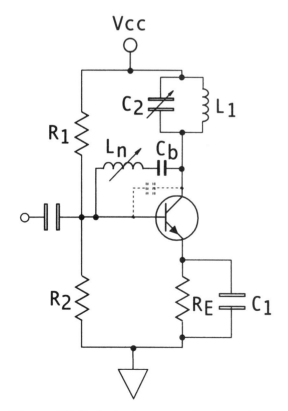

Figure 10-8. *Inductor neutralization circuit of L_n and C_b.*

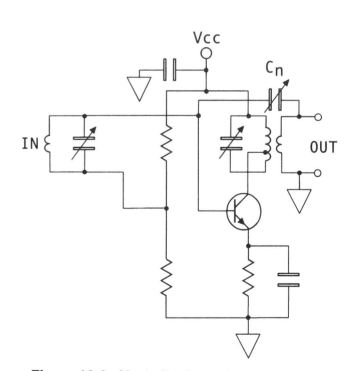

Figure 10-9. *Neutralization using an adjustable feedback capacitor C_n.*

To neutralize a final amplifier tube transmitter: Place 70% full power into the resistive load and look at the cathode current. A dip in this cathode current will occur at the same time as maximum output power. If not, then alter the neutralizing capacitor until cathode current decreases and power out increases.

A *losser resistor* is an inferior way to neutralize a stage, but may be found in some amplifiers. These resistors are placed at the base lead, are from 10Ω to 1kΩ, and cause a voltage drop, which creates a loss that counteracts *small* values of regeneration.

Amplifiers can also be neutralized by *inductive neutralization* (*Figure 10-10*), with part of the amplifier's output inductively coupled back to the input.

Another form of inductive neutralization, but for a push-pull stage, is shown in *Figure 10-11*. L_1 feeds degenerative voltage back to the transistor's input for neutralization after tapping a small voltage from L_2. Push-pull amplifiers may also have two neutralization capacitors, which must be adjusted at the same time to maintain equal capacitance and feedback values.

With modern transistors, due to their small values of interelectrode capacitance, it is possible to run amplifiers up to 500 MHz that are un-neutralized and that will not oscillate.

Even though common-base amplifiers are far more stable than common-emitter circuits during high-frequency operation, some may still need neutralization at certain very high frequencies to increase their stability and prevent oscillations.

The input capacitance of a testing device or probe (a DMM, for instance, may add up to 1000 pF to a circuit) can cause a sensitive circuit to oscillate. As well, probing an oscillating amplifier may stop any oscillations in an unstable amplifier, creating confusion during the troubleshooting process.

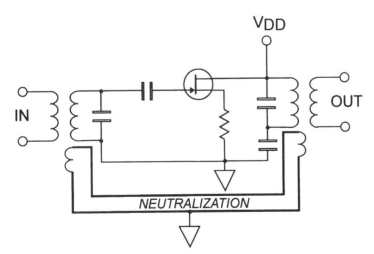

Figure 10-10. Neutralization by induction.

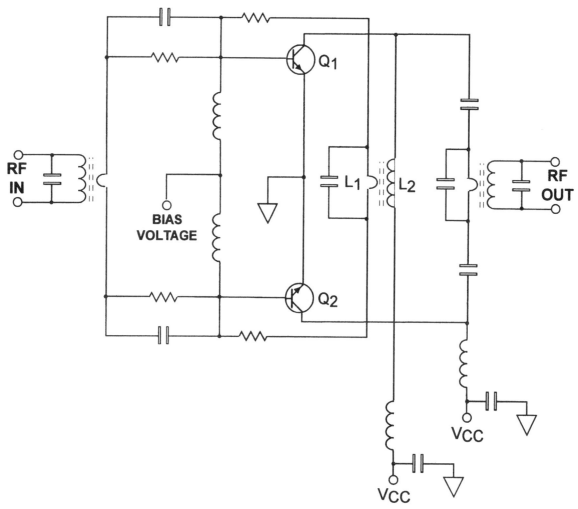

Figure 10-11. *Inductive neutralization of a push-pull stage.*

10-5B. Parasitic Oscillations

Parasitic oscillations are caused by stray capacitances and inductances, created by coupling capacitors, coils, leads, or active devices, forming tuned circuits within an amplifier, making the amplifier act as an oscillator. These oscillations are at a higher or lower frequency than the amplifier's design frequency. In fact, the parasitic oscillations can reach into the UHF spectrum, even if the transmitter's operating frequency is 30 MHz. The frequency of the oscillations depend on the value of the stray capacitances and inductances.

Low-frequency parasitics are usually aggravated by a transistor's increasing gain as the frequency is lowered. This produces low-frequency instability in an amplifier, since any low frequencies naturally have a higher gain than higher frequencies in any active device. Low-frequency parasitics can destroy an active device very quickly.

Parasitic oscillations cannot be treated by a neutralizing capacitor or inductor. The cure is to lower the Q of these undesired parasitic circuits with added low-value resistances in series with the inductors; or use ferrite beads in the grid or base and the plate or collector to suppress any VHF or UHF parasitic oscillations (VHF parasitic oscillations in a high-frequency (HF) amplifier sounds like a hiss or hash noise superimposed on the desired signal); or in the circuit of *Figure 10-12*, both Z_1 or R_1 are acting as *parasitic suppressors* (any component used to eliminate parasitics is referred to as a parasitic suppressor); or utilize a small value choke in series with the grid or plate circuit of a vacuum tube RF power amplifier (to apply any reactance, if not so designed by the OEM, to the output of a (typically) low output impedance RF power transistor's collector can have disruptive consequences, and is not recommended); or extend the plate lead of the power tube for added inductance.

Low-frequency parasitics can be suppressed by a low-pass filter added between the collector and the base of the offending amplifier, which feeds a progressively low-frequency degenerative out-of-phase signal back to the input, thus decreasing low-frequency gain.

A word of caution: It is strongly recommended that a technician should never add additional components or modify a transmitter or receiver without the expressed permission of the supervising RF engineer.

Use a spectrum analyzer with a sniffer probe to detect parasitics. Any unexplained frequencies can indicate parasitics. Also, when checking AM trapezoidal patterns on an oscilloscope, any unexplained non-linearities in these patterns can mean parasitics.

In a well designed transmitter little or no parasitics are present, but after repair (such as tube replacement) or modification, they can arise. Simply confirming that lead dress and length are as designed and, if not, reworking to specs, will cure parasitics 80% of the time.

A transmitter with parasitics will waste power and can create internal arcing, overheating, blown fuses and relays, destruction of capacitors, tubes (caused by increased plate and grid currents) and transistors, and can produce erratic transmitter tuning and the creation of interfering (spurious) signals.

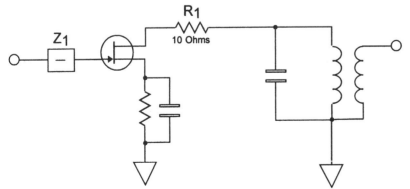

Figure 10-12. *Parasitic suppressors in-circuit.*

10-6. Receiver Troubleshooting

Troubleshooting AM and FM receivers, as well as their associated AGC and specialized circuits (such as squelch, frequency synthesizers, and battery saver circuits), can appear to be quite daunting. However, if a certain logical procedure is followed, most problems can be determined and repaired quickly.

The reason behind total failure of a receiver to perform its function is usually much easier to discover than the sometimes subtle loss of selectivity, sensitivity, dynamic range, or S/N.

10-6A. *Receiver Interference, Distortion and Noise Troubleshooting*

Interference and noise in a receiver can be caused by:

Inadequate selectivity, since wide bandwidth equals more natural noise pickup and increases the chance of adjacent channel man-made interference; by regeneration (positive feedback) in an audio amplifier, causing low-frequency noise-producing oscillations referred to as *motorboating*, which is a put-put sound that can also occur as a consequence of a low battery in some portable radios (if an audio amplifier begins to hum or whistle then it may be breaking into oscillations due to bad capacitors).

Noise on the power line feeding the receiver, producing hash-type noise interference (cured by the use of a low-pass power line filter).

RF transmission line noise pick-up induced by damaged or poorly shielded transmission lines (utilize only high quality shielded and undamaged lines).

A poor ground, causing a high-impedance path to ground for stray RF and noise, and intermittent ground, creating high noise levels, or *ground loops*, producing low-frequency noise pick-up through induction caused by power transformers, switching power supplies, brush type motors, etc. (employ proper size ground straps of short length to an adequate ground, well bonded to the grounding rods, and remove all ground loops and bond to a single ground point).

An improperly shielded or leaky receiver case, allowing interfering signals into the internal high-gain amplifiers stages, causing intermodulation distortion (which then produces many noise generating and non-harmonically related new frequencies), and desensitization (creating extra noise due to the need to increase gain in RF and audio amplifiers in an attempt to overcome the desensitization).

A low-frequency hum (at 60 or 120 Hz) caused by faulty filtering or regulation of the power supply, or poor shielding of the power supply's power transformer, that feeds the affected receiver.

Bad (cold) solder joints, creating a high-resistance connection where noise is dropped across.

Unbalanced push-pull amplifiers in the audio section.

Intermittently bad connections, such as faulty cable or socket connections, which can produce popping or scratching noises (physically tap areas that are suspect for any increase in noise output).

Bad or failing components, which can create noise that is usually sensitive to physical impact or temperature variations (tap suspected components, or freeze components with freeze spray — if noise level changes, the component just tapped or frozen is the culprit).

A receiver with squelch, since a bad squelch can cause a lack of muting when no signal is being received — thus amplifying any noise present.

If there is distortion in a receiver when listening, first check the speaker — the paper cone could be torn. Check for crossover distortion (*Figure 10-13*) in any push-pull amplifiers. Crossover distortion is produced by a lack of true Class B operation caused by improper bias. As well, any Class A amplifier that reaches either saturation or cutoff produces amplitude distortion, which can be caused by a change in its bias, either through DC bias or AGC component changes; or power supply variations; or an excessively high input signal. If this amplitude distortion occurs only during strong signals, AGC is bad, or an output transformer is becoming saturated and is unable to pass any increase in signal level, creating the distortion. Also, bad LC or RC components in a filter network can create frequency distortion, which attenuates some frequencies more than others outside of its design parameters. And frequency distortion can be produced in an AM or SSB receiver if the LO drifts off frequency, or begins FMing, or is mistuned, causing *sideband cutting* to either the lower or upper sideband (which can be viewed on a spectrum analyzer as one sideband lower than the other). A damaged transistor may likewise introduce distortion, as can positive feedback through the interelectrode capacitance between its collector and base (See *Self-oscillations*).

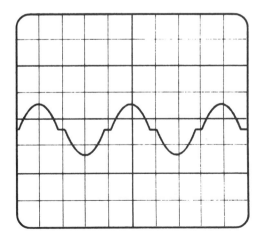

Figure 10-13. *Crossover distortion as viewed on an oscilloscope.*

10-6B. AM Receiver Troubleshooting

Turn on a radio at high volume. If there is only white noise, which is circuit generated (mainly by the RF sections), even when tuned through its entire band, then the audio and RF amplifiers are probably good. Check local oscillator or PLL for proper frequency and amplitude output, then mixer, IF, and detector.

If a receiver is not receiving at the indicated dial frequency after signal injection, then the LO or PLL is probably off frequency. If the LO is oscillating, then examine antenna tuning capacitor for warped or shorted rotor or stator, or if the RF antenna coil is open. An LO may also become loaded down, causing a frequency shift, if the next stage (usually a buffer or a mixer), becomes defective. Also, a low supply voltage caused by a bad regulator can cause the LO to shift in frequency or become unstable.

If the receiver is completely dead: Confirm that the proper voltage is reaching the power supply's transformer. Check fuse and voltage output of power supply (a power supply regulator may have internal protection circuits that shut down the regulator if temperatures or currents go too high due to a shorted load. It may appear that the power supply is bad when it may actually be a bad load causing the regulator or power supply protection circuits to shut down the entire supply).

If power supply output is low, power supply leads should be disconnected from the receiver and tested, as a receiver fault could be pulling down supply voltage, or current limiting could be kicking in too early in the PS regulator circuit, furnishing low current levels to the receiver. If the output voltage or current level of the PS is still bad after disconnecting, then troubleshoot power supply.

If all PS voltages and currents are acceptable, then increase the receiver's volume and use half-split signal injection method by injecting an audio signal at the beginning of the audio stages (customarily at the volume control with squelch *off*). If no signal is heard, then confirm that the squelch circuit itself is operating properly; then troubleshoot down the audio stages to the speaker. If a signal is heard at the speaker after injection at the volume control, then inject a modulated RF signal of the correct (dial) frequency into the antenna. If there is now no speaker output, then move the injection point, changing frequency as required, toward the audio stages, until a tone is heard. The stage before that point is the defective stage. When the faulty stage is localized, then troubleshoot to the component level.

If the output is distorted: Insert an audio tone at the first audio stage, then check for distortion at the appropriately terminated output of the last audio stage with an oscilloscope. This distortion will usually be in the form of clipping of the peaks of the audio waveform. If these are present, then the bias levels of the audio stage may be faulty. If clipping of the signal occurs before the detector, the AGC could be defective, or there may be very bad IF amplifier DC bias levels. Distortion in amplifiers can also be caused by a *prior* stage feeding an excessively large input signal into the amplifier under test.

If a receiver stage has a low output: The bypass capacitor across the amplifier's emitter resistor may be open; or a coupling capacitor to a next stage or circuit may be open; or AGC may be defective; or the active device of the amplifier could be faulty. Possibly amplifier self-oscillations are occurring, creating a strong AGC reaction to what may only be a weak input signal. Finally, check the alignment of the receiver.

Almost all receivers have AGC circuits that limit gain in the IF or RF section. AGC will have to be disabled by cutting the AGC line, or by injecting a reverse *bucking* voltage into the AGC line, or by turning off the AGC with a panel switch (when so outfitted), if the AGC and the amplifiers that the AGC controls need to be isolated from each other to segregate the bad circuit. However, if the DC bias to the amplifiers is being completely or partially supplied by the AGC at all times as a normal condition of circuit operation, then clamp the AGC line to an appropriate DC voltage and continue troubleshooting. Also, with any receivers having poor sensitivity, check that all AGC voltages are correct. Defective AGC may also completely or partially shut down one or more good RF or IF stages. Also keep in mind that it is sometimes difficult to work with receivers equipped with AGC, since AGC is a loop — a bad IF signal output causes incorrect AGC voltages; while a faulty AGC can cause incorrect IF output.

If the receiver is drifting or has shifted in frequency, then check its regulated bias supply for voltage variations to the LO (when a transceiver, receiver, or transmitter starts drifting past tolerance levels, it is usually the fault of instability in the power supply); or if the LO is varactor diode controlled, also check for a bad varactor. Older receivers and transmitters that use ferrite cores may shift in frequency due to the drying out of the ferrite, causing its permeability to vary from its when-new value. A cracked crystal will also cause frequency shift. Certain crystal-controlled oscillators may drift lower in frequency with age, and may have to be read-justed on a yearly basis due to oxidation of the silver plating on the face of the crystal. Certain higher-frequency amplifiers may shift in frequency at low power when switched from a high power setting. This is ordinarily a design fault, but can be a bad ground, bad base bias, or bad transistor.

Always replace all removed shielding, as the shielding of stages or components prevents or reduces interaction between stages that can cause undesired oscillations, degeneration, or IMDs. Coils, tubes, and capacitors, as well as interstage leads, are susceptible to electro-magnetic-field coupling, causing interference. Some old, poorly designed, or improperly re-paired equipment may need extra or supplementary shielding to attenuate this coupling.

SSB troubleshooting is similar to standard AM except for the need to confirm that the BFO is on frequency and outputting a clean and sufficiently high amplitude signal into the detector. If not, then noisy, off-pitch, or a lack of audio output can result.

10-6C. FM Receiver Troubleshooting

See above, except for the following differences:

Many FM receivers do not use AGC in the IF sections due to their built-in limiters, which automatically restrict amplitude variations for the FM detectors; FM receivers, nevertheless, may use AGC to limit overloading in their RF amplifiers. Also, do not be mislead into thinking that the last IF stage or two of an FM IF section has defective DC bias, as this stage is sometimes overdriven to act as a limiter for the detector. Furthermore, if an FM receiver drifts, then the AFC (or PLL) section of the local oscillator stage is faulty; or the P.S. regulator is failing, causing frequency instability; or the LO crystal may be damaged.

In addition, handhelds with *Private Line* must be unsqelched by the proper tone injected into the front end, or no output into the speaker is possible.

10-7. Transmitter Troubleshooting

Transmitter troubleshooting requires much more caution than working on receivers. Transmitters can <u>generate very high voltages and currents,</u> which can prove fatal to whomever comes in contact (if tuning a high-powered transmitter, switch the power switch from *high* to *low,* if the transmitter is so equipped).

In fact, some high-powered transmitters will not allow the transmitter to be internally inspected or repaired with the power on; <u>all access doors may have safety interloc</u>ks that shut power down as soon as a panel is opened.

As well, the technician must deal with a device that radiates a signal that can, if improperly repaired or aligned, interfere with other RF communications — much to the intense displeasure of your employer and the FCC.

By the appropriate and cautious use of oscilloscopes, DMMs, spectrum analyzers, and SWR/wattmeters, most transmitter troubles can be quickly isolated.

Caution: <u>*Touching the finals of a transmitter or its antenna while in operation will cause RF burns, sometimes very deep and/or very fatal. Also, do not be at ground potential when probing any high-voltage and high-current circuits.*</u>

10-7A. *Common Transmitter Problems and Troubleshooting Techniques*

As mentioned above, transmitters, due to their inherently higher voltages and currents present during operation than those existing in receivers, can be dangerous to troubleshoot in all but the low-powered (QRP) types. However, the larger transmitters generally have external panel voltage and current meters that can help the technician to isolate most problems down to a stage, while some even have switchable (into a desired circuit) oscilloscopes, signal monitors, and SWR/wattmeters. This allows the transmitter to be troubleshot while under power.

In smaller transmitters, and almost all transceivers, it is appropriate, if one follows proper safety precautions, to internally signal trace down to the bad stage while under power.

Whatever the transmitter, first check for a proper signal output by monitoring the frequency accuracy (there should be no modulation or CTCSS present during this check), signal cleanliness (harmonics, IMDs, parasitics), RF power level, and modulation percentage. Examine for the correct power supply voltage (large transmitters may have a separate power supply for *each* power amplifier stage), and lack of AC ripple to each stage, as well as any low or distorted levels in the audio amplification stages.

A large commercial high-powered transmitter should also be checked when first installed; when any changes are made; or *at least* once a year. The tests required are: power output, frequency, and modulation percentage.

If a high SWR is shown on the SWR bridge, look for a mistuned transmitter final; a broken or mistuned matching network or antenna; a damaged RF cable; or a corroded RF connector.

Even if the SWR does not seem excessively poor, it can appear to be better than it is if the transmission line is lossy. Since the SWR is commonly sampled at the transmitter output and the transmitter power must travel down the lossy line and then return (if not properly terminated), then the reflected power, and the forward power through the line, is consumed as heat. This makes the reflected power lower when it reaches the SWR bridge at the transmitter's output, and reduces the *ERP* (Effective Radiated Power) below what it should be.

If one RF amplifier stage in a transmitter is slightly detuned, then phase modulation will occur. This is due to the detuned amplifier tank presenting either a capacitive or an inductive reactance to the signal, creating phase modulation, causing possible distortion at the remote receiver.

As well, general rapid-carrier frequency instability can be produced by poor power supply regulation, creating a master oscillator that instantly alters frequency due to a varying supply of power (*dynamic instability*), sometimes only during modulation — or it can be created by actual mechanical vibrations causing momentary changes in component values. This is referred to as *frequency shift* and can be detected by attaching a frequency counter to the output of the unmodulated transmitter and checking for oscillator instability by vibrating the offending transmitter or by stressing the power output of the common supply.

FMing (also called *incidental FM* or *residual FM*) is produced by another source (such as noise or AM) modulating the carrier, creating a generally rapid and fleeting, but sometimes continuous, unwanted frequency modulation. If riding on a desired AM signal, incidental FM may show up on a spectrum analyzer display as one sideband being of a higher amplitude than the other (*Figure 10-14*), or as a slight horizontal jitter of the carrier on an oscilloscope display (*Figure 10-15*).

Excessive (out-of-tolerance) *frequency drift* is a gradual change in frequency over time, caused by defective, or badly designed, oscillators or frequency control circuits (or a blocked or damaged cooling fan(s) causing heat build-up, changing the internal capacitance of the transistors and the values of the LC components).

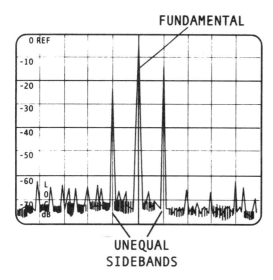

FUNDAMENTAL

UNEQUAL
SIDEBANDS

Figure 10-14. *Incidental FM as viewed in the frequency domain.*

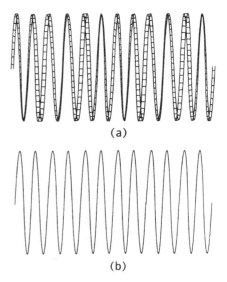

(a)

(b)

Figure 10-15. *Oscilloscope waveform (a) with incidental FM; and (b) without.*

A transmitter's driver and finals are where most faults occur due to their high power, and thus high heat, output. If the transistor finals of a transmitter are burnt out, first suspect the cause as high SWR. If tube driven, then suspect worn-out tubes.

When RF transistors, tubes, components, or leads are replaced during service, then a retune may be necessary due to the probable changes in the reactances of the sensitive RF circuit.

If a transceiver does not transmit or receive, but there is some white noise during receive, check for a bad PLL (the frequency readout on the transceiver may not indicate an error with the PLL).

10-7B. Transmitter Tube Final Troubleshooting

In a defective tube circuit the same troubleshooting techniques apply as above except, in view of the high powers always involved with a modern tube transmitter, also check for arced tube pins and sockets (causing a constant or intermittent open, or a high-resistance contact), broken tube pins and, of course, bad capacitors, burnt resistors, and bad tubes (failed vacuum tubes can account for up to 70% of tube circuit failures).

To check for tube failures: If the tube's heater is not glowing or the tube's envelope is not extremely warm (*Caution*: Many tubes become *very* hot) then the heater element, or its power supply circuit, is defective. Remove the tube and test the filament with an ohmmeter for continuity.

While powered, check for a leaky tube, frequently evidenced by a green, blue or purple haze inside the tube (a soft tube) causing excessive plate current and erratic amplifier operation (a dim dark purple glow is generally normal). Also, if a vacuum tube's *getter* (a little cup within the tube containing alkali to help evacuate the tube) is white or red, then the vacuum tube has lost its vacuum.

Check for the appropriate tube amplifier bias. If the bias goes out in some old and unprotected tube circuits, excessive currents will flow and the tube may be destroyed.

Also, cooling fans are needed in most power tube circuits due to the extreme heat generated under normal operation. If these fans cease to work or become blocked, the tube will be damaged or destroyed.

Moreover, well-used tubes will naturally decrease their current output. This is caused by the cathode's coating flaking off or becoming pitted (*cathode depletion*) with use or excessive current flow.

Confirm that the electrodes inside the tube have not become shorted. If the electrodes are not visible, employ an ohmmeter to verify a lack of continuity from electrode to electrode (cathode-to-suppresser grid and beam electrode-to-cathode shorts are normal).

Vacuum tube miscellany:

If internal tube sparking is observed, this can indicate excessive voltages, loose elements, or parasitics.

Grid and plate leads must never be placed in close proximity to each other (unless so designed) due to the mutual coupling created; causing regenerative feedback, and thus oscillations.

A vacuum tube (VT) transmitter should be warmed up (filaments lit) for five or more hours before aligning the transmitter after replacing a tube. Tubes can be purchased *preseasoned* (or *burned in*) before replacement use in a transmitter, which stabilizes their electrical characteristics.

Since a tube, if overheated, may become soft (lose some of it vacuum) due to the release of gasses present in the metals of the cathode, heater, or plate, there are temperature-sensitive paint, crayons, tapes, and decals available to mount on the tube to monitor its temperature.

10-7C. Transmitter Harmonics

If excessive harmonics are present in the output of a transmitter, then confirm proper operation of the transmitter: Establish that all stages are properly biased and tuned and that none are being overdriven (which causes flattopping and thus harmonic generation), and that all low-pass filters and/or wavetraps are set and in operation. Make sure that suitable grounding of all appropriate devices (such as the transmitter, test equipment, and external filters) is in place. Confirm that any Faraday screens, which are sometimes installed between the final RF power amplifier transformer's primary and secondary to lessen capacitive coupling of harmonics to the antenna, are also grounded. If a transmatch is in use check for proper tuning (to attenuate any frequencies not of the fundamental).

Harmonic distortion can manifest itself as multiples of the carrier and/or the modulating frequencies: Nonlinearity within an *audio* amplifier stage, or modulation (AM) of over 100%, can result in harmonics of the *modulating* signal; overdriving the RF oscillator or RF amplifiers can result in harmonic generation at the *carrier* frequency.

10-8. Common Tests

The following are important tests that allow the technician to confirm proper operation of RF and AF equipment and their associated components, and as an aid in troubleshooting and alignment.

Caution: Never hook up test or other equipment as shown in Figure 10-16, *since ground loops are formed.*

A ground loop causes low-frequency magnetically induced interference to be picked up by this closed loop. The main culprit is commonly a 60 Hz power transformer transferring its hum to the equipment.

Minimize loops by grounding to a single point, thus decreasing noise-driven induced voltages by minimizing the loop area.

Warning: Do not normally use a scope to measure voltages across a component, or a short to ground, or circuit instability may result. The scope probe's ground should be connected to the unit-under-test's common ground (if the UUT is isolated by a transformer).

When the voltage drop across a component must be measured, then employ a DMM or VOM instead, or use special differential probes (See the *Oscilloscope* section in this chapter for more information on this type of measurement).

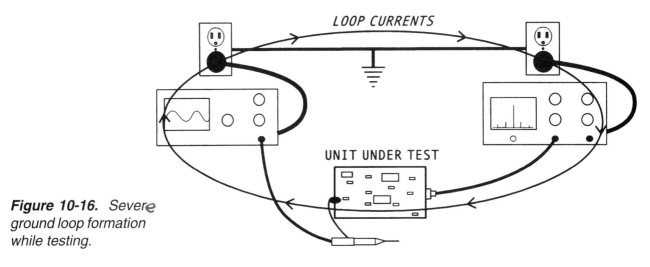

Figure 10-16. *Severe ground loop formation while testing.*

10-8A. Transistor In-Circuit Test

A simple in-circuit good/bad test to confirm the basic operation of a transistor (*Figure 10-17*): Short the base to emitter — the transistor should turn off (voltage across the emitter-collector increases). If it does not, then the transistor is probably bad. If the transistor does turn off, then parallel a resistor across R_1 (which decreases R_1's value, which increases forward bias) — the current through the transistor should increase (voltage across the emitter-collector decreases). If the current does increase, then the transistor is probably good.

Table 10-1 furnishes possible in-circuit voltage levels for single and double stage 12V-powered RC CE amplifiers for different component defects. This may help in tracking down a faulty component in an ordinary Class A amplifier (shorted and opened component values assume complete, and not partial, shorts and opens).

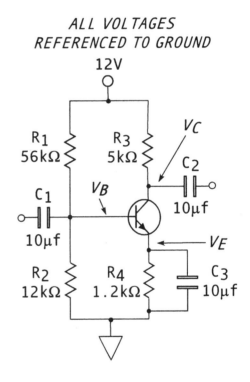

ALL VOLTAGES REFERENCED TO GROUND

12V Vcc

BIAS COMPONENTS		V_B	V_E	V_C
NORMAL		2.0	1.5	5.8
OPEN R1		0	0	12
OPEN R2		3	2.5	2.5
OPEN R3		.7	.175	.18
OPEN R4		2.1	2.1	12
OPEN C1, C2, C3*		2.0	1.5	5.8
SHORTED C1	TO GROUND:	0	0	12
	FROM NEXT STAGE:	3.25	2.7	2.7
SHORTED C2	TO GROUND:	.5	0	0
	TO NEXT STAGE:	2.0	1.5	3.25
SHORTED C3		.55	0	.05
TRANSISTOR		V_B	V_E	V_C
OPEN CB JUNCTION		0.68	1.75	12
SHORTED CB JUNCTION		2.7	2.2	2.7
OPEN EB JUNCTION		2.1	0	12
SHORTED EB JUNCTION		0.1	0.1	12
SHORTED CE JUNCTION		2.1	2.3	2.3

*An open C3 would cause a lower AC signal out;
an open C1 or C2 can cause a total lack of AC out.

Table 10-1. DC voltages for troubleshooting a CE amplifier.

10-8B. Transistor Out-Of-Circuit Test

If in-circuit tests indicate that the transistor may be damaged, remove the unit and perform an out-of-circuit test.

Even though advanced transistor testers exist, they are rarely needed. Most tests can be performed with a basic ohmmeter.

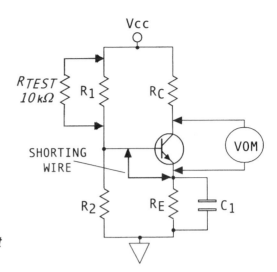

Figure 10-17. *In-circuit test of a transistor.*

First confirm that the red, or positive, lead on the ohmmeter is actually at positive potential, as many are not. By placing the ohmmeter's leads as indicated in *Table 10-2*, with the ohm's scale set at R x 100 and, if so equipped, the low-power ohms position *not* chosen, you will quickly obtain relative good/bad indications of a transistor's function. Even though any LOW indication will vary tremendously in ohmic value from transistor type to transistor type, the HIGH readings should be almost infinite.

When testing a transistor or a diode with a DMM, select the *Diode Check*, or there may not be enough voltage to turn on the semiconductor junction, resulting in an erroneous reading. The Diode Check function measures the approximate forward voltage drop of the semiconductor's junction — usually in mV or V.

Sometimes there may be a *damper diode* and a base resistor built into power transistors (a damper diode is used to prevent ringing or oscillations), which may indicate a bad transistor that may actually be fine. The damper diode is between the collector-emitter, so a transistor test between the C-E may test low in one direction (instead of high at all times, as in a normal transistor). When there is a damper diode there will also be a resistor between the base-emitter. This will show up as a nearly zero voltage junction drop, even though the power transistor is still quite good.

Table 10-2. *Relative indications for a good NPN transistor.*

E	B	C	RELATIVE OHM VALUES
+		-	HIGH
+	-		HIGH
	-	+	HIGH
-	+		LOW
-		+	HIGH
	+	-	LOW

10-8C. N-Channel JFET Out-Of-Circuit Test

To test for a good/bad JFET, set the ohmmeter to R x 100, connect the negative lead to the gate, and the positive lead to the source or drain. The ohmmeter should show almost infinite resistance. If a medium or low resistance reading is indicated, than the device is leaking.

To further test the JFET, reverse these connections to forward bias the device. The ohmmeter should indicate between 500 to 1,000Ω. Then set ohmmeter to R x 10 K, attach positive lead to the drain and the negative lead to the source. Hold the drain lead; then touch the gate. The forward resistance from the source to the drain should decrease.

10-8D. MOSFET Out-Of-Circuit Test

Attach any lead of the (low voltage) ohmmeter, which must be set to the highest ohms scale, from the gate to the source or drain. Unless infinite resistance is indicated, the MOSFET is leaky or shorted. *As always, be careful of static when handling MOSFETs.*

10-8E. Capacitor Out-Of-Circuit Test

Make sure the capacitor is discharged by shorting both leads. Using an ohmmeter, attach one lead to each side of the capacitor. This charges the capacitor through the ohmmeter's internal battery. Reverse the leads. The ohmmeter's needle should now rise quickly, then fall as the capacitor is discharged through the meter. This indicates a good capacitor. If there is a low constant resistance across the capacitor, then it is shorted — however, keep in mind that an electrolytic capacitor may have only about 0.5 MW of resistance and still be normal. If there is an infinite resistance reading with no rise and fall of the needle, then it is opened. Reliable only on capacitors larger than 100 pF.

10-8F. SCR Out-Of-Circuit Test

To test low-current SCR's for proper function attach an ohmmeter, set to the R x 1 scale, as shown in *Figure 10-18*. Take reading across the SCR with the polarities as shown. The resistance should be almost infinitely high. Now short the gate to the anode and the resistance should drop to under 50Ω. When the short from gate to anode is removed, the resistance reading should remain the same.

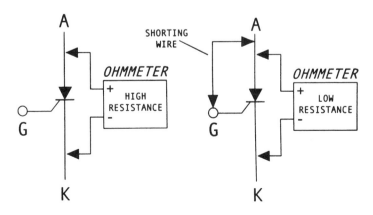

Figure 10-18. *Testing procedure for an out-of-circuit SCR.*

10-8G. Diode Out-Of-Circuit Test

To verify a properly functioning diode, utilize this diode test procedure. Employing an ohm-meter, set to the R x 100 range and forward bias the diode by attaching the ohmmeter's probes to the diode's leads — but do not touch fingers to leads or probe tips (bodily resistance will affect reading). When forward biased, an ohmmeter indication of approximately 600Ω is normal for silicon diodes, less for germanium, and can be appreciably more for some high-voltage types. Any values that significantly vary from these amounts indicate either a shorted or opened diode. When reverse biased the resistance readings should be, for a silicon diode, almost infinite (up to hundreds of megaohms); and for germanium anywhere between 100 kΩ to 1 MΩ or more. Any values substantially less indicate a leaky or shorted diode.

Many ohmmeters do not have a positive voltage at the plus lead, or a negative voltage at the negative — always test, then mark, leads for proper polarity. Also confirm that the ohmmeter has enough forward voltage to turn *on* the diode. When using a DMM the Diode Check function can be selected, which measures the approximate forward voltage drop of the semiconductor's junction — usually in mV or V.

10-8H. Zener Voltage Out-Of-Circuit Test

The simple diode ohmmeter test above will indicate opens, shorts and leakage, but not a Zener's proper Zener voltage (V_Z). To test this Zener voltage, reverse bias and connect as per *Figure 10-19*. Supply the diode's rated Zener current by increasing the power supply's output voltage until the Zener's current rating is reached. Vary this value above and below the rated Zener current while monitoring its voltage drop and current. (Normal Zener currents are approximately 50 mA for a 1 watt 5.1V Zener and around 20 mA for a 1 watt 12V Zener — the higher the Zener voltage, however, the lower the Zener current that the device can withstand, given the same power rating). If the Zener voltage remains relatively constant, and as rated, while the input voltage is varying, then the Zener is operating properly.

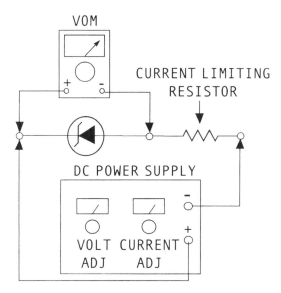

Figure 10-19. *Test setup for checking a Zener diode's V_Z.*

10-8I. AM Modulation Percentage Test

First, a word of caution: Never place a transmitter's signal directly into either an oscilloscope's or spectrum analyzer's input. The oscilloscope's input has a impedance of 1 MΩ, which would reflect almost all of the transmitter's power directly back into the transmitter's finals, possibly destroying them, while the spectrum analyzer's front end would be destroyed by any power level over 1W entering its input. However, the spectrum analyzer can be hooked up to the transmitter's output by feeding a properly attenuated transmitter signal into the analyzer's input, with the amplitude of its vertical amplifier set to logarithmic (See *Spectrum analyzer*).

The necessary attenuation for the spectrum analyzer can be accomplished by using a *step attenuator* (*Figure 10-20*), which is capable of varying the attenuation in preset steps either by manually turning a knob, or setting switches, or by electronic control. If the power output of the transmitter is beyond the safe operating range of the step attenuator, which is commonly the case, then employ a fixed-value *power attenuator* in back of it — while adjusting the needed value of attenuation with the step attenuator.

Small coaxial in-line attenuators are also available that can supply various levels of fixed attenuation (*Figure 10-21*) of up to 30 dB, with a maximum safe power dissipation of 25W. 2W low-powered models can supply up to 60 dB attenuation.

To test the percentage of AM modulation: Set the microphone gain, if so equipped, to its center position. Inject the audio modulation at a desired amplitude into either the microphone or directly into the microphone's input jack. Confirm that the output of the transmitter is appropriately terminated by an antenna, or by a dummy load capable of handling the transmitter's full output power. The percentage of AM can then be checked either with an oscilloscope or with a spectrum analyzer.

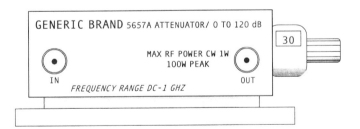

Figure 10-20. *A popular step attenuator.*

Figure 10-21. *A common coaxial in-line fixed attenuator.*

When using the oscilloscope or spectrum analyzer employ an RF pickup loop or small antenna (*Figure 10-22*) placed close to the transmitter's antenna, its dummy load, or near to the transmitter's final tuned circuit. A spectrum analyzer can also have the transmitter's output fed into its 50 W input if proper series attenuation is employed (See *Caution* above), while the oscilloscope's probe can be attached in parallel to the transmitter's load, or a small portion of the transmitter's power can be tapped off and fed into the scope through a *T* connector or line sampler.

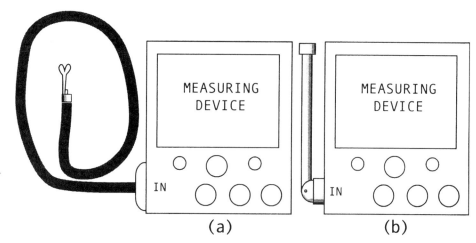

Figure 10-22. *Setup for observing an RF signal: (a) A pickup loop; or (b) a small antenna.*

When working with the spectrum analyzer, read the signal's sideband and fundamental amplitudes on the CRT display, take the *difference*, in dBs, between the two, and use the graph of *Figure 10-23* to accurately interpret the modulation percentage. As an example, utilizing the analyzer display of *Figure 10-24*, the sidebands are shown to be exactly 30 dB down from the carrier. Glancing at *Figure 10-23* and finding the -30 dB on the lower horizontal dB sideband level scale, and reading up and across to the vertical percent modulation scale, we find the modulation percentage to be 6% (the vertical scale is logarithmic, and not linear).

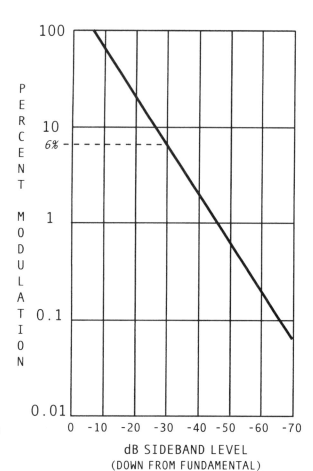

Figure 10-23. *Graph employed to calculate the AM modulation percentage when using a spectrum analyzer.*

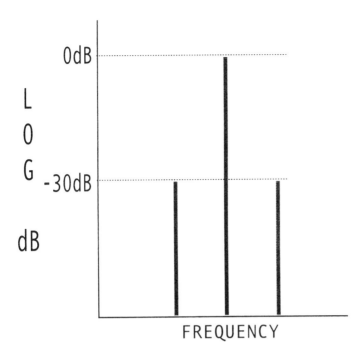

Figure 10-24. *A spectrum analyzer display of an AM signal at 6% modulation.*

When employing an oscilloscope, use the formula:

$$\frac{V\max - V\min}{V\max + V\min}\,100$$

to easily calculate the modulation percentage. For example, in *Figure 10-25*, if V_{max} equaled 5V and V_{min} equaled 1V, then the modulation percentage is calculated as:

$$\frac{5V - 1V}{5V + 1V}\,100 = 66.6\%$$

The oscilloscope can be a very low bandwidth type, since only the low frequency modulation envelope is of interest, if the AM output is tapped off of a dummy load (through a coaxial *T* connector) and inserted directly into the scope's vertical deflection plates. If a large AM transmitter is equipped with a panel-mounted oscilloscope and set to display *trapezoidal* modulation patterns, these can be employed to read the modulation percentage (See Appendix).

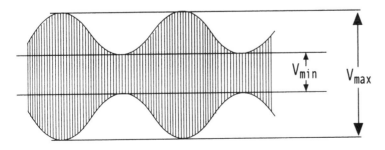

Figure 10-25. *V_{min} and V_{max} of an AM modulated waveform in the time domain.*

10-8J. FM Deviation Test

To find the *approximate* deviation of an FM signal on a spectrum analyzer, locate the side-band where the amplitude begins to drop and *continues to drop* (*Figure 10-26*). The frequency spread between the carrier and this sideband is the deviation.

An *FM deviation-meter*, if available, will supply a simple, fast and accurate deviation reading.

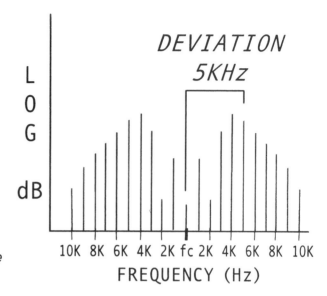

Figure 10-26. *Finding the approximate frequency deviation of an FM signal.*

10-8K. Harmonics and Overmodulation Test

The harmonics and excessive bandwidth created by single-tone AM transmitter overmodulation will show on a spectrum analyzer display as very high amplitude sidebands pairs (LSB and USB, both within 6 dB of the carrier), with the generated harmonics evenly spaced from each other in multiples of the *modulation* frequency (*Figure 10-27*). If voice or music were the baseband signal, the harmonic frequency generation would be far more complex.

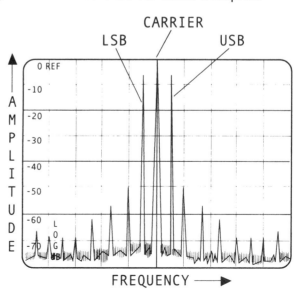

Figure 10-27. *Single-tone AM over-modulation as seen on a spectrum analyzer.*

In addition, watch for transmitter carrier harmonic generation at least out to the third harmonic (*Figure 10-28*) by widening the frequency span on the spectrum analyzer and confirming that harmonic suppression is within specifications for the equipment under test.

If placed in the field without a spectrum analyzer, harmonic generation and suppression can still be measured with other test equipment. At some distance from the transmitter site, while utilizing a *tunable field-strength meter* with the output fed to a *dB meter*, take a dB reading of the fundamental frequency and its second harmonic by tuning the field strength meter to these frequencies and notating the results. The difference between the two readings is the amount of harmonic attenuation.

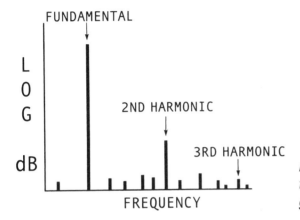

Figure 10-28. *Check at least out to the third harmonic for excessive harmonic generation.*

10-8L. Amplifier Bandpass Test

By the use of a *sweep generator* and oscilloscope, the bandpass of an amplifier or circuit can be determined. The setup is shown in *Figure 10-29*. As the sweep generator sweeps through a desired band of frequencies, the bandpass response of an amplifier or tuned circuit is displayed on the scope's CRT.

Figure 10-30 shows typical narrowband and wideband IF-section receiver responses. The bandwidth of either is measured as the frequency values between the half-power points (or at the points where the waveform's voltage falls to 70.7 percent of its peak amplitude) by counting the horizontal graticule divisions between these two points.

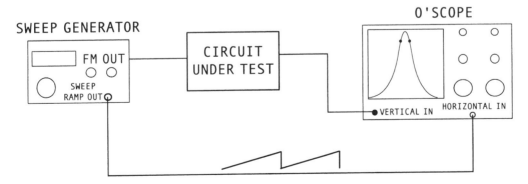

Figure 10-29. *Viewing the bandpass of an amplifier with a sweep generator and oscilloscope.*

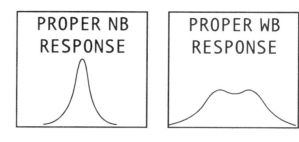

Figure 10-30. *Appropriate IF-section frequency responses.*

An alternate setup is shown in *Figure 10-31*. Utilizing an oscilloscope's internal trigger as well as a simple function generator that can be manually tuned from just below to just above the bandpass of the circuit, similar results can be obtained. When the point of resonance of the circuit under test is reached, the oscilloscope will display a peak indication of the frequency generator's frequency. By measuring the frequency on either side of this peak at the calculated half-power points, an approximate measurement of the circuit's bandwidth can be found.

Another amplifier test, referred to as *square-wave analysis*, is employed to discover if an audio amplifier has the proper linear frequency response within design parameters. Begin by feeding a 1 kHz squarewave pulse train into the input to the amplifier under test, at a amplitude that will not introduce distortion. Use an oscilloscope or spectrum analyzer placed at the output of the appropriately terminated amplifier and compare the waveforms of *Figure 10-32*. Since a squarewave produces odd harmonics, feeding a squarewave into the amplifier's input indicates any possible frequency distortion. A relatively clean output verifies that the amplifier is passing at least up to the ninth harmonic without excessive attenuation or overamplification of certain frequencies.

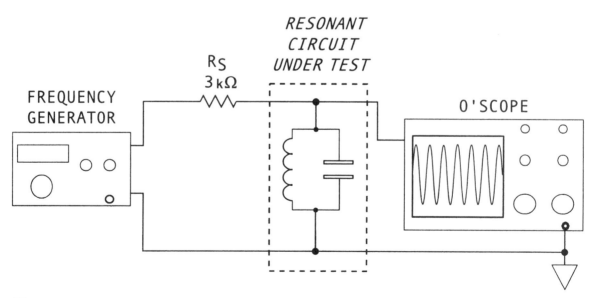

Figure 10-31. *Finding the passband or resonant frequency of a test circuit.*

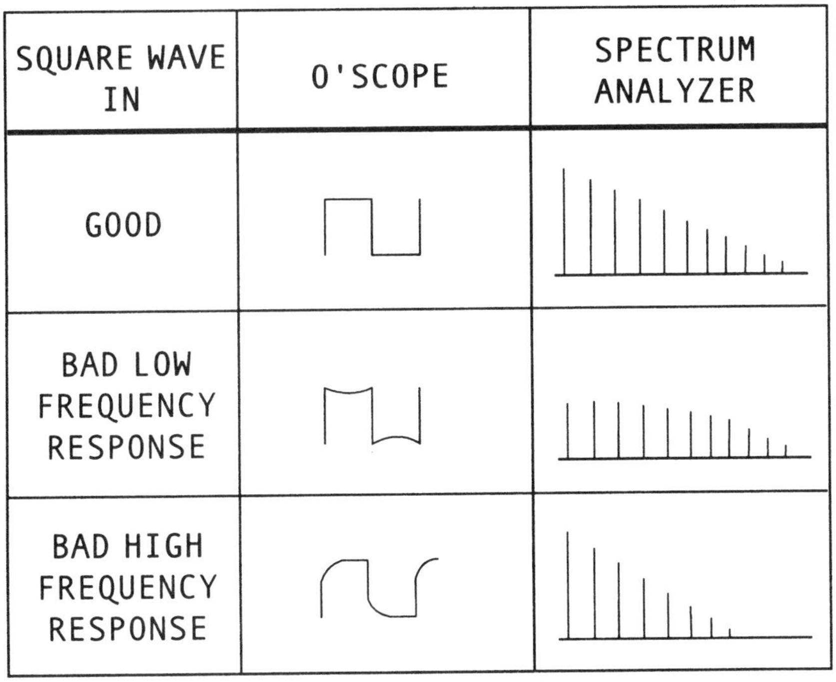

SQUARE WAVE IN	O'SCOPE	SPECTRUM ANALYZER
GOOD		
BAD LOW FREQUENCY RESPONSE		
BAD HIGH FREQUENCY RESPONSE		

Figure 10-32. *The square-wave analysis with possbile test results for an audio amplifier.*

10-8M. Phase Angle Comparison Test

To compare two phase angles, both must have the same frequency and be sinusoidal waves. Using a dual-channel oscilloscope, feed each signal into a separate channel. Measure the time between the two signals (T_2, or the time between points A and B, *Figure 10-33*). Then, measure the time period of one complete waveform, or T_1. Apply the formula:

$$\frac{T_2}{T_1} \; X \; 360°$$

to obtain the phase angle between the two signals.

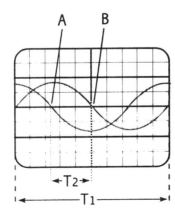

Figure 10-33. *Phase measurement between two voltages.*

10-8N. Resistance In-Circuit Test

Switch off all power to the circuit under test and discharge all capacitors. When making in-circuit resistance tests, confirm that all semiconductors will be reverse biased by the ohmmeter voltage — or false readings will be obtained if the PN junction is forward biased *on*. As an alternative, use the low-power setting on an appropriately equipped ohmmeter so that any forward biased junction will not be turned *on*. Also, utilize only the R X 100 or R X 10 ranges on standard ohmmeters when testing semiconductor circuits, as this lowers the ohmmeter's short-circuit current. This will prevent any possible transistor damage if its PN junction should be biased *on* unintentionally.

10-8O. Receiver Sensitivity Test

To check for correct AM receiver sensitivity at its rated S/N ratio, inject an RF generator's signal into the receiver's input, preferably by removing the antenna and directly attaching a coax cable from the generator (*Figure 10-34*). This generator's signal output should be at the receiver's rated minimum sensitivity level (frequently 1 uV for HF receivers; 0.16 uV for VHF receivers — if the receiver is suspected of having lower than rated sensitivity, increase the initial voltage input), and at the *exact* dial frequency along with 30% 1 kHz modulation. With the receiver unsquelched and the volume at a reasonable amplitude, read the output, in dB, at the properly terminated speaker output with a voltmeter set to the decibel scale. Then remove the modulation at the RF generator and read the output in dB (the receiver's no signal condition). The difference is the S/N ratio, and is ordinarily at least 10 dB. In this case, the specification would be: Sensitivity 1 uV/10 dB (S+N)/N. In other words, the 1 uV signal is 10 dB above the noise floor. Certain obligatory audio output powers (such as 1 watt) may be specified while measuring this sensitivity and S/N ratio.

The 12 dB SINAD test for FM receivers is similar, except it uses a 1 kHz tone with 3 kHz deviation from the frequency generator. Turn off the receiver's squelch. Place a SINAD meter (an *audio analyzer*) across the receiver's audio output. Now, lower the RF signal generator's level until the SINAD meter reads 12 dB. Note this RF signal level. This reading is the 12 dB SINAD sensitivity in uV or dBm.

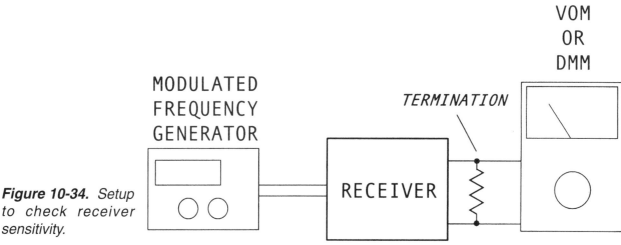

Figure 10-34. *Setup to check receiver sensitivity.*

Another FM sensitivity test is the 12 dB (or 20 dB) quieting threshold reading: With no RF signal, adjust the receiver's audio output until the dB meter (such as a VU meter) reads 0 dB (squelch OFF). Now, increase the unmodulated signal generator output until the dB meter drops to -12 dB (or -20 dB). Measure the signal generator's output in uV or dB (usually around 0.3 uV for 12 dB quieting).

If employing a standard voltmeter or an oscilloscope, then calculate AM sensitivity by:

$$20LOG_{10} \frac{V_{signal}}{V_{noise}}$$

and FM SINAD by:

$$20LOG_{10} \left(\frac{V_{signal} + V_{noise} + V_{distortion}}{V_{noise} + V_{distortion}} \right)$$

10-8P. Receiver Frequency Response Test

A frequency response test is performed on a communications receiver by hooking a modulated generator to the input of the receiver, while monitoring the output at the speaker with a DMM, oscilloscope, or audio frequency analyzer. The modulation at the generator is varied between 300 Hz and 3 kHz. Up to a 3 dB drop or less in audio output at the speaker between these two frequencies is considered acceptable for most receivers.

10-8Q. SSB PEP Transmitter Test

Attach an oscilloscope across (in parallel with) a properly terminated transmitter's antenna output (*Figure 10-35*). Feed equal-in-amplitude two-tone audio signals, usually at 500 Hz and 2,400 Hz, directly into the microphone or the microphone's panel input, since SSB must be modulated to produce any transmitter output for a PEP measurement. Increase, equally, these audio amplitudes until the output SSB signal is just under flat-topping on the scope. Then obtain the PEP reading of the displayed waveform by using the formula:

$$PEP = \frac{V_{RMS}^{2}}{R}$$

with the *R* being the value of the termination (ordinarily 50Ω).

If a peak-reading RF wattmeter is available, this can be employed to indicate the peak (maximum) envelope power when voice is used to modulate the transmitter, since a standard wattmeter will only indicate the average power of the signal, which may be as low as one-fourth of the PEP value over time. In other words, the peak-reading wattmeter senses the highest maximum (peak) value of the modulated SSB waveform, and displays this as the PEP reading — while not averaging in the lower amplitude peaks, which would significantly decrease the reading and give no indication of the power in the peaks of the modulated RF signal, but only the average power over time.

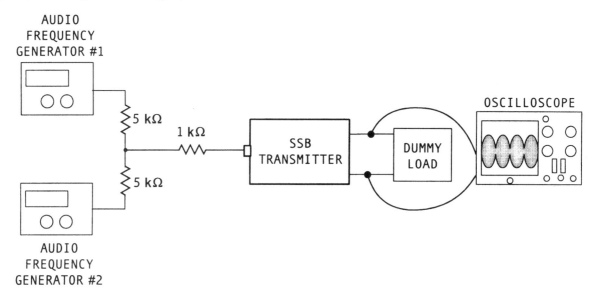

Figure 10-35. *Setup for SSB PEP test.*

10-8R. SSB Transmitter Frequency Test

To check the frequency of a SSB transmitter, modulate the transmitter with a 1 kHz tone injected into the microphone, or directly into its MIC input jack, and make a properly termi-nated frequency counter measurement of the transmitter's output (or place a pickup loop or small antenna into the BNC connector of the frequency counter to make an indirect but safe measurement). USB/LSB switch should be set to USB, with any *TX RIT* (or *TX clarifier*) control switched out or set to 0. Subtract 1 kHz from the indicated reading. The resultant is the f_r of the transmitter.

10-8S. AM/FM Output Power Test

Place a wattmeter in-line between the transmitter and its antenna or dummy load, or place an oscilloscope across a proper noninductive output resistance, such as a dummy load. Switch to transmit and read the wattmeter indication or, if working with the scope, use the formula:

$$P = \frac{V^2_{RMS}}{R}$$

to calculate the transmitter output power. Do not modulate while the reading is being taken.

10-8T. AM/FM Transmitter Frequency Test

Set up as in *Figure 10-36*, confirming that the attenuator is reducing the power to the counter's 50Ω input to safe levels — however, verify that there is adequate power to acquire an accurate reading. If the counter's Zin is 1 MΩ, place an antenna or pickup loop on the frequency counter's BNC input while the transmitter is feeding an antenna or dummy load. Switch to transmit with zero modulation applied. Read the frequency indication on the counter display.

Figure 10-36. *Setup for AM/FM transmitter frequency test.*

10-8U. Ohmmeter Click Test

The ohmmeter click test can be used to quickly troubleshoot a communications receiver, in the field, with a simple VOM:

Power on receiver.

Do not ground negative VOM lead.

Set VOM to ohmmeter's highest range.

Touch the positive lead to the collector or base of an amplifier stage for an instant. If a click is heard at the receiver's speaker, then the stage or stages all the way to the speaker are good.

Work back from the last audio stage until no click is heard at the receiver's speaker. That stage is bad.

10-9. Test Equipment

Test equipment, such as multimeters and oscilloscopes, are standard on any test bench, while spectrum analyzers, frequency counters, frequency generators, and logic probes are considered a little more specialized. These pieces of test equipment, and the knowledge to effectively operate them, are now regarded as mandatory in the field of RF.

10-9A. RF Probes

A word about RF probes. Shielded cable passive attenuator (high-impedance) *low-capacitance probes*, usually found on oscilloscopes, are very common and are built to have the least effect on the signal to be measured. They increase the oscilloscope's input impedance and decrease the scope's input capacitance, and are designed not to distort the input signal within the probe's rated bandwidth (commonly up to about 50 MHz for 10X probes; 20 MHz for 1X probes). Low-capacitance probes are important to lessen shunt capacitance (which would attenuate high-frequency signal amplitudes), reduce pickup of stray fields (which would create interference), diminish capacitive loading (which would cause distortion of non-sine waves), and decrease circuit loading (which would cause incorrect voltage indications and create circuit instability).

RF and demodulator probes are operated with VOMs or oscilloscopes and are built to convert the RF signal into either an RMS or a peak-reading DC voltage (depending on the model), allowing the low-frequency VOM or scope to furnish relatively accurate voltage indications (the demodulator probe produces a steady DC voltage when RF is present, and AC when AM is present). Some are reliable up to 0.25 GHz. Many RF and demodulator probes do not necessarily give the true voltage of the circuit under test, but only a relative indication.

Passive divider probes are available that can function up to 6 GHz with extremely low input shunt capacitances of around 0.25 pF.

Input adapters are available to convert the 50Ω input impedance common to spectrum analyzers into 1 MΩ, allowing the analyzer the capability to carry out high-impedance (1 MΩ) measurements, similar to an oscilloscope. However, they are commonly good up to only 100 MHz and cost about \$1,000. Higher-frequency active probes that supply 1 MΩ of input resistance and a bandwidth of up to 3 GHz are also available for spectrum analyzers for about \$2,500. These devices are utilized because the spectrum analyzer's low 50Ω input impedance would excessively load down the circuit-under-test while probing circuits during troubleshooting.

Most spectrum analyzer probes can be used in low powered circuits only, such as receivers and the low-powered sections of a transmitter. They are damaged above relatively low input voltages (1.5 to 2 V_{RMS}), and are ordinarily equipped with a 10:1 voltage divider to increase this maximum voltage to about 20 V_{RMS}.

If such spectrum analyzer probes or adapters are not available, a simple probe can be manufactured (*Figure 10-37*) that will exhibit to any circuit-under-test a 5 kΩ load for about \$2, which gives a 1/100th voltage reduction of the signal under test (C_1 supplies DC blocking).

Special RF 50Ω input oscilloscopes use special probes called *low-capacitance* or *low-impedance* probes. They have a bandwidth of up to 8 GHz, with an impedance of 500Ω, giving a 10:1 voltage ratio. There is no choice of different probe tips due to the high frequency of operation. These low-impedance probes do not need compensation, but must only probe circuits that were designed to work into 50Ω at very low power levels.

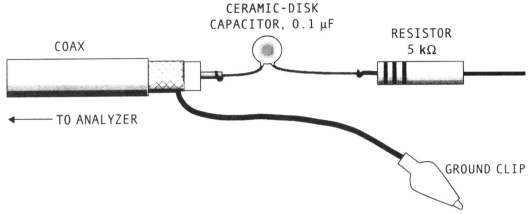

Figure 10-37. *A handmade spectrum analyzer probe.*

Active FET probes are available for the oscilloscope and are capable of very high bandwidth operation and very high prices (up to $3,500). They have very low input capacitances of 0.6 pF or less. Commonly available active FET oscilloscope probes provide 1:1 voltage ratio, impedance of 1 MΩ, and a bandwidth of 1 GHz. Some high-end FET probes are 10:1, with an impedance of 2 MΩ, and a bandwidth of 2.7 GHz. Models are available that work into either 50Ω or 1 MΩ oscilloscope inputs, so they can also work for spectrum analyzers.

All active probes require a power source, either supplied from the oscilloscope or from a separate supply, and are easily damaged during over-voltage measurements. As with most test equipment, the higher the frequencies being measured, the lower the voltages that the probes can withstand.

Caution: *If high DC levels are present during probing, use a blocking capacitor in series with the probes (if the probe is not already so equipped) to protect the signal generator, spectrum analyzer, or frequency counter from damage.*

10-9B. The Multimeter

Multimeters, as with all test equipment, are available with varying features, qualities and prices.

A digital multimeter (DMM) is used far more than the older analog type of VOM. The DMM is capable of measuring AC and DC voltages and currents, as well as resistance values, with high accuracy. Many are equipped with auto ranging, dBm measurement abilities, a diode tester, a continuity buzzer, and even transistor gain and capacitance measurement capabilities.

The analog VOM type is, however, superior to the DMM for keeping track of rapidly changing current and voltage level trends. Also, when working near an active transmitter, DMMs can be severely influenced by the RF, causing erratic operation. Under these circumstances, an analog model must then be used.

A DMM has its highest accuracy and speed when not set to *auto-range*. They may also inject noise from their digital circuits into sensitive circuits, creating erratic test results.

The input impedance of a multimeter is measured in Ω/V of sensitivity. If using an analog meter, do not settle for less than 50,000 Ω/V so that loading is kept down to negligible levels in most circuits. When working with a digital model, or an FET-input analog model, any good quality unit will have 10 MW input impedance or better.

Keep in mind that most VOMs will only give accurate readings when the AC voltage being measured is at 20 kHz and under for better analog units, or up to 100 kHz for most quality DMMs.

Another type of meter has gained popularity in the last few years; the *true RMS* meters. They will accurately measure non-sinusoidal waveforms, unlike average-measuring DMMs, which can be off by up to 50%.

10-9C. The Spectrum Analyzer

In RF, the ability to operate and interpret the results of a spectrum analyzer is essential. While an oscilloscope is ordinarily used more often, there are occasions when a time-domain reading cannot give all the information necessary. Since a spectrum analyzer reads in the frequency domain (a spectrum analyzer displays a range of frequencies-versus-amplitude), it is capable of presenting all the frequency components that shape a wave, as well as display an entire spectrum of diverse signals. A signal that seems clean on an oscilloscope may appear to be anything but when observed on a spectrum analyzer's CRT, since harmonics must be at least 20 dB or less from the fundamental to be seen reliably as distortion of a sine wave on the oscilloscope's CRT. Also, very high frequency oscilloscopes are expensive and quite rare, while the *average* spectrum analyzer is good to one gigahertz.

A spectrum analyzer can examine signal levels, bandwidths, and frequencies; determine noise and signal-to-noise ratios; measure IMDs and harmonics, percent modulation, FM deviation; detected spurious signals; and align transmitters and receivers.

Different designs for spectrum analyzers have been used, but the most common is the super-heterodyne class. This type of spectrum analyzer can be visualized as nothing more than a normal superheterodyne receiver that visually displays all frequencies within its passband, rather than outputting an audio signal, by sweeping, or tuning, its local oscillator through a range of frequencies. The amplitude of these frequencies are displayed in the vertical, while the scanned frequencies are displayed in the horizontal (*See Figure 10-27*), with the highest frequency to the right of the CRT and the lowest to the left.

The amplitude of a signal is normally shown in a logarithmic fashion, in decibels (most spectrum analyzers are referenced to dBm (0 dBm=1 mW)). If all incoming signals were displayed in a linear manner, the usable dynamic response of the analyzer would be quite low.

Consider that if input signals differed by just 60 dB, one incoming voltage would differ from the other from one thousand times; small signals would get lost, while large signals would overdrive the display. Any meaningful comparison between the two signals would be almost impossible.

There are many exotic uses for the spectrum analyzer, but the application that the technician is most concerned with is the spurious content of a transmitted waveform. A transmitter must be tuned for the highest spectral purity attainable, since harmonic, IMD, and parasitic output must be kept to a minimum — or severe interference with other stations will occur (as well as creating inefficient transmitter operation).

To operate a basic spectrum analyzer, refer to *Figure 10-38*. *First, a warning*: Any high input levels, whether they are DC or AC, may instantly destroy the analyzer's front end. Too large an input signal at the first mixer of a spectrum analyzer can result in distortion, spurious signal products, and/or damage to the first mixer. External attenuation *must* be employed for any possible high amplitude signals entering the analyzer's front end (*Figure 10-39*, see *Attenuators*). A spectrum analyzer's input usually cannot exceed + 20 dBm or + 30 dBm, which is 2.24V to 7.07V across 50Ω.

When viewing a weak signal, and a strong signal is being picked up by the spectrum analyzer (even many MHz or GHz away), signal attenuation must not be decreased to view the weaker signal. The strong signal would still destroy the spectrum analyzer's wideband front end (increasing the spectrum analyzer's IF gain, if possible, may be acceptable).

Low-powered transmitters (one watt or less) may be attenuated solely with the analyzer's *INPUT ATTN* (or *input level* or *reference level*) knob set to +30 dBm initially. If working with such low-powered transmitters, the transmitter and analyzer can be connected directly through a coax cable (the transmitter would typically take a BNC or PL-259 connector, while the analyzer would normally take an *N* connector, though some are outfitted with BNCs).

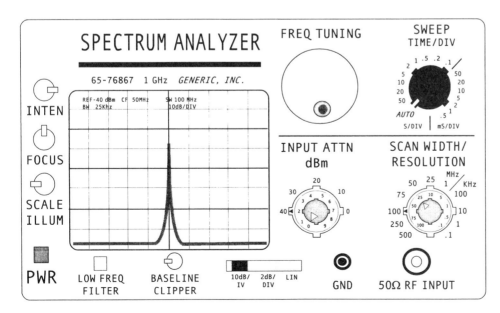

Figure 10-38. A no-frills spectrum analyzer.

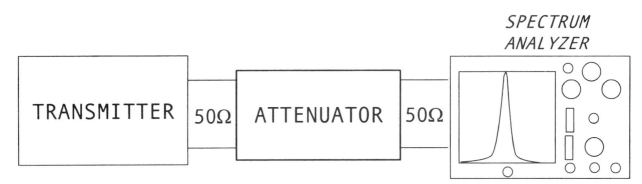

SPECTRUM ANALYZER

Figure 10-39. *Proper spectrum analyzer setup for testing a transmitter.*

The *INPUT ATTN* knob, as the name indicates, will attenuate the input signal — normally in calibrated discrete steps of 10 dB, with a fine adjustment in 1 dB steps. In most analyzers, this input attenuation value is now our starting reference point for all amplitude measurements which, on a spectrum analyzer, is the very top line of the CRT display, and is referred to as the *reference level*. For example, if the input attenuation is set to 0 dB, then the top line of the display is at 0 dB — any input signal is now referenced to that line, and anything below that line is of a negative dB value. Obviously, due to the analyzer's sensitivity, it would be judicious to start with the maximum attenuation first, and work down, while viewing the entire frequency spectrum by using the *SCAN WIDTH* knob at its widest setting (generally marked as *Max.*). This is to insure that there are no high-amplitude signals further up or further down in the frequency spectrum that can destroy the analyzer's front end.

The *SCAN WIDTH* (or *frequency span*) knob adjusts the frequency range of the sweep across the frequency spectrum to be displayed on the CRT. In the case of the analyzer of *Figure 10-38*, anywhere from a width of 100 Hz to 500 MHz can be scanned, or swept.

Many analyzers, though, are calibrated in *SCAN WIDTH/DIV* (or *FREQ. SPAN/DIV*), and to obtain the swept spectrum range of the entire display, simply multiply the number of horizontal divisions (ten) by the number across from the SCAN WIDTH/DIV dial indication.

When using a narrow *frequency span* setting on a spectrum analyzer, some spectrum analyzer's internal oscillators may drift, causing the signal to apparently drift. Better analyzers phase-lock the internal oscillators to stabilize them, decreasing this false drift.

The internal bandwidth of the analyzer itself is adjusted by the smaller inset *RESOLUTION* (RBW, or *Resolution BandWidth*) knob, which alters the filter selectivity (bandwidth) of the IF stages of the analyzer. This setting is used when two sinusoidal signals are very close together in frequency — the narrower the setting, the easier it is to separate these two signals. As well, small amplitude signals can be grabbed out of the noise grass by narrowing the RBW setting. An excessively narrow resolution setting, however, will create erratic responses to be displayed on the CRT, since any analyzer instability will create a false or distorted display. As a rule, it should be set to *coupled* or *auto*.

The importance of a properly adjusted RBW can be shown in *Figure 10-40*: With a wide RBW the side frequencies are lost, while with the narrow setting they can be clearly viewed. Furthermore, as the RBW filter is narrowed, then the noise amplitude on the noise floor will also, of course, diminish, since the wider the bandwidth the more noise (external signals and internal analyzer noises will both be affected).

The *FREQ TUNING* (or *center frequency*) knob allows the technician to center the frequency of interest on the display (similar to a radio's tuning knob), while choosing the width of the scan on either side of this frequency with the SCAN WIDTH knob. In some analyzers and in some modes of operation, the frequency control is referenced to the left side of the CRT screen, and is referred to as the *starting frequency*.

The vertical (amplitude) range of the display is adjustable by the *log/lin* switch, indicating 10 dB/DIV, 2 dB/DIV, and LIN (linear). By switching to the 10 dB/DIV range, each vertical division will now equal a value of 10 dB. This setting is used for normal measurements with signals that are expected to have high dynamic range and amplitude. The 2 dB/DIV setting is for more precise, low amplitude measurements. However, due to the logarithmic nature of spectrum analyzers, amplitude measurements, even on the 2 dB/DIV setting, are not very accurate. Switching to LIN will give the amplitude an (ordinarily) RMS linear response for reading very low amplitude values, with a low dynamic range, of any input signals.

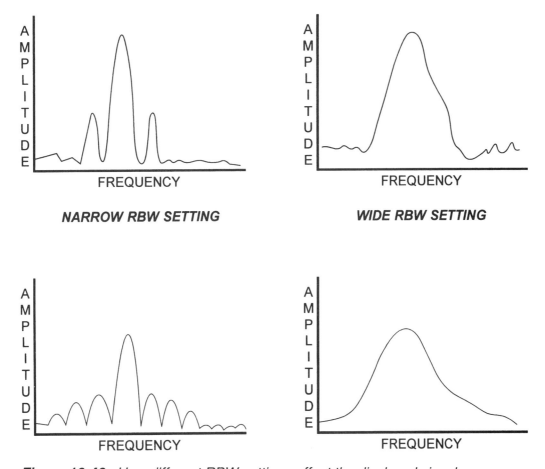

Figure 10-40. How different RBW settings affect the displayed signal.

The *BASELINE CLIPPER* control is used to clip-off the bottom of a signal to remove any annoying noise (grass) present in the lower part of the CRT display.

The *LOW FREQ FILTER* (or variable *video filter*, or *VF*) is a low-pass filter that smoothes out the display and rids it of excessive noise hash. The video filter may also reduce the amplitude of video modulation and short duration pulses, which is not desired. As well, use of the video filter may demand a slower *sweep rate*. Most spectrum analyzers allow the choice of several video bandwidths. The default bandwidth for the video filter should be the same as the *RBW filter* selection, or switched off if not needed, or set to *auto*. Selection for the appropriate video filter can be made by panel switch or CRT on-screen menu.

On most units an adjustable *SWEEP* (or *TIME/DIV*) knob is also available to regulate the sweep speed of the trace across the CRT. If the sweep time is set at too fast a speed, the apparent amplitude of the signal will decrease and a shift will occur in the signal's apparent frequency (the RBW filters will not have time to charge, resulting in inaccuracies, while if the sweep speed is set too slow the CRT display will flicker. This may be overcome if *Display Storage* is available. With this feature, the spectrum displayed on the CRT can be saved in the analyzer's memory by pushing a *Save* button, which saves the screen in storage for future inspection). As needed, set the sweep knob for a flicker-free and stable display or, typically, switch to *auto*.

Generally, most of the analyzer's settings (such as BW, RBW, dBm, ATTN) as made by the operator are arrayed at the top of the CRT for easy and quick reference.

Figure 10-41 reveals a basic digitally controlled spectrum analyzer commonly found in many shops. Most have the capability to store and recall favorite frequencies, SPAN/DIVs, REF LEVELs, RBWs, VERTICAL SCALEs, SWEEP SPEEDs, etc, which greatly speeds up the troubleshooting process, especially of certain repetitive alignment and testing tasks.

Adjusting the frequency tuning on this type of analyzer can be accomplished in several ways:

Rotating the circular analog-type FREQ wheel.

Pressing the digital FREQ button, then pressing the adjacent up/down *arrow keys*.

Pressing the FREQ button and directly selecting the frequency of choice by using the KEYPAD buttons, then choosing either GHz, MHz, kHz, or Hz.

Pushing the proper MENU button and choosing the frequency from an on-screen menu.

SPAN/DIV, REF LEV, RBW, and SWEEP speed can also usually be adjusted similarly to the above.

As well, picking between ZERO or MAX SPAN (see below), selecting the VERT SCALE values, or adjusting the INTENS, FOCUS, or BASE CLIP, can be accomplished by pushing the appropriate button and pressing the adjacent up/down arrow keys, or by flipping switches or rotating knobs, depending on the analyzer.

The MAX SPAN button informs the spectrum analyzer to sweep from 0 Hz to the analyzer's maximum frequency (in this case 0 Hz to 1.8 GHz). The spectrum analyzer can also be set to stop *auto-scanning* (auto-scanning is considered NORMAL MODE), and be set to ZERO SPAN mode. This means the analyzer can be tuned by the FREQ (*Center Frequency*) knob. In this mode the spectrum analyzer operates much like a receiver with switchable bandwidths. An AM or FM station's audio can even be detected and fed to the analyzer's internal speaker. As well, any spectral lines passing through the IF filter will cause a level CRT display, which is sometimes called the *selective voltmeter function*. By manually changing the spectrum analyzer's sweep control, you now have a representation of the signal's envelope passing through the RBW filters (similar to an oscilloscope with adjustable bandwidths).

This analyzer is also equipped with a *Tracking Generator* (TG) output, which greatly simplifies many tests. The TG is integral with the analyzer (some may be standalone units) and outputs a frequency sweep in step with the analyzer when in NORMAL MODE, or outputs a CW signal at the spectrum analyzer's center frequency when in ZERO SPAN MODE. An analyzer with a tracking generator can check filter bandwidths, amplifier frequency response, and SWR. A tracking generator can be hooked to the DUT as shown in *Figure 10-42*.

A movable *marker* indicator on the CRT is also available on some analyzers (such as *Figure 10-41*). This marker can be placed, ordinarily by arrow cursor buttons on the analyzer's front panel (MKR), at any spot along the CRT. The signal markers may be a dot, cross, or intensified section of a waveform on the display. It enables the operator to quickly read the frequency and amplitude at any desired location on the CRT, which is then displayed either on the CRT face or on a separate panel meter.

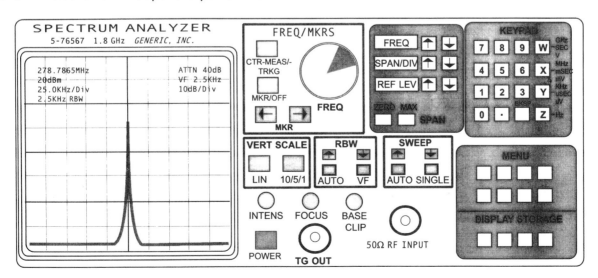

Figure 10-41. *A no-frills digitally controlled spectrum analyzer.*

CTR MEAS (*CENTER MEASURE*) control will allow the spectrum analyzer to measure the frequency of the signal peak nearest the center of the screen, or of the signal nearest the signal marker. Depressing the TRKG (*TRACKING*) button causes the CENTER MEASURE function to repeat continuously, which centers a drifting signal in the center of the CRT for better viewing.

The DISPLAY STORAGE section allows the operator to store multiple waveforms in different registers to be recalled later. Even though not shown in the basic analyzer of *Figure 10-41*, most new digital analyzers can temporarily store the spectra as displayed on the CRT, but still allow it to be updated with each new sweep, by choosing PEAK (which plots only the peak values of each sweep), or by choosing MAX/MIN (which plots both the minimum and maximum values). This data is temporarily stored in the *Waveform Memory*. Pressing the SAVE button will save the on-screen spectra in RAM for future examination.

A MAX HOLD feature can enable the operator to capture the maximum signal amplitude encountered during a series of sweeps. Pressing the SAVE button will then save the screen in memory. MIN HOLD is analogous.

When a digital spectrum analyzer is in *Digital Storage Mode*, a WAVEFORM SUBTRACTION switch can be used to compare two spectra which may or may not be identical. If they are, then they will cancel each other on the CRT — if not, it will be distinctly visible in frequency or amplitude.

Most modern spectrum analyzers have frequency counters built in, and are used with the *markers* and CTR MEAS function. Many are accurate to +/-60 Hz, and may also read out accurate amplitude measurements.

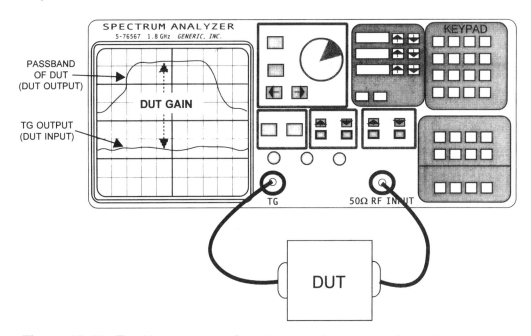

Figure 10-42. *Tracking generator/spectrum analyzer setup for testing.*

A FREQUENCY RANGE control is found on microwave spectrum analyzers. This allows the operator to choose a range of frequencies that the very wideband microwave analyzer must display (example: 0-1 GHz, 1-2 GHz, etc.). The control is similar to band switching on a radio.

When the *digital waveform registers* are turned off on a digital spectrum analyzer, the analog trace is displayed, which has the benefit of being able to catch faster changing waveforms, while the trace is also able to be swept at a much faster rate.

Most of these more advanced analyzers have built-in floppy or hard drives for permanent storage of all displayed information.

If taking any absolute amplitude measurements with a spectrum analyzer, such as the amplitude of a carrier, rather than *ratios* between a signal and its harmonics, it must be remembered to compensate for the amount of external attenuation employed, and adjust your amplitude readings accordingly. Also, since most RF spectrum analyzers have a 50Ω input impedance, there will be an impedance mismatch between a 75Ω output transmitter and the50Ω input of the analyzer, with the normal loss of power and inaccurate amplitude readings (there are impedance matching couplers available, with an internal DC blocking capacitor, to compensate for this mismatch).

To sustain absolute amplitude calibration of an input signal on the analyzer's CRT screen (so that any input signal has a definite and unchanging value), no matter what the analyzer's input attenuation setting may be set for, two methods are commonly employed to compensate.

The first: When setting the input attenuation knob, the indicated *reference level* (*RL*, the value of the top line of the CRT's graticule) will automatically change value to compensate for this variation in input signal strength. As the analyzer's input attenuation is increased, the displayed signal will decrease in amplitude, while the reference level's value at the top of the graticule will increase in value (in step with the panel-mounted input attenuation knob's setting). This allows the operator to calculate the signal's true power by starting at the reference level, and counting the divisions down to the peak of the signal.

The second method: With some newer analyzers, as the input attenuation is increased, the top of the displayed signal will always remain at the same amplitude, but the baseline rises (the analyzer's IF amplifier gain increases to compensate for the increase in input attenuation), while the reference level itself also remains unchanged. This method automatically compensates for any input attenuator settings by allowing an input signal to remain at the same relative location on the CRT's graticule, and in relation to the reference level, no matter what input attenuation settings are employed. There will now be two different dB measurements near the RL of the CRT. The RL in dBm (or dBmV), which still has not changed; and the ATTEN = #dB text, which has changed.

In other words, both of the above methods assure that if a -10 dB signal is placed at the input to the analyzer, then the CRT will allow the operator to observe instantly that the amplitude of this signal is -10 dB, whether this signal is heavily attenuated by the input attenuator or not.

Upconverting superheterodyne analyzers have what is referred to as *local oscillator feedthrough*. The internally generated first LO signal reaches the analyzer's display and causes an indication that appears to be a high amplitude signal. Since this signal is always at 0 Hz, it is a useful indicator of where the technician is looking in the frequency spectra, as well as allowing more accurate evaluation of low-frequency signals, since we now have a precise reference point from which to begin measurements. Also, any signals found to the left of this 0 Hz area on the spectrum analyzer display are just mirror images of the true signals to the right of 0 Hz, and should be ignored.

Although different FCC rules apply as to the maximum spurious emissions allowed to be generated (as a very general guide, FCC regulations state that there must be no more than 50 mW of spurious signal output from a transmitter), depending on the transmitter's model year, frequencies generated, output power levels, etc., the following examples will give you a starting idea of the spectrum analyzer displays that will commonly be encountered.

It is recommended that the *scan width* of the analyzer be set wide at first to catch any high amplitude harmonics, or other spurious responses, that would cause interference to other stations from the transmitter under test.

The *Figure 10-43* display of an unmodulated carrier from an AM transmitter shows a good spectrum out to the fourth harmonic, with the second harmonic 60 dB below the fundamental. *Figure 10-44* illustrates a much narrower analyzer sweep of a very good AM signal modulated with a single 1 kHz tone at approximately 50% modulation (the analyzer's amplitude is set to *Log*). The signal shows that the sidebands have equal amplitude, indicating no *inci-*

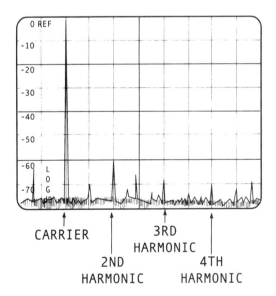

Figure 10-43. *An unmodulated carrier with properly attenuated harmonics.*

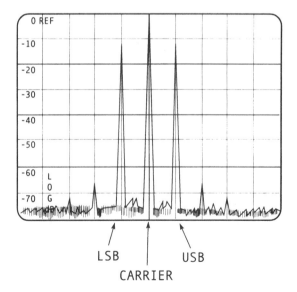

Figure 10-44. *A good single-tone AM waveform.*

dental FM, and all harmonics of the original modulating waveform (harmonic sidebands), are at least 60 dB below the fundamental, with no *parasitics* visible. *Figure 10-45* presents a poor AM display with overmodulation causing severe spurious responses. *Figure 10-46* presents an excellent SSB two-tone signal with the carrier and the opposite sideband, in this case the LSB, properly suppressed, with little visible intermodulation distortion (IMD). An inadequate two-tone SSB display showing extensive IMDs is exhibited in *Figure 10-47* (such IMDs in SSB would make the voice sound gravely). IMDs should, as a general rule, be at least 40 dB below the PEP output. The display of *Figure 10-48* demonstrates an unmodulated FM signal with all carrier harmonics well under the legal limit. *Figure 10-49* displays a good FM single-tone modulated spectral exhibit. Depending on the modulation index, FM displays can vary drastically from each other and still be completely acceptable (See *FM fundamentals*).

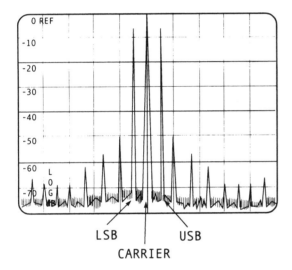

Figure 10-45. *A single-tone AM over-modulation waveform.*

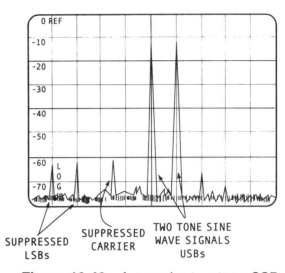

Figure 10-46. *A superior two-tone SSB signal.*

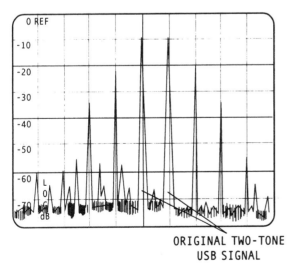

Figure 10-47. *A two-tone SSB signal with massive IMD levels.*

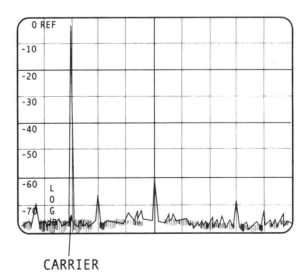

Figure 10-48. *A great unmodulated FM carrier with little harmonics present.*

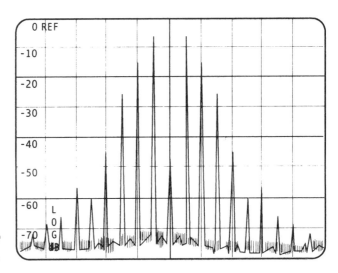

Figure 10-49. *A single-tone FM signal showing sidebands.*

Short instructions for use:

Warm up all equipment and the device-under-test (in this case a transmitter).

Place attenuators, or use a line sampler, between analyzer and transmitter (*Note:* At least 20 dB of added attenuation is needed for a 100 watt transmitter. Check an unknown signal's amplitude before feeding it to the fully attenuated spectrum analyzer by using an oscilloscope or an RF voltmeter. The signal should not be over 2.24 volts RMS across 50Ω for a spectrum analyzer set to +20 dBm. Conversely, in order to view a very low amplitude waveform on a spectrum analyzer, decrease the input attenuation (if safe), decrease the RBW filter and, if still needed, switch in the Video Filter).

Set analyzer preliminarily as follows:

    a. *INPUT ATTENUATION* to maximum (+30 dBm).
    b. *SWEEP SPEED* to auto.
    c. *LOG/LIN* switch to 10 dB/DIV.
    e. *SCAN WIDTH* to maximum (Max.).
    f. *RESOLUTION* to auto.
    g. *BASELINE CLIPPER* completely counterclockwise.
    h. *VIDEO FILTER* (*LOW FREQ FILTER*) off.
    I. *INTENSITY, FOCUS, SCALE ILLUMINATION* knobs adjusted as desired.

Switch on transmitter to transmit.

Adjust analyzer's *INPUT ATTENUATION* for desired CRT signal amplitude.

Center signal of interest with *FREQ TUNING* knob.

Narrow *SCAN WIDTH* as desired to view signal or its harmonics.

Engage *BASELINE CLIPPER* and/or *LOW FREQ FILTER* as desired.

Take measurements of signal and any spurious responses.

10-9D. The Oscilloscope

The oscilloscope is an indispensable tool for all who work in electronics. It is a device that allows the operator to view the shape of any type of waveform in the time domain, as well as measure its amplitude, phase, and frequency. Any artifacts or anomalies present in the waveform or distortions of the waveform will be quickly detected.

Most oscilloscopes have 1 MΩ input impedances, while some specialty scopes also have a low-powered 50Ω input for certain RF measurements.

The oscilloscope of *Figure 10-50* is a basic no-frills dual-trace oscilloscope. The *TIMEBASE* (the horizontal amplifier) adjusts the sweep speed, either in milliseconds or microseconds per division, of the electron beam that traces out the signal being viewed on the face of the CRT. This horizontal axis exhibits the signal in the time domain. The faster the sweep, the smaller the time slice that is being displayed, thus allowing a higher frequency to be viewed.

The *CH 1* and *CH 2* knobs adjust the sensitivity of the vertical amplifiers of each channel's input signal. The vertical axis displays the amplitude of the signal, typically in volts and mV. Since the values given around each knob decrease as this knob is rotated in the clockwise direction, any input can be increased or decreased in amplitude to adjust the height of the signal on the CRT. These values, as in the horizontal amplifier, are per division. Thus any value read must be multiplied by the amount of graticule divisions the signal's maximum positive and negative amplitudes reach for a peak-to-peak voltage reading.

The *AUTO/NORM* button can be switched to AUTO to set the trigger to automatically begin a sweep at the same relative part of the waveform each time (such as when an input signal crosses zero volts, going from negative to positive), allowing for a stable display that does not float across the CRT. If the button is pushed to *NORM*, triggering must be manually adjusted by the *LEVEL* knob, which allows the operator to adjust the triggering to occur at any point on a waveform. This is normally employed with complex waveforms that the auto trigger mode cannot handle. Triggering on the positive or negative-going edge of a waveform can be chosen by depressing the ± button.

The *TRIGGER* selector switches in different filters in-line with the trigger circuits to supply more reliable triggering at different frequencies. The *AC* setting is utilized to block any DC components, and can be employed for most signals; the *DC* setting will pass DC or very slow AC, as well as being the recommended setting for most digital input signals; *HF* filters out low-frequency fluctuations; *LF* filters out high-frequency interference and noise; *LINE* is adopted

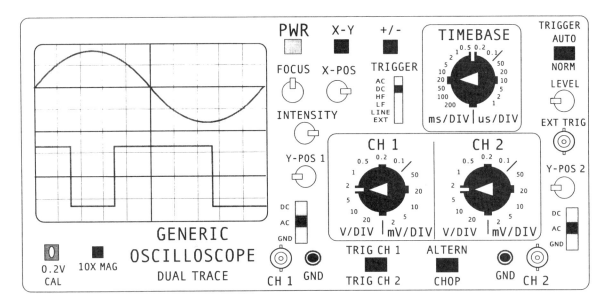

Figure 10-50. *The dual-trace oscilloscope.*

to supply triggering at power line frequencies; *EXT.* works in conjunction with any trigger signal at the *EXT. TRIGGER* BNC input.

FOCUS and *INTENSITY* adjust the CRTs electron beam (trace), while the slider switch marked *DC, AC, GND* adjust the input coupling for each channel into the vertical amplifiers. With *DC* coupling both DC and AC signals are passed to the vertical inputs; *AC* coupling blocks the DC, as well as low-frequency alternating current, and passes the AC; *GND* places chassis ground at the input, allowing for a reliable no-input reference to adjust the traces.

X-POS adjusts the position of both horizontal traces on the face of the CRT; the two *Y-POS* knobs adjusts the position of each vertical trace.

CH 1 and *CH 2* BNC connectors accept the standard 10:1 (or 100:1) attenuator probes (*Figure 10-51*), or any other suitable probe (See *Probes*). Through the probe the input signal is fed into the vertical amplifiers.

The *10x MAG* button magnifies a portion of the signal on the CRT by ten times for closer observation of the waveform.

The *0.2V CAL* connector outputs a 0.2 V_{P-P} 1 kHz square wave for compensating (adjusting) the 10:1 attenuator probes. (Probes may have both a high and low-frequency adjustment or merely a single adjustment — either way, the RC time constant of the probe must be adjusted for accurate, undistorted measurements. Simply set INPUT COUPLING to DC, and turn the internal probe trimmer(s) until a perfect square waveform is obtained on the CRT, with no droop or overshoot).

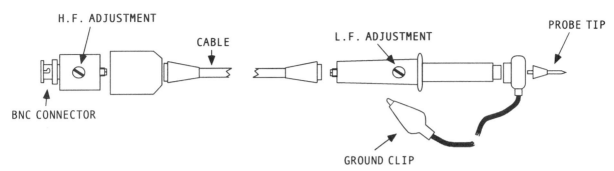

Figure 10-51. *Disassembled oscilloscope attenuator probes showing adjustment points.*

TRIG CH 1/TRIG CH 2 pushbutton chooses between Channel 1 or Channel 2 for triggering (when using both channels). *ALTERN*, also used when working with both channels, selects between *alternately* completely sweeping the first channel from left to right, then sweeping the second channel completely, then the first, etc. If the sweep speed must be very slow, then the first sweep will begin to fade before it can be refreshed by a new pass of the electron beam. When this occurs, *CHOP* mode should be adopted. Each trace is now chopped up into many equal sections; the beam switches between the two traces at a very high speed, not allowing either to fade.

The *X-Y* measurement button is pressed to display Lissajous patterns, which is an old method for comparing phase or frequency between two similar signals.

Even without a spectrum analyzer, harmonic generation can be roughly detected with a basic oscilloscope, but only if the harmonics are at relatively high levels and within the bandwidth of the scope (the harmonics of a waveform are, of course, much higher than the fundamental signal being measured). If the sinusoidal wave, such as the unmodulated carrier of a transmitter, is deformed in any way (*Figure 10-52*), then harmonic products are present in the signal.

Apart from having the ability to view the condition of a signal in the time domain, the oscilloscope gives the technician the capability to measure the amplitude and frequency of a signal. Since oscilloscopes naturally display peak and peak-to-peak voltages, any RMS measurements must be calculated. Also, because 10:1 attenuator probes are typically used to decrease loading on the circuit under test, any amplitude measurements must be multiplied by ten (some scopes have a setting on the vertical amplifier knob that will take this attenuation into account). To measure the amplitude of an input signal, find the number of vertical divisions between the positive and negative peaks (*Figure 10-53*), multiply this by the volts/DIV setting. Then, if operating a 10:1 probe, multiply the voltage obtained by 10. This is the peak-to-peak amplitude of the signal. To find the RMS voltage divide by two to obtain the peak voltage, then multiply the peak reading by 0.707.

Frequency can also be *approximately* determined by multiplying the number of horizontal divisions from the beginning to the end of a single wave by the time-base setting, which supplies the time period of one wavelength of the input signal. Then, use the formula $f = 1/t$ to obtain the signal's frequency.

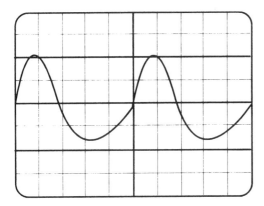

Figure 10-52. *Distorted sine wave, indicating possible harmonics.*

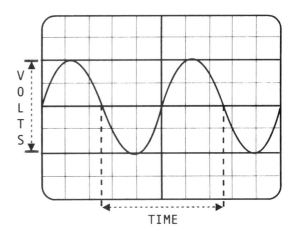

Figure 10-53. *Frequency and amplitude measurements with an oscilloscope.*

In order to safely measure the voltage across a single component in a circuit with an oscilloscope (*differential mode*), instead of making a ground referenced measurement to a circuit's test point: First switch Channel 1 to INVERT (which reverses the signal as compared to Channel 2); set both channels to the same V/DIV value, and press ADD (which combines, or adds, the two signals from Channel 1 and 2). Then connect both probes to the same end of the component under test and adjust the variable gain of one channel for a null on the screen. This equalizes both channels for a differential reading. Place the probe for Channel 1 to one lead of the component, and the Channel 2 probe on the other component lead. Try grounding or not grounding the scope to the circuit's ground (with ground there may be some hum). This will give the difference between the two probe's voltages, which is the same as the voltage across the component (or the *differential voltage*). When scoping in this differential mode Channel 1 is acting as the negative, while Channel 2 is acting as the positive, lead of a DMM. Both channel's V/DIV must be changed equally when adjusting the vertical sensitivity. Not an effective technique for high-frequency readings due to delay matching problems.

It is usually safe to measure the voltage drop directly across a component, with a scope ground placed on one side of a component, and the probe on the other side, when using a free-floating (battery operated) scope. However, if the circuit is for low-voltage RF (especially in RF receiver circuits), a ground clip may add too much capacitance and destabilize it.

With the popular digital oscilloscope, or DSO (*Digital Storage Oscilloscope*), an operator is able to view a waveform that occurred a few seconds ago, allowing stubborn glitches and anomalies to be tracked down. DSOs are higher bandwidth than analog scopes, and are rapidly replacing analog scopes in many applications. The good DSOs have very fast sample rates, with an internal floppy or removable hard drive that stores setup info, screen shot graphics, and waveform data storage. They are capable of reaching up to 500 MHz, or up to 50 GHz with the sampling types.

Short instructions for use:

Warm up all equipment, as well as the unit-under-test.

Set scope preliminary as follows:

a. *V/DIV* (vertical amplifier) fully counterclockwise.
b. Adjust *FOCUS* and *INTENSITY* as desired.
c. ± set to +, *TRIGGER* to AC and AUTO.
d. Input *COUPLING* to GND, set *X* and *Y POSITION*, switch *COUPLING* to *AC* to block any DCV (use DC coupling for non-sine waves or distortion can result).
e. Confirm that *10x MAG* and *X-Y* buttons are *not* engaged.
f. *TIMEBASE* to center position.

Place the *ground clip* of properly compensated 10:1 attenuator probe on case or chassis of device-under-test (chassis must not be hot — use isolation transformer if so).

Probe test points and check for proper amplitudes and waveforms while adjusting the *TIMEBASE, VERTICAL AMPLIFIER, TRIGGER* filtering, and *input coupling* as desired.

For two-channel operation, follow appropriate instructions above for the new channel, set switches to *TRIG CH 1 or TRIG CH 2* and *ALTERNATE* or *CHOP*, as desired.

10-9E. The SWR/Wattmeter

Rather than manipulating Ohm's law to determine the power output of a transmitter, such as V^2_{RMS}/R if into a pure resistance, a far easier and faster method is to utilize the wattmeter, or the combination SWR/wattmeter. SWR/wattmeters measure the reflected and forward power, as well as the SWR. The unit is placed in series between the transmitter and its transmission line, or can be placed at the end of a line or transmitter as an end termination form of wattmeter.

While there are different types of wattmeters on the market, the example in *Figure 10-54* is referred to as a *directional single-needle* type.

Cross needle meters are common also, allowing the operator to read both forward and reflected power, as well as SWR, at a single glance. These meters have two needles and separate forward power, reflected power, and SWR scales. One needle indicates reflected power, the other points to the forward power value — and the area where the two needles cross is the SWR.

To operate the single-needle SWR/wattmeter of *Figure 10-54*, the transmitter is connected at the *IN* (or *trans*) port, with the antenna coupled to *OUT* (or *ANT*) port. To obtain a power measurement, the *POWER/SWR* button is set to *POWER*, the *HIGH/LOW* button is set to the appropriate power range expected out of the transmitter (on some meter models, choose HIGH for transmitters up to 2,000 watts; LOW for transmitters up to 200 watts), and the *FWD/REF* button is set to *FWD*. The AM or FM transmitter is then keyed on, indicating on the appropriate high or low *WATTS* scale the RMS output power of the transmitter.

A wattmeter that is capable of reading the PEP is needed for most SSB power measurements, since a SSB transmitter has no carrier and must be modulated to obtain an output signal. This modulation will cause the SSB signal to no longer follow the normal relationship of P_{AVG}=PEP, as would be the case in a power measurement for a CW carrier wave of AM and FM transmitters. With two-tone and voice modulation, the average power will vary from the true PEP; and *average* power, over time, is what a regular wattmeter is built to measure. This is why a SSB transmitter cannot employ a normal wattmeter, but must use one with a built-in sample-and-hold amplifier and RC circuit. This circuit will charge the meter's internal capacitor to the peak voltage of the SSB signal (using a fairly long time constant), and read this voltage as V_{RMS}^2/R, which is the PEP. This allows the technician to observe the true peak voice and two-tone output power of the SSB signal (See *SSB PEP transmitter test*).

To read the SWR present on the transmission line, switch *POWER/SWR* to *SWR*, switch *FWD/REF* to *FWD*, transmit and, at the same time, use the *CALIBRATION* knob to set the needle to CAL (calibration). Cease transmission. Switch the FWD/REF to REF, and transmit. The needle will point to your standing wave ratio on the *SWR* scale. An indication of 1.1:1 to 1.5:1 is considered excellent, a 1.6:1 to 2:1 is acceptable, while anything over 2:1 should be looked into by checking the antenna, the transmission line, and their connections for broken or damaged parts.

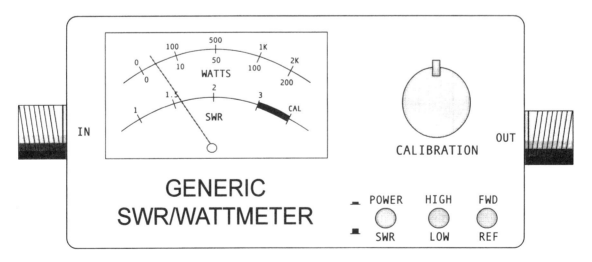

Figure 10-54. An RF SWR/wattmeter.

10-9F. The Frequency Counter

RF frequency counters are available from the very simple, all the way up to models with every bell and whistle imaginable. The highest frequency range and the counter's accuracy are also tremendously variable. However, in virtually all modern applications, an inaccurate frequency counter has little more use than as a paper weight. The type in *Figure 10-55* has all the functions needed in 99% of real-world applications.

It is always vital to attenuate signals from a transmitter into a frequency counter. Excessively powerful signals can damage the unit, even though most are equipped with Zener input protection. The maximum power that a counter can accept goes down as the frequency is increased. All professional counters come with a derating curve with these maximum input values graphed. Typically, a maximum safe input voltage will be around 150V at 1 MHz.

A very safe method to obtain a frequency reading from a transmitter is to attach either a "rubber ducky" or a small telescoping antenna to the counter's BNC input. Then, key the transmitter into an antenna or dummy load and take the measurement.

To operate a standard RF frequency counter, view *Figure 10-55*. The *LOW-PASS* button switches a low-pass filter into the counter's input when taking readings of low-frequency signals (under about 100 kHz). This removes any high-frequency components that may give an erroneous or unstable reading. The *10x ATTENUATOR* attenuates any input signals by ten times to save the front end from destruction from high amplitude signals. The three *FUNCTION* switches are pressed to read the *FREQUENCY* of an input signal, the *PERIOD* of a waveform, or the number of *EVENTS* that occur. The *RESET* button is used to clear the display and start the count again. *TRIGGER LEVEL* allows the operator to control where on the waveform triggering takes place. Only adjust if proper triggering does not occur when in the "*0*" (or *NORM*) position.

The device is used by first warming up all equipment as well as the device-under-test. For correct measurement accuracy the transmitter should be terminated in its proper output impedance, with the signal then inserted into the input of the counter. If the transmitter is of sufficiently low power, a direct coaxial connection from the transmitter into the counter can be used if Zin is 50Ω. Many counters have both a 1 MΩ BNC and a 50Ω BNC input, while others may have an input impedance of 1 MΩ at 100 MHz and below, and a 50Ω input impedance at 100 MHz and above. Either way, switch the *LOW-PASS* filter off, the *10X ATTENUATOR* on, the *FUNCTION's FREQUENCY* button pressed, and the *TRIGGER LEVEL* set to *0*. Key transmitter.

If the input frequency is above that which the frequency counter is designed to measure, a red *OVER* indicator will light. If the frequency counter shows double the proper frequency expected, or the readings are unstable, then there may be signal noise, improper signal shapes or signal amplitudes. Most frequency counters can also be fooled if there is more than one frequency, or there is modulation present, at a test point.

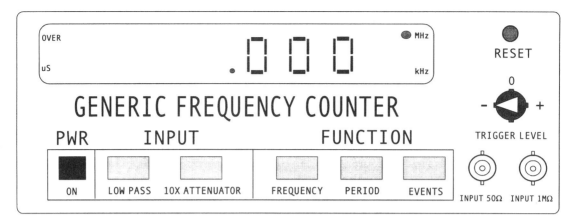

Figure 10-55. AN RF frequency counter.

10-9G. The Signal Generator

The signal generator, with integral *function* and *RF generator* (*Figure 10-56a*), is also an irreplaceable piece of test equipment for troubleshooting, aligning, and maintaining electronic equipment all the way from audio to RF (Note: most RF and function generators are separate devices — the illustration combines both for simplicity. Function generators are typically available up to a maximum frequency of 2-50 MHz (depending on cost), while RF generators are available into the GHz range).

The signal generator is used extensively for signal injection and signal tracing in electronic devices to check for improperly operating circuits and to measure sensitivity, selectivity, bandwidth, etc. A function generator is a *signal generator* that can output different waveforms; while an RF generator is a high-frequency sine-wave signal generator.

The *FUNCTION* section has the controls to choose the type of output waveform desired. In the generator of *Figure 10-56a* we have a choice of outputting a sine wave, a square wave, a triangle wave, or pulse waveforms (employed for digital circuits). Pulse levels are further selectable to be compatible with TTL (5V) or adjustable for CMOS circuits by the *CMOS AMPLITUDE* knob. This pulse output is taken from the *PULSE OUT* BNC. The *AMPLITUDE* knob sets the level of the AC output voltage, while the *DC OFFSET* adjusts the DC offset level from 0 VDC to some negative or positive value. The *DUTY CYCLE* of complex waves is capable of adjustments from 1:1 to 10:1. The *RANGE* buttons are used to choose the appropriate frequency ranges. Within the range chosen, the *TUNING* knob is varied to output the desired frequency, as indicated in the LCD or LED display (display readings on most generators should not be considered as accurate enough for most uses — employ a frequency counter to confirm the generator's display reading).

Since some type of modulation is needed for many RF tests, a *EXT MOD/INT MOD* switch is included. When *INT MOD* is chosen, the generator outputs a signal with internally generated 1 kHz amplitude or frequency modulation, depending on the position of the *AM/FM INT MOD* button. When *EXT MOD* is selected, a source of external audio frequency modulation can be input into the *EXT MOD* BNC.

To use, warm up all equipment and the unit-under-test. Set AMPLITUDE almost completely counterclockwise (to minimum), set *RANGE* to the proper value, place *DC OFFSET* to *0*, DUTY CYCLE to 50%, *FUNCTION* set to the desired output waveform, and adjust the *TUNING* knob to the preferred frequency. Connect a coaxial cable between the *OUTPUT* BNC and the input of the unit-under-test; or attach a rubber duck or other type of antenna to the generator's *OUTPUT*; or attach a probe from the generator to the point of injection of the unit-under-test — whichever is appropriate. Adjust *AMPLITUDE* for proper output, choose *EXT MOD* or *INT MOD* or no modulation. If *INT MOD* is chosen, then select either the internal *FM* or *AM* modulation.

The more modern fully synthesized RF signal generator of *Figure 10-56b* can have very high frequency resolution values (down to 0.1 Hz), with high frequency stability (almost drift free), and no band switching needed. A large choice of internal modulation schemes (AM, FM, PM, SSB, BPSK, FSK, QAM, etc.), and the capability to sweep over a large frequency range, is usually provided.

Most of these digital-type signal generators allow the user to change frequency either by rotating a large analog-type TUNING dial; or by pressing FREQUENCY and pushing the adjacent up/down arrow keys; or by depressing FREQUENCY and directly choosing the frequency desired through the keypad keys. The carrier AMPLITUDE, or the percent or amplitude of the MODULATION, is similarly chosen. The type of internal modulation of the RF carrier can be selected by pressing MODE, and then the appropriate keypad choices (modulation schemes are marked under each keypad number key).

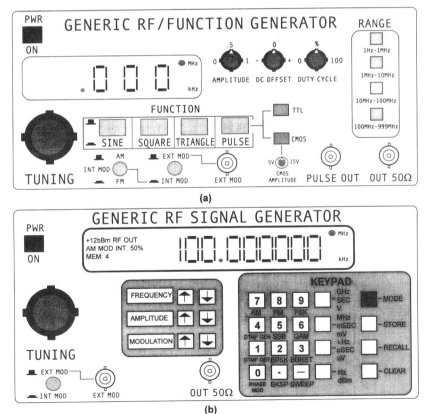

Figure 10-56. (a) An RF/function generator; (b) a fully synthesized RF signal generator.

Favorite front panel setups (frequencies, amplitudes, modulation types and percentages, etc.) can be saved in memory by depressing *STORE*. Choosing *RECALL* will then configure the generator at the touch of a button.

Any type of exotic modulation scheme can be inserted into the EXT MOD input.

Some RF signal generators (*Figure 10-57*) may employ a large analog knob with a *dial pointer* for manually choosing the frequency (frequency TUNING section), and RANGE buttons (marked A, B, C, D, as well as in MHz) for selecting a certain band of frequencies. The dial face itself is calibrated for each band, such as band A, B, C, and D. Amplitude of the output signal can be controlled by the RF AMPLITUDE knob, while the percent of FM or AM (1KHz) modulation is controlled by the MOD AMPLITUDE knob. An exact frequency can be easily selected by the FINE TUNING knob.

To use this low-cost analog model RF frequency generator, first set the RANGE buttons to the desired band of frequencies (such as range B, which is 1 to 10 MHz). Then rotate the dial pointer TUNING knob to choose the frequency within this bandwidth (such as 1.8 MHz) to output from the BNC. The frequency chosen with the dial pointer must be read from the proper lettered scale, such as scale "B". Finely adjust the frequency with the FINE TUNING knob.

Keep in mind when changing the power output of older or low-cost RF generators that the output signal may be pulled off frequency, necessitating re-setting. Many generators may also have excessive and unacceptable frequency drift over time.

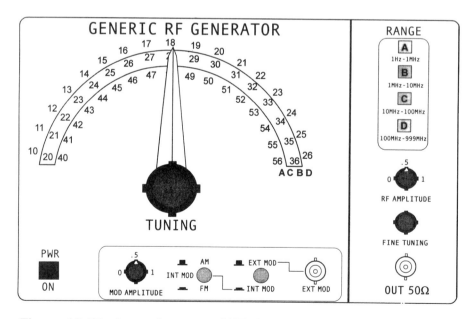

Figure 10-57. *An analog type of RF signal generator.*

10-9H. The Logic Probe

The *logic probe* (*Figure 10-58*) is a basic tool for quickly testing most digital circuits encountered in RF work.

A logic probe consists of *HI* and *LO* LEDs, which indicate when a signal is a 1 (HI) or a 0 (LO). A TTL/CMOS switch is included so that the probe will recognize which input voltage levels are appropriate for the type of logic under test. When the *MEMORY/PULSE* switch is set to *PULSE* detection and the logic probe is receiving a 50/50 duty cycle pulse train, such as a clock input at under 100 kHz (upper frequency limit depending on model), then both *HI* and *LOW* LEDs will light with equal brightness and the *PULSE* light will blink. Over 150 kHz only the PULSE light will blink. If the duty cycle is lower, then the *LO* LED will be brighter than the *HI* LED and the *PULSE* light will blink.

If the MEMORY/PULSE switch is set to *MEMORY*, then a single pulse, which might normally be missed, will light the PULSE light. This LED will remain lighted until reset by switching the MEMORY/PULSE switch to PULSE.

If LEDs are not lighted, then there is a bad level or open circuit present, or you could be testing *tristatable* bus lines that are in their third state, which is high impedance. In this case a lack of standard levels is normal.

Clip the operating power to the probe by simply attaching any positive 5 to 18 V_{DC} to the red lead, with the black lead going to circuit ground. (Read logic probe specifications to confirm particular DC operating voltage span for the probe in use). When attaching DC supply power to the probe when testing CMOS circuits, connect to the CMOS power supply (Vcc) circuit to obtain accurate HI/LO indications (a *LOW* is indicated by a voltage lower than 30% of Vcc; a *HIGH* is greater than 70% of Vcc — most probe circuits use Vcc as the reference voltage).

When checking digital circuits with a logic probe, make sure the probe's maximum operating frequency is *above* the frequency under test.

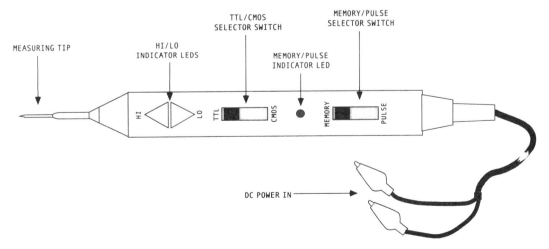

Figure 10-58. *A common logic probe for troubleshooting digital circuits.*

10-10. Soldering

Even though it is second nature to most technicians, it may be worthwhile to reiterate proper soldering practice. Both through-hole and SMD (Surface Mount Device) soldering and desoldering techniques will be covered. While the former is, once learned, very easy, the latter is usually such a complex task that many technicians find it very difficult to perform proper SMD soldering or desoldering — with, or without, special tools.

Follow ESD (Electrostatic Discharge) procedures when handling static sensitive components — always ground the soldering iron tip when working with ESD-sensitive components, even if Zener input protected. If a grounded iron is not available, at least touch the tip to ground before soldering to discharge any built-up potential.

Most modern soldering tips are plated with nickel and do not normally need to be filed, while the old copper tips rapidly become pitted and must be filed to maintain their shape.

Rosin should be removed from solder joints or it may absorb moisture and degrade joints. Also, heated residue of rosin, if not cleaned off a PCB, can short two or more high-impedance traces within sensitive circuits. As well, a badly cleaned board may leave low-resistance areas all over the PCB, leading to erratic operation.

Solder that has been reheated several times will crystallize and lose its strength.

The pistol-grip type of heavy-duty soldering irons may induce destructive currents in semiconductors, and should always be avoided for light duty soldering.

10-10A. Through-Hole Soldering

When soldering through-hole PCB components, use a 25W soldering iron with a tinned chisel tip. Tin by melting a small amount of solder on the tip to aid in efficient heat conduction between the iron's tip and the component's lead. As shown in *Figure 10-59a*, warm up the trace pad on the PCB and, at the same time, the component lead. Now, as in *Figure 10-59b*, place the rosin flux solder against the opposite side of the pad and component lead. Let the pad and the component's lead melt the solder, and not the direct tip of the iron. Complete melting should take no more than two seconds if the iron is at the correct temperature. The solder will then flow around the leads. Remove tip from board.

After the solder joint cools it should have a shiny silver appearance. If it appears to be dull, a cold solder joint has formed and should be reheated to reflow the solder around the connection.

Keep the iron's tip clean (by wiping the tip on a moist sponge frequently) and well tinned for reliable soldering.

For desoldering of through-hole components, three options are available. The first is a manually operated spring-loaded desoldering vacuum pump. Heat the solder joint with a soldering iron, placing the pump's tip into the molten solder, and releasing the pump's trigger. This sucks up the molten solder into a reservoir (which should be safely emptied periodically). The desoldering bulb is a simple version of this concept.

The second technique employs a desoldering wick. Place wick over joint to be desoldered and arrange iron on top of wick. As the solder melts it is "wicked" up into the finely woven copper braid. Wick comes in rolls — use one section, snip off used part and discard.

The third method uses an electric desoldering station that melts and vacuums up the solder, in a single semi-automatic step, through a hollow tip and into a filter that must be replaced occasionally.

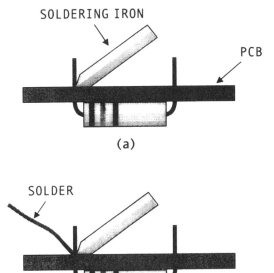

Figure 10-59. Proper soldering technique: *(a) Heat pad and lead; (b) melt solder against pad and lead.*

10-10B. SMD Soldering

Various methods have been developed to solder and desolder surface-mount devices for repair purposes. The following procedures are proven, and do not require special equipment (which is frequently unavailable and of questionable value), except for *J* leaded devices.

To remove two-terminal SMDs, such as capacitors, resistors, inductors, and diodes, from a PCB: Quickly heat both ends of the surface mount device by rapidly alternating from one end to the other (*Figure 10-60*), then *twist* and pull up on the SMD with tweezers (*Caution*: SMD's are generally epoxied to the PCB) after the end connections are molten. Clean off excess solder on PCB with solder braid. There are also accessory soldering tips available that will heat both ends simultaneously, facilitating removal.

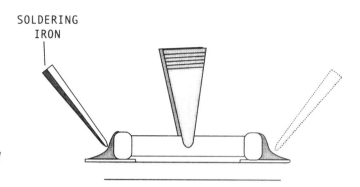

Figure 10-60. Removal of a two-terminal SMD.

Follow this procedure to replace two terminal SMD's on a PCB: Melt a dab of solder on *one* pad; place SMD over both PCB pads (*Figure 10-61a*); push down on SMD as you heat the solder dab, while not touching the end termination of the component with the iron (or the delicate device can be damaged). The end of the SMD will now sink through the melting solder dab to the copper pad. Remove heat to the connection after 3 seconds (any longer and component and/or board damage can occur). Complete operation by carefully soldering other end by heating *just* the copper pad (and *not* the component's termination), and placing the solder against the pad (*Figure 10-61b*) — allowing the solder to melt and flow into and around the end termination contact of the SMD. When set, the solder fillet should not significantly extend beyond the top of the SMD and must have a shiny, smooth appearance — with no apparent visible thermal damage to the device (such as cracks in the body or delamination of the end termination contacts).

Integrated circuit SMDs must employ other techniques for desoldering and soldering due to their multiple leads. The removal and replacement of SMD ICs of the gull-wing leaded SOIC's (Small Outline Integrated Circuits) and QFP (Quad Flat Packs), as shown in *Figure 10-62*, are feasible without special equipment. *J* leaded ICs are considered too difficult to perform without special equipment.

To desolder a gull-winged SMD, grasp the number 1 IC lead with tweezers. Heat the IC's lead at the point where it contacts the PCB pad; twist up on lead after the solder liquifies. Continue until all leads are desoldered. Twist and pull up on the IC's body to break epoxy bead. Timing, practice, and patience are vital — or damage to the PCB is a certainty.

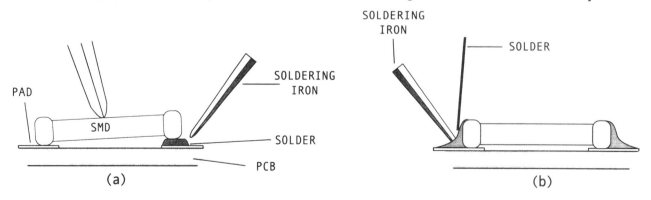

Figure 10-61. *Soldering two-terminal SMDs to a PCB: (a) Soldering first terminal; (b) soldering second terminal.*

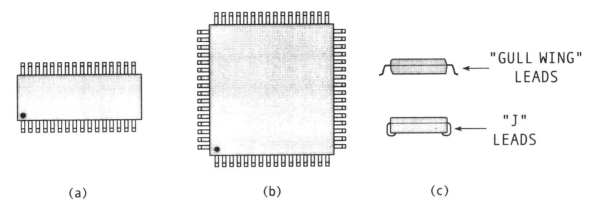

Figure 10-62. *IC SMDs: (a) SOIC; (b) QFP; (c) Gull wing and "J" lead types.*

Another great IC SMD desoldering method is by melting a large amount of solder on one side of an IC's leads. This melts all of the solder at the same time on all leads, allowing the technician to lightly lift up an entire side of an IC. Repeat for the opposite side. Also fantastic for three- and four-lead package removal.

To place a new SMD IC: Clean solder from the copper pads with a wick and iron. Remove any old epoxy and check pads for delamination from PCB. Place SMD IC over the pads (with number 1 pin aligned with the number 1 pad!) and solder one of its pins in place (never pick up SMDs with the hand: Use tweezers or a special vacuum pencil, or the leads may become bent or the IGFETs inside the IC damaged). Confirm proper placement of all pins over the pads. Solder remaining pins to pads. Check for solder bridges and any incomplete joining of IC lead to pad by inspecting all work under a magnifier.

The use of .015 inch (sometimes hard to find) *eutectic* (63-37 tin-to-lead) rosin core solder and a 25W 1/16 inch or less diameter soldering iron tip will make your *SMT* (*Surface Mount Technology*) soldering chores a little easier.

A promising new method for removing SMDs is the patented Chip Quickä SMD removal system. A special liquid flux is spread on the device to be removed. A unique low-temperature desolder/alloy is melted onto the SMD's pins with a soldering iron, combining with the original solder. The SMD will now pull off the board without damage to the copper pads. Complete by cleaning up the area with soldering iron and braid or other desoldering technique. Wipe off the PCB area with cleaner solvent. Install new SMD chip as above.

10-11. Digital Troubleshooting

Digital electronics is now integrated into many aspects of RF technology: Digital display metering, frequency tuning, automatic channel scan and memory, onboard clocks and timers, digital modulation methods, spread spectrum communications, digital multiplexing, cellular communications, *packet* radio, digital signal processing (DSP), etc.

A common use of digital circuits in modern communications radios employs dedicated micro-processors, or single-chip microcontrollers, to accurately and smoothly control transceiver and receiver tuning. These digital ICs, at the request of the radio operator, send data commands to the frequency synthesizer to change frequency, while driving the radio's display. They are ROM or EPROM based with their instructions burned in, with RAM registers available for user programming (such as favorite frequencies available — at the touch of a button) and as temporary registers for internal program utilization.

Most digital ICs, in general, are extremely reliable and almost never fail.

What appears to be a microprocessor failure is usually elsewhere. To troubleshoot, first make sure the IC is getting the proper V_{cc} (within 0.25V) with no noise and very little ripple (ordinarily not more than 1/50 of the amplitude of V_{cc}). Confirm that ground is not floating due to a bad solder connection or broken trace (there should be zero volts when referenced to system ground — if not, then the ground line is floating).

Check all clock inputs for proper frequency, amplitude, and signal cleanliness. Inspect for proper I/O levels while unit is under operation. If input levels are bad, trace back for shorts or opens on the traces or components. When a pin or line should be HIGH but is found to be LOW, check for a short to ground. Make sure no pins are just *floating* — generally noise will be seen on the oscilloscope, or the pin will be in a high or intermediate voltage state when it should be low.

Confirm that the microprocessor resets when V_{cc} is cycled off and on by checking the reset pulse at the proper pin (do not depend on the power switch since the microprocessor may still be receiving its supply voltage — disconnect the power cord for several minutes to discharge the charged backup capacitor).

If any output levels from the IC are not correct, then lift the bad pin after checking all I/O ports for proper operation. Now check pin for proper levels. If level is good then examine trace paths "exiting" from pulled pin for shorts, or for any shorted coupling caps, etc.

Rotate the radio's tuning knob. The logic probe should begin pulsing at the proper output pin (serial) or pins (parallel) to the PLL chip programmable divider input and to the frequency display or its display driver.

If pin levels are still wrong after lifting lead, and all other pins have been checked for proper levels, then the microprocessor should be replaced.

A few words about *Tristatable* (also referred to as *three-state*) devices, since a chip so equipped can appear to be floating or open and may create confusion during digital troubleshooting.

Tristatable switches are necessary for most modern microprocessors to access their support chips, such as ROM, RAM, and I/O ports on a common bus line. The tristatable switch allows only one chip to output its digital word on the bus at a time. If there were no tristate outputs on a chip, then the chip outputs would not be isolated from the common bus: As soon as a chip's output went high, it would be shorted out by a low, which is usually ground or zero volts, allowing large currents to flow — with no way of telling whether the output was a one or zero. Tristatable switches, on the other hand, are present as part of the output/input of any supplementary memory and I/O chips of a microprocessor, as well as in the microprocessors themselves.

Chips outfitted with tristate bus outputs can supply a normal high (5V for TTL) and low voltage (0V), as well as a high-impedance state, which acts as an open to the common bus line. In other words, the microprocessor must tell its support chip that it wants to read or write to that chip, or its outputs are switched "off". If chosen, the output would be switched *on*, allowing either a 1 or 0 bit to be sent into or out of the support chip and on to the common bus lines. As soon as that bit (or an entire word — one bit per line) is sent, the tristate switch is turned off and another is turned on so that that chip can also communicate.

Only an external RAM, ROM, or I/O chip that receives an active low from the microprocessor into its *chip select* input will open that chip's tristate lines, allowing that chip to speak with the microprocessor. This permits other devices to output their signal, at different time slices, unto the common bus lines without interfering with each other and nullifying the need for each output to have its own line.

The tristate level will be floating and is usually between 0.8 and 2.4 volts on an oscilloscope or DMM, though bad levels will be indicated on a *logic probe's* LEDs.

Some dedicated processors are single-chip types, with the ROM, RAM, and I/O all on one chip, and may or may not have tristatable output/input considerations. However, output enabling control pins, as well as address and data pins for supplementary memory and I/O chips, may be present — but most may go unused.

One of the things to keep in mind when troubleshooting high-speed digital circuits is that of the limitations of the test devices: When measuring digital (square wave) signals with an oscilloscope, the oscilloscope's vertical amplifier must be rated at over ten times the frequency of the square wave under test. This is so that all the square wave's harmonics can be passed in order to produce an undistorted test output (a 20 MHz scope will only accurately measure a 2 MHz square wave). If using a logic probe, confirm that the probe's upper operating frequency limit is not passed by the digital signal, or all readings may be incorrect.

10-12. Electromagnetic Interference

Considering electromagnetic interference (EMI) can sometimes simulate equipment failure, either continuous or intermittent, the technician should be familiar with its identification and cure. As well, the technician may be required to track down the interfering device and, hopefully, eliminate or at least attenuate the problem, either at its source or at the receiver. The technician must also be able to identify and cure any EMI that may be exiting any transmitters in the form of RFI (Radio Frequency Interference).

Electromagnetic interference is generated by all forms of devices, and can adversely effect AM, SSB, CW and FM receivers, television sets, biomedical apparatus and many other types of equipment. EMI can cause anything from white noise in the receiver to a total annihilation of the desired input signal.

Sources can be spurious generation, or even proper and clean fundamental transmissions, from any type of RF transmitter. EMI can also be generated from the operation of computers, light dimmers (thyristor circuits), fluorescent and neon lights, arcing in defective transformers or internal combustion ignition systems, and in all brush-type electric motors, to name just a few. In fact, anything that produces electrical sparking or fast current steps from zero to some value (or from some value to zero), will generate electromagnetic radiation up to very high frequencies. This radiation can be transmitted through the air or through the AC power line for extreme distances.

If the transmitter under test has been accused of creating RFI, then examine the transmitter's output directly using a spectrum analyzer (See *Spectrum analyzer* and *Attenuators*). If all right, then attach a sniffer probe (*Figure 10-63*) or antenna to the analyzer, checking around the entire transmitting installation, for any spurious outputs. If found, confirm that the transmitter is well grounded and tuned, making sure that ground straps are not excessively long and not radiating RF, and that the transmitter cabinet is not radiating harmonics through open inspection doors (or even empty screw holes). As well, check that the antenna and its feed line are undamaged, and that all connectors are tight and clean. Establish that the transmitter is not generating parasitics or feedback oscillations. Check for master oscillator stability during modulation. Confirm that any low-pass filters, isolators or wavetraps used for harmonic or nearby transmitter frequency (transmitter-generated IMD) suppression, if present, are functioning as designed. Sometimes brute-force line filters will have to be employed to filter out any RFI from entering or leaving the transmitter through any power cables.

If the on-site transmitter is sending a clean signal, yet TVI (Television Interference) and RFI complaints continue, the complainant's radio receiver or television itself may not be properly shielded or filtered against fundamental overload or low-level harmonics for use in an EMI filled world. If the transmitter is in the shop, it can be seen that a complete check of this nature, as outlined above, is impossible.

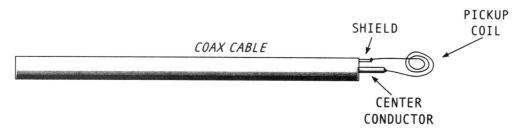

Figure 10-63. *A sniffer probe.*

If the technician is on-site and suspects EMI instead of equipment failure while troubleshooting a receiver, and the receiver is blanking out intermittently, or there is excessive noise generated at its output such as a popping, whining, cracking, buzzing, or pulsing sounds, etc., then investigate for nearby noise-producing devices. As mentioned above, check for light dimmers, brush-type electric motors of all kinds, arc welders, adjacent clean high-powered transmitters (which can create fundamental overload and powerful, but legal, harmonic amplitudes) or low-powered but spurious generating transmitters. Also look for passing noise-generating vehicles (standard and electronic ignitions creating electrical arcs); nearby computers (especially older types) or microcontrollers (standard in many kinds of electrical and electronic consumer equipment); starter-type fluorescent lights; high-voltage neon signs (the longer the run of neon, the higher the possible voltage); and any device that uses an oscillator or frequency synthesizer, such as poorly shielded or defective television sets and radio scanners (TVs may emit RFI every 15.734 MHz, which is the TV's high-voltage flyback transformer frequency).

The RFI problem can be abated by utilizing a line filter at the receiver under test, or filtering its input with a high-pass or low-pass filter to attenuate any lower or higher-frequency interference, or by trying out a bandstop filter to notch out the offending frequency. Confirm that the interference is not entering any of the receiver's cabinet openings and that the cabinet, if so designed, is properly grounded, with no ground loops present (some devices were not intended to have a grounded cabinet — check for "hot-chassis" transformerless or autotransformer designs). Check that the feed cable from the antenna is undamaged. Use 0.01 μF ceramic-disk capacitors from the receiver's speaker to ground, and/or place ferrite beads around the speaker leads, or wrap any connecting cables around a ferrite rod or toroid (to attenuate any interference picked up by these cables). If necessary, replace unshielded cable with shielded type.

Of course, if it is expedient and possible, attempt to neutralize any EMI-generating sources at the point of origination by requesting the device's owner repair the defective equipment, such as arcing power-line transformers or faulty transmitters. Or, if the offending device is on-site with the receiver and your company so authorizes, nullify the aberrant unit's EMI output: Use bypass capacitors on electric motors by adding 0.01 μF ceramic-disk capacitors from each commutator brush to ground or, if the brushes are worn and the commutator is dirty, then clean and repair. For neon signs, place 10 kΩ resistors in series with each supply lead. With fluorescent lights, confirm that the ballast and light tube are operating properly — if not, replace ballast or tube (older units employ starters and must be outfitted with bypass capacitors). Replace or bypass offending light dimmers. For EMI-producing computers, place split ferrite toroids around all cables exiting the central processing cabinet of a computer (as close to the cabinet as possible). When feasible, utilize AC line filters on the offending equipment to prevent EMI from entering the common power line.

If unfamiliar with the electrical equipment above, advise owner to have offending devices repaired or replaced by a competent electrician.

If receiving complaints that a new or old mobile receiver or transceiver installation is acquiring interference, confirm that the power leads to the units are well shielded or filtered, that the suppresser ignition wires and spark plugs are installed and undamaged, or that any aftermarket suppresser kits are installed and functioning. Make sure that all shields for the spark plugs, wiring, and other ignition devices, if so installed, are present. Establish that ground-strap bonding points (for maintaining proper ground) are still functioning. Verify that the alternator is not creating a *whine*. Examine the alternator's connection integrity to the battery, while also confirming that any EMI filter present in its power lead is operating.

If the onboard vehicle microprocessor or microcontroller is suspected of creating EMI, check with vehicle manufacturer (transmitting from a modern vehicle may also cause adverse and unexpected results to any unshielded microprocessor controlled circuit, such as the ABS breaking and ignition systems).

10-13. Important Troubleshooting Notes

The following notes are important to remember when troubleshooting RF circuits.

Aluminum invisibly oxidizes quickly in contact with air, forming an insulation layer. Always use toothed-type metal washers and/or sandpaper at connection points to aluminum chassis during a new connection or a reconnection. As well, anodized aluminum chassis do not conduct electricity, as the anodization process creates an extremely thin layer of insulation. This must be scraped off to supply a proper electrical connection to the chassis.

Use only nonconductive tuning tools for adjustable capacitors and inductors — if not, the added reactance will shift the proper tuning point of the circuit during alignment.

Transformers employ color codes for their leads, but a replacement part may use a different color code. Also, the same type of *transistor* may have a different lead arrangement than that replaced.

Resistors: Some metal film resistors are flame retarding and when changed out must be of the same type. When working on older equipment, observe that carbon resistors may increase their resistance by 100 percent with age. Do not use wirewound resistors, which have a lot of inductive reactance (unless wound noninductively), and cannot normally be adopted at RF. In sensitive circuits replace resistors with the *exact* same type: Carbon composition resistors have a bad TC (*Temperature Coefficient*), and will vary resistance with temperature, while thin film resistors have a good temperature coefficient, for example. A very high value resistor ($R=10^8$ or above) can appear to have low resistance if fingerprints or dirt are across it, since current will be shunted around the resistor.

In a large rotor-stator air-tuned capacitor there is a trimmer capacitor for each stage of the capacitor. These are used to adjust the minimum capacitance when the plates are completely unmeshed, and must be properly adjusted or tuning calibration will be inaccurate.

Mylar caps have replaced paper caps and due to Mylar's relatively high absorption losses (energy lost in the dielectric) at higher frequencies, as well as inductive effects caused by their typical foil-wound construction, they must be used for lower-frequency applications only. Noninductively wound Mylar and paper capacitors are also available.

Ceramic-chip capacitors have the highest operating upper frequency range, with ceramic disk capacitors just under the chip type. Mica and ceramic capacitors are employed at RF due to their low inductance and low absorption losses. Ceramic capacitors may also have a *variable temperature coefficient* (an increase or decrease in capacitance depending on temperature). N### equals a negative temperature coefficient (NTC); P### equals a positive temperature coefficient (PTC); NPO (or *COG*) equals no temperature change. The numerical values (represented here by #) given after the N or P designations are in parts-per-million per degree Celsius change in *capacitance* from the reference temperature of 25°C. Confirm that any replacement capacitor has the same temperature coefficient (oscillators may use either NTC or PTC capacitors to correct for frequency drift), as well as the same voltage rating (replacing a capacitor with a rating too low and the dielectric can be punctured, causing continued circuit failure in the future). When AC is placed across high-Q ceramic capacitors they may exhibit the piezoelectric effect, and may hum or buzz. In some circuits (such as power supplies) it can be vital that the same capacitor with the same ESR rating be chosen when replacing a bad component. If not, the capacitor may overheat due to the increased ESR and series impedance, causing it to fail prematurely.

Electrolytic capacitors, especially the aluminum type, should have close to their voltage rating placed across them to maintain their oxide film, and thus their capacitance. They must never have a large AC voltage placed across their terminals, unless there is a larger DC voltage to maintain the necessary polarity as marked on the component (there are also nonpolarized electrolytic capacitors but, due to their internal structure being composed of two opposite polarity electrolytics back to back as well as two diodes, their capacitance per package size is less than a standard electrolytic). Use only fresh aluminum electrolytics as replacements to avoid any with dried-out or leaking electrolyte. *Tantalum* electrolytics, however, have virtually unlimited shelf life. Most electrolytics have poor tolerance, and a leakage current of around 0.2 mA per µF of capacitance (this is usually not a consideration because of its common application in filtering, bypassing, and coupling at low frequencies).

Never overlap when winding a wire around a binding screw, or premature failure of the wire will occur (as the screw is tightened the top wire turn will press against the bottom turn, causing instant damage).

When troubleshooting large transmitters ground the hot side of the power supply (after turning off the supply) to assure your safety.

Reflected impedance may cause detuning of a prior stage as you tune each stage down the line of a receiver or transmitter. Retune any sensitive stages.

Insulate all probes up to 1/8 inch of the probe's tip to prevent expensive and embarrassing equipment shorts, especially in today's densely packed PCBs.

If a glass fuse looks good but it is actually open, then thermal shock brought about by age and heat may have caused it to fail. Replace and see if it fails again. On the other hand, a severed fuse wire means approximately 2 times its rated current flowed through it, while a fuse with blackened or silvered discoloration means a total short circuit with high current flow.

Keep in mind that a stage may appear bad due to a weak output, yet the next stage may be pulling the signal down, or an amplifier may have a bad output due to a prior stage feeding it an excessively high input signal.

If an inductor or transformer is excessively saturated, even once, it may change the magnetic properties of the core, affecting sensitive circuits in a negative way.

When testing an amplifier in CW mode that is meant for speech (which is typically an average of 15 dB below the peak levels of CW), use a cooling fan across the output transistor's heat sinks — or the transistors may be damaged by the excessive heat buildup.

With many wideband push-pull amplifiers, if one transistor is damaged then both should be replaced with *beta-matched* transistors.

Unexpected voltage at the input of a stage might indicate that the coupling capacitor from the prior stage is leaking.

CHAPTER 11
Support Circuits

Any circuit that is not absolutely necessary in receiving or transmitting a signal of a basic receiver or transmitter can be seen as a support circuit. Many support circuits are so invaluable that a modern transmitter or receiver could not adequately function without them, such as speech processing, mixing, AGC, and power supply circuits.

Other circuits, which are not discussed here, could be termed as "bells and whistles", and are strictly for added operator convenience and are not indispensable for reliable and accurate communications.

11-1. Speech Processing

Speech processing modifies the audio signal in amplitude, frequency, or both, before it is inserted into the modulator of a transmitter. A special form of compression, called ALC, affects the RF, instead of the audio, of a transmitter.

Overmodulation in AM causes *splatter*. Splatter is created by the additional harmonics produced in the baseband signal, which further modulates the carrier, creating extra sideband components and affecting a much wider bandwidth than desired (See *Overmodulation*). This produces adjacent channel interference, as well as making the primary AM signal less understandable. Signal processing, such as *speech compression,* can prevent this from occurring.

Compression amplifies normally up to a set level, but then begins to reduce gain by 1 dB for every 2 dB of signal increase. Simple speech compression limits the maximum amplitude of the AM or SSB audio, while *dynamic compression* also increases the low-amplitude audio signals.

Compression can be very effective in increasing the effective output power of a transmitter: If an SSB voice transmission goes above half of the CW power peaks (with the CW power peak arrived at during transmitter tuning), then severe distortion and splatter will result. However, with compression, this value can safely rise to almost three quarters of the CW test value. This maximizes the average power output of the transmitter while also limiting maximum modulation to 100% or under.

Speech compression functions similarly to standard AGC, but is placed in the audio stages (*Figure 11-1*).

Compandoring is related, and can be used in some SSB communication systems. It compresses the speech amplitudes at the transmitter (*compression*) and then expands them at the receiver (*expanding*), as shown in *Figure 11-2*. This allows for a high dynamic range with a better signal to noise ratio.

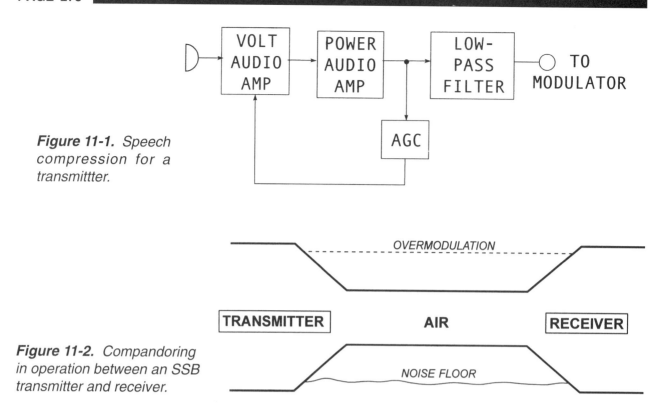

Figure 11-1. *Speech compression for a transmittter.*

Figure 11-2. *Compandoring in operation between an SSB transmitter and receiver.*

FM (and some AM and SSB) transmitters may employ a *speech clipper* (*Figure 11-3*). The clipper slices off any voice signal that is above a set design amplitude. This clipping action creates harmonics, which are removed by a low-pass filter at the output of the audio section. The low-pass filter is also used to limit the maximum frequency excursions of the modulating waveform. A similar circuit, called an *audio clipper*, is shown in *Figure 11-4*, and uses back-to-back diodes to send a degenerative feedback for any audio peak voltage over a certain preset level, which is governed by the value of these Zener diodes. R_2 sets the gain of the circuit when clipping is not occurring.

Another type of compression (sometimes called *RF compression*) is employed in SSB, AM, and a few FM transmitters, and is placed in the intermediate and radio frequency stages. It is referred to as *ALC* (*Automatic Level Control*) and is also basically a standard AGC circuit, but located in the transmitter section. Automatic level control manages the gain of the transmitter's IF by applying a voltage to the gate, grid, or base of an IF amplifier. This is accomplished by taking the final's plate or collector voltage and rectifying and filtering it into a DC control voltage, then sending it to the IF amplifiers.

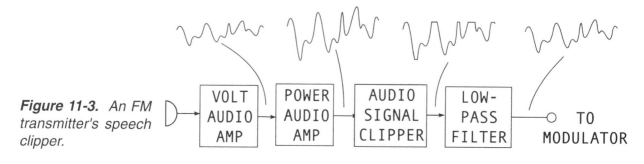

Figure 11-3. *An FM transmitter's speech clipper.*

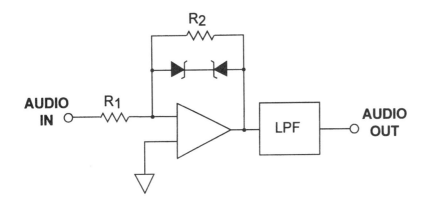

Figure 11-4. *An audio clipper circuit.*

The ALC circuit of *Figure 11-5* works by rectifying the output signal from the final amplifier through C_1 and D_1, which is filtered through C_2 and R_2, and applied to the *gain controlled amplifier*, lowering gain as required. The *threshold bias* is set so that no compression occurs before a preset level has been reached.

ALC will lower excessively high amplitude output levels, but amplify medium and low level signals normally — in other words, a form of compression.

Since the modulation frequency in AM, SSB, and FM transmitters governs the transmitted bandwidth, some way to limit the maximum baseband frequency must be used. Any one of a variety of audio low-pass filter designs can be placed in the audio section for this purpose.

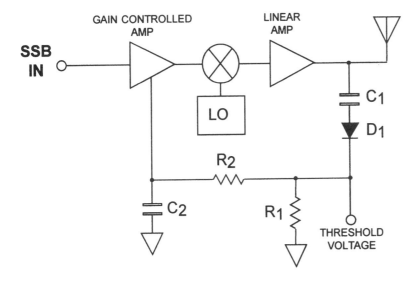

Figure 11-5. *An ALC circuit.*

11-2. Mixers

Mixers are devices utilized to produce a sum and difference frequency from an input of two unlike frequencies. This process is referred to as *frequency conversion*, and is employed in both transmitters and receivers to decrease or increase a sinewave or complex signal's frequency. If the sum is used, it is referred to as *up-conversion*; if the difference is used, it is called *down-conversion*.

Many AM and SSB transmitters use mixers to convert up to a higher frequency for transmission. In superheterodyne receivers a mixer is sometimes referred to as the first detector, and is utilized to convert a received signal to a lower frequency, called the intermediate frequency, or the IF. The conversion process, called *heterodyning*, is adopted for receivers because lower fixed frequencies are much easier to efficiently amplify than higher varying frequencies. This increases sensitivity and selectivity to obtain a narrower and higher gain response in the receiver, since the IF amplifiers can be tuned, filtered and optimized for a single low frequency.

A receiver obtains its IF through this processes: Any received input signal *beats* (heterodynes) with the local oscillator frequency in the mixing element, which can be a diode, transistor, tube, or FET run in the non-linear portion of its bias. For active devices, this bias is usually Class AB. However, many modern RF mixers prefer to employ single- or dual-gate MOSFETs as their non-linear conversion elements, due to their low mixing noise, lower IMDs, and wide dynamic range.

This beating commonly produces, at the mixer's output, the signal frequency, the local oscillator frequency, and the sum and difference frequencies of the signal frequency (with any attendant sidebands) and the local oscillator frequency. The new desired translated frequency, consisting of the carrier and any sidebands at the (generally) difference frequency, is filtered at the output of the mixer and sent through fixed tuned IF amplifiers. All other unwanted frequency components, such as the sum of the signal and LO frequencies, as well as the original signal and LO frequencies, are removed by the passband selectivity of the IF filters.

There are different classifications of mixers. The output of *unbalanced* mixers, such as that described above, are f_S, f_{LO}, $f_S - f_{LO}$, $f_S + f_{LO}$, with very little isolation between the mixer's input and output ports, resulting in signals which can radiate or interact with any other input or output port.

A mixer that is *singly*-balanced significantly attenuates either the original signal *or* the LO, and passes the resulting mixing products on to its output.

A *doubly*-balanced mixer (DBM) furnishes great IF-RF-LO port isolation while outputting *only* the sum and difference frequencies of the signal and LO — making the job of filtering a far simpler task.

Amplitude modulated and single-sideband transmitters cannot use simple frequency multiplication circuits as are employed in FM circuits (to increase the signal's frequency after modulation). AM and SSB must apply the heterodyning techniques described above — or major distortion of the delicate modulation envelope would result.

11-2A. Single-Ended FET Mixers

In the *single-ended FET* mixer of *Figure 11-6*, the RF input signal is dropped across the tuned input tank and coupled to the FET's gate, while the LO output is dropped across R_S and into the source lead. The mixed signal is then taken from the drain and placed into the output tank, which is tuned to the desired IF frequency. All other mixing, signal, and LO frequencies are discarded by the tuned tank. The IF is now induced across the secondary of the output transformer and on to the IF amplifiers.

Dual-gate MOSFET mixers, such as that shown in *Figure 11-7*, can also used, with the LO output being injected into the first gate and the RF signal into the second gate. The difference frequency, equal to the IF, is removed from the secondary of the tuned output transformer of T_1 and sent on to the IF strip for further amplification and filtering.

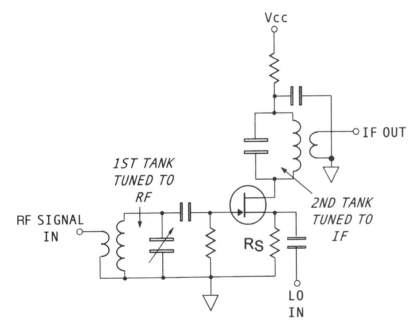

Figure 11-6. A single-ended FET mixer stage.

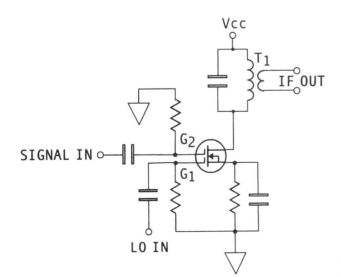

Figure 11-7. A dual-gate MOSFET mixer stage.

11-2B. Single-Ended Bipolar Transistor Mixers

The *single-ended transistor* mixer of *Figure 11-8* injects both the signal and the LO into the shared base, where they are mixed by the non-linear Class AB biased transistor. All mixing products, as well as the original signal and the LO frequency, are in the output — but only the IF is passed to the next stage by virtue of the selectivity of the output tank circuits.

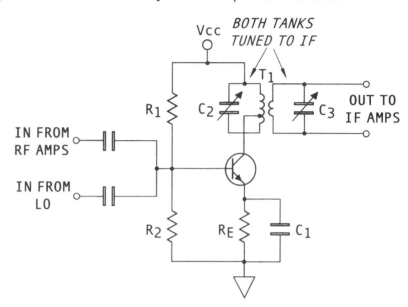

Figure 11-8. *One type of single-ended transistor mixer stage.*

11-2C. Diode-Ring Mixers

The doubly-balanced mixer of *Figure 11-9* uses a *diode ring* to perform frequency conversion of the input signal. The LO switches on and off the diodes, which has the RF frequency *in* them, mixing the two signals.

Although diode mixers have a conversion loss (the output amplitude is less than the input amplitude), they have the advantage, as stated previously, of only outputting the sum and difference frequencies and attenuating the original carrier and LO frequencies, while offering IF-RF-LO port isolation (which is of primary importance in eliminating any LO frequency from being radiated from the antenna) — as well as a general increase in mixer stability.

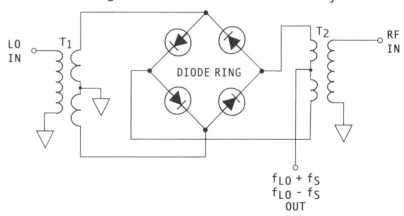

Figure 11-9. *A doubly-balanced mixer stage with diode ring.*

This type of mixer can be operated at VHF and above by using *hot-carrier* (AKA *Schottky barrier*) diodes, or other such high-frequency diodes, due to their low noise and high conversion efficiency.

Doubly-balanced mixers are available in a module, with the diodes and transformers already balanced in a single package.

Other diode mixers are available that use either a single diode (*Figure 11-10*), or two diodes (*Figure 11-11*). However, they are not doubly-balanced, and must be supplied with a very high amplitude LO output for proper mixing. They are, nevertheless, low in cost and need few components.

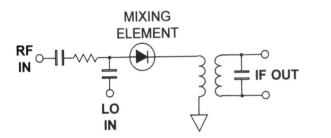

Figure 11-10. *A single-diode mixer.*

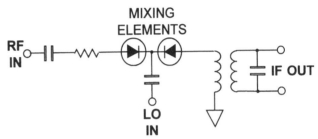

Figure 11-11. *A double-diode mixer.*

11-2D. Autodyne Converter

Autodyne converters are utilized in low-cost receivers that operate below 30 MHz. They employ a single transistor for combined mixing and local oscillator chores (*Figure 11-12*). The LO frequency is selected by T_2 and C_4; the IF by the resonant tanks consisting of T_1, C_2, C_3.

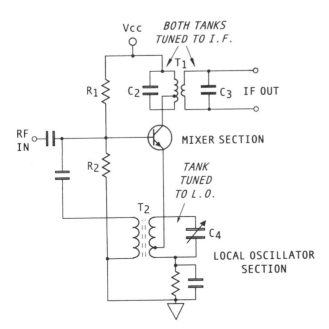

Figure 11-12. *An autodyne converter: A combined mixer and local oscillator.*

11-2E. On-Chip (IC) Converter

On-chip converters boast on-board voltage regulation, a local oscillator, and a mixer (*Figure 11-13*) — all in a single integrated circuit. By adding a few external discrete components the circuit is complete. Some types of IC mixers are usable up to and beyond 0.5 GHz. The LO frequency is set by an external crystal, or an LC tank, and the IF is chosen by a ceramic, crystal, or other filter method at the mixer's output.

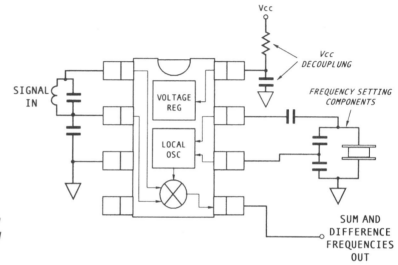

Figure 11-13. *An IC with local oscillator mixer and voltage regulator.*

11-3. Automatic Frequency Control

AFC (Automatic Frequency Control) is adopted to maintain the local oscillator of older, or some newer but low-cost, receivers on frequency without employing frequency synthesis.

Any receiver that can be tuned over a wide range using LC oscillators needs some method of remaining on frequency even with internal and external temperature changes (which would create frequency drift that is unacceptable in modern equipment). A superheterodyne receiver with local oscillator drift converts the incoming signal to an IF that is not exactly at the IF amplifier's center frequency — thus causing signal attenuation and distortion of the received signal.

The AFC of the FM receiver of *Figure 11-14* can detect the FM from the last stage of the IF section with a separate demodulator, or a slightly altered circuit configuration can tap into the receiver's own detector. Looking at *Figure 11-14*, the AFC's demodulator will vary its output voltage in step with any drifting of the intermediate frequency, which is caused by LO drift, and sent to a low-pass filter, which filters the output of the demodulator into a DC control voltage, adding to or opposing the varicap's rest frequency bias. This varicap has been placed across the tuned circuit of the LO. The varicap, and its bias control voltage from the filter/demodulator, has a certain amount of control over the frequency of the LO. If the local oscillator drifts off frequency, the demodulator will vary from its resting output voltage and feed a correction voltage to the varicap, which then changes its capacitance, forcing the LO back on frequency.

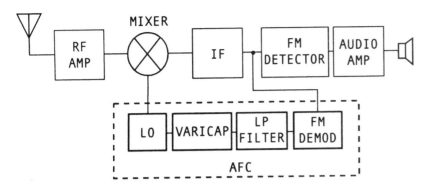

Figure 11-14. *An AFC circuit for an FM receiver.*

Most modern equipment uses the phase-locked loop (PLL), which does not need AFC and is crystal steady and tunable over a broad range.

11-4. Automatic Gain Control

AGC, also referred to as AVC (*Automatic Volume Control*), is utilized in virtually all modern receivers. Without AGC strong signals would overdrive the receiver and cause distortion, while weak signals would be virtually undetected by the listener. The dynamic range of any receiver would be extremely low.

AGC functions due to a certain simple but important transistor characteristic: Increasing the transistor's collector current increases the transistor's gain.

An easy way to increase the flow of current in the collector is to increase the forward bias at the base, since increasing base current increases collector current. A point is eventually reached where this ability levels off and then gain begins to decrease if collector current is increased any further (*Figure 11-15*).

However, it must be emphasized that this base and collector current is direct current that increases the gain, and not the signal current. This current is produced by a *DC bias voltage* impressed at the base of the transistor by the AGC circuit. Indeed, in some AGC amplifiers the only DC bias, which fixes the Q-point, is set strictly by this AGC voltage.

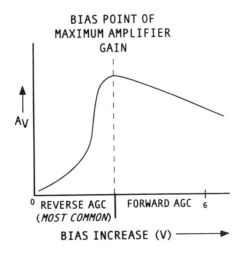

Figure 11-15. *Graph of base-bias voltage versus amplifier gain for AGC.*

As stated above, increasing collector current increases the transistor's gain, yet a point is reached where this ability tapers off. Gain then begins to decrease if collector current is increased further. Because of this ability to increase and decrease gain by increasing or decreasing collector current, it only follows that there are two ways to apply AGC: *reverse* and *forward* AGC. Reverse AGC generally controls the gain of the IF section, while forward AGC commonly controls only RF amplifiers. Reverse AGC lowers gain by decreasing the bias on the base, which decreases collector current, and thus gain. Forward AGC increases forward bias on an amplifier, which increases collector current, and thus reduces stage gain.

The AGC voltage can be tapped off of the last IF stage (*Figure 11-16*) or, in some receivers, after detection by the detector (*Figure 11-17*). In *Figure 11-16* the IF signal is tapped, amplified, rectified, filtered to obtain a DC voltage then, in some receivers, further amplified — and applied to the base of the 1st IF amplifier through an *AVC bias line* (actually just a wire or trace on a PCB board). In many cases AGC is also administered to the second IF stage as well.

When the AGC is taken after the detector as in *Figure 11-17* (a technique frequently used in AM radios), a separate AGC rectifier is not needed — operation is basically the same as above. FM receivers frequently tap the signal amplitude for the AGC circuits from the main discriminator.

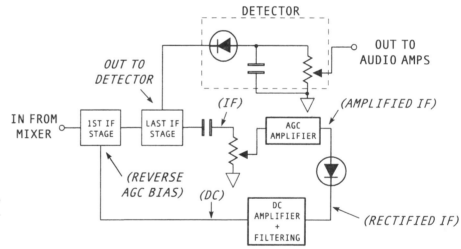

Figure 11-16. A popular AGC circuit using the IF signal.

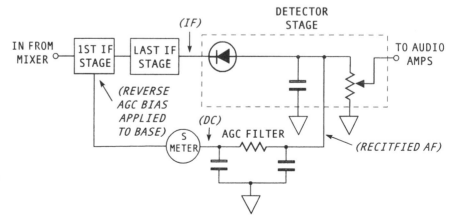

Figure 11-17. An AGC circuit using the detected audio signal.

In *Figure 11-18* the IF Class A amplifier shown utilizes reverse (negative, for NPN transistors) AGC to subtract from the base's normal positive bias. As a strong signal is received, the rectified and filtered negative (DC) AGC voltage is applied to the base of the transistor. The stronger the received signal the more negative the AGC voltage. This voltage opposes the positive bias voltage at the base (which biases the transistor at its Q-point, allowing a set amount of collector current to flow). As the base becomes less positive, less DC collector current flows, which lowers the transistor's gain. When AGC is applied to an NPN transistor, typical measured values at the base will be +1 for minimum gain and +3 volts for maximum gain.

For forward AGC, the positive (for NPN transistors) DC bias is increased at the base as a strong signal is received, which increases base current, which increases collector current, which decreases gain. The RF transistor amplifier of *Figure 11-19* utilizes this type of automatic gain control.

Entire IF sections are now obtainable on a single integrated circuit. Most of these IF ICs have pins available for manually setting the AGC level with an external potentiometer. This will adjust the amount of AGC amplification desired (very little AGC for distant stations — a lot for strong local stations).

As well, regular AGC will still lower gain even when weak input signals are received — *delayed AGC* corrects this common problem by not supplying any gain reduction until a set signal amplitude is reached. This assures that low-level signals will receive maximum amplification. The threshold input signal voltage needed to activate AGC is adjustable by attaching a resistor or trimmer to the proper IC pin.

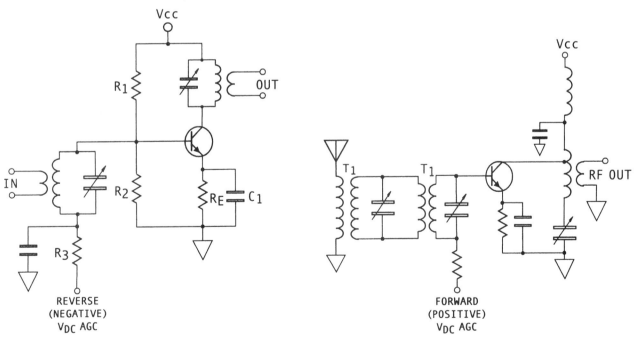

Figure 11-18. *An IF amplifier employing reverse AGC.*

Figure 11-19. *An RF amplifier with forward AGC.*

Accessory pins may also be available for feeding AGC-generated voltages into *tuning indicators*. Considering that the greater the input signal amplitude the greater the generated AGC voltage, these indicators can consist of a simple LED that becomes brighter when a maximum signal strength is reached.

The AGC output voltage may also be fed to a voltage or current meter, referred to as a *signal strength indicator*, and suitably calibrated in *relative signal strength units*, to show to the operator the intensity of the received signal.

Dual-gate MOSFETs are also commonly used in AGC-controlled IF amplifiers (*Figure 11-20*). The MOSFET's second gate can be fed a positive AGC voltage, which increases stage gain, or a negative voltage, which decreases gain.

Another form of AGC (*Figure 11-21*) exploits a single diode. When a low signal is received, CR_1 is off and offers a high resistance to ground. However, when a large enough signal is received, CR_1 is biased *on*, acting as a shunt resistance to ground for the greater IF signal. This circuit is rarely used for AGC, but can be found as a *limiter* by employing two back-to-back diodes.

An FET's AGC action works in a similar manner to that of the BJT, but need only supply a reverse bias *voltage* to the gate-source terminal, but no current, and thus no power, due to the FET's almost infinite DC input impedance (IxE=P; so 0xE=0W). A BJT's AGC must supply both voltage and current, and thus power. This allows for a simpler and lower-cost FET AGC circuit. FETs also have a gain control capability over an input signal covering a 20 dB range in common source configuration, and up to an amazing 50 dB in common gate.

See also Section 6-2 for *SSB AGC* on a description of popular single-sideband automatic gain control schemes.

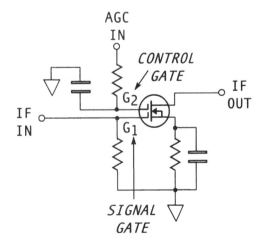

Figure 11-20. A dual-gate MOSFET IF amplifier with its second controlling stage gain.

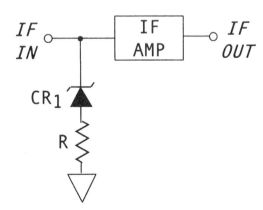

Figure 11-21. Simple single-diode IF AGC.

11-5. Squelch

Squelch, or muting, circuits are found in many AM and FM communication receivers. Squelch *removes* the annoying static that occurs when the radio is not receiving a signal, since gain is then normally at its highest (See *AGC*), as well as saving battery power (an unsquelched receiver may use 100 mA, while a squelched receiver may use only 16 mA).

In a common squelch circuit, the audio amplifiers are disabled in some way to remove any noise from reaching the speaker. Certain audio amplifiers are disabled by applying a reverse bias to the base of the stage, others completely cut off the supply voltage to these audio sections, while still others simply prevent the noise energy from ever reaching the audio amplifiers by either shorting the static to ground, or blocking it with a series-pass transistor. Most exploit the DC output of the AGC circuits to detect whether or not there is any RF input signal present.

In the squelch circuit of *Figure 11-22* AGC operation occurs as follows: When there is a low or absent input signal level, the negative AGC bias is decreased, and is almost nonexistent (allowing the IF stages to be biased at maximum amplification). This AGC level is wired into the AGC IN of the squelch amplifier and on to the *squelch gate*, which is basically a transistor switch and is normally biased ON when low input signals are present. The low negative AGC level is not high enough to turn OFF this squelch gate, so the noise signal is shorted to ground through the low resistance of the ON transistor before it can even reach the audio amplifier stage; as well as completely cutting off the audio transistor. Thus, no audio signal reaches the speaker. However, when an RF signal of the proper amplitude is received, the AGC now generates enough voltage to cutoff the squelch gate, allowing the detected audio to pass into the audio amplifiers to be amplified; and without being shorted to ground.

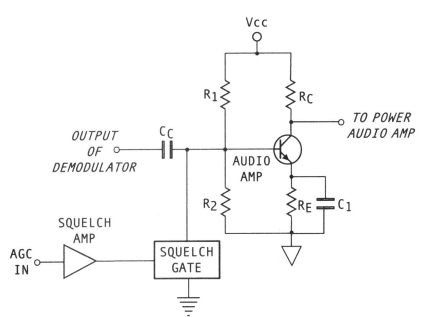

Figure 11-22. A squelch circuit using AGC for signal detection.

Other squelch circuits use a bandpass filter tapped into the IF section (*Figure 11-23*) to pass the high-frequency noise (above 6 kHz) into a static (noise) amplifier, and on to be rectified and filtered to a DC control voltage. The control voltage is applied to a squelch gate which, if no signal is being received (always a high noise condition), is then biased into conduction, shorting the detector output to ground. This removes the audio signal from the input to the first audio amplifier, thus quieting the audio output. Any desired (or undesired) audio modulation received would instantly attenuate the high-frequency noise by replacing it with the 3 kHz or lower audio frequencies, which would shut off the squelch gate, allowing the audio to reach the audio stages and be amplified.

Other methods employ IF integrated circuits (*Figure 11-24*). The IF IC outputs a voltage level from one of its pins that corresponds to the received signal strength and into a *comparator*, which is adjustable for the preferred squelch threshold by the level control R_1. If Input 1 is below the reference level as set by the voltage divider of potentiometer R_1 at Input 2, which sets the level of squelch desired, then the comparator swings close to its positive rail (+Vcc), switching ON the squelch gate, shorting the collector of Q_1, the voltage audio amplifier, to ground through diode D_1. This removes the audio signal from the input to the next stage, the power amplifier, which quiets the speaker.

Other more advanced IF ICs have built-in squelch circuits (and a 1st audio amplifier) that simply require an external potentiometer attached to the appropriate pin to adjust squelch level.

These are just a few of the many circuits developed to mute a receiver.

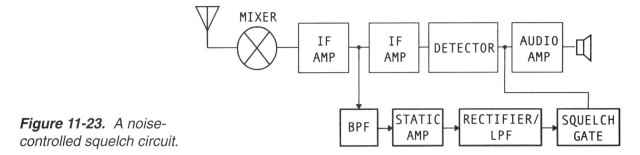

Figure 11-23. A noise-controlled squelch circuit.

11-6. Power Supplies

Since batteries can only supply current for a limited period of time, the majority of RF equipment runs off of the AC mains through power supplies (PS). Most power supplies convert this AC mains voltage into the desired DC levels needed to furnish power to the electronic circuits at a constant and regulated voltage. Banks of lead-acid batteries, however, are utilized extensively for backup power when a failure in main's power delivery occurs.

The average single-phase power supply consists of a transformer (for voltage conversion), a rectifier, a filter, a regulator, and ordinarily some form of protection circuits to safeguard both the PS and its load.

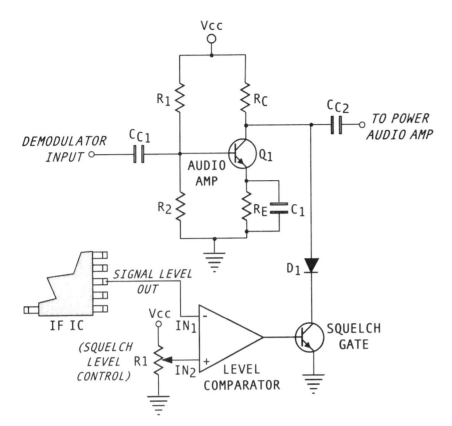

Figure 11-24. *An adjustable squelch circuit utilizing an IF IC's signal level output and a comparator.*

A complete power supply is shown in *Figure 11-25*. It contains the transformer needed to convert the AC mains voltage of 120 V$_{RMS}$ into any desired voltage. The bridge rectifier circuit then converts the AC into a pulsating DC, which is then filtered by the filter network into a steady DC. The *three-terminal regulator* holds the output voltage steady, while C$_3$ suppresses any output oscillations, as well as aiding in regulation (some high-powered amplifier's power supplies are not regulated due to the high power dissipation levels that would result due to the regulation dissipation). R$_B$, the bleeder resistor, is used to draw a fixed current, helping to stabilize the output voltage while the power supply is on; and draining dangerous voltage levels from the filter capacitors when the PS is shut down. Regulated low-voltage power supplies may not use bleeder resistors. Each of these circuits will be examined in more detail below.

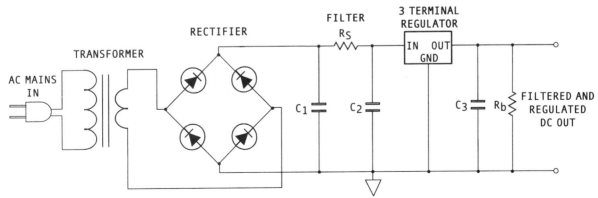

Figure 11-25. *An entire regulated linear power supply.*

11-6A. Rectification

There are three common ways single-phase (household) AC power can be converted into pulsating DC power, which is the first step in the power supply chain of obtaining a smooth DC output voltage, and is referred to as rectification.

Half-wave rectification (*Figure 11-26*) supplies a difficult-to-filter 60 Hz output, but at a peak voltage that is equivalent to the incoming AC peak voltage from the power line or transformer, while requiring few components (a single diode).

Full-wave rectification (*Figure 11-27*), delivers an easy to filter 120 Hz output, but at only half the AC voltage peak of the input (because of the transformer's center tap), while utilizing an expensive center-tapped transformer.

Bridge rectification (*Figure 11-28*) gives an easy-to-filter 120 Hz at the full input AC peak voltage levels, and is by far the most common rectifier in modern, quality equipment.

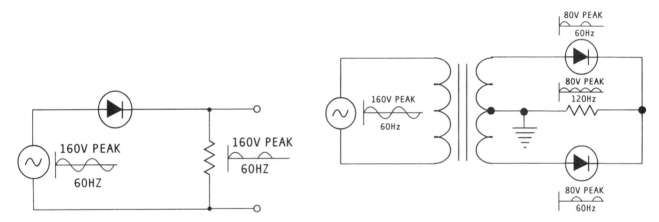

Figure 11-26. *A half-wave rectifier.*   **Figure 11-27.** *A full-wave rectifier.*

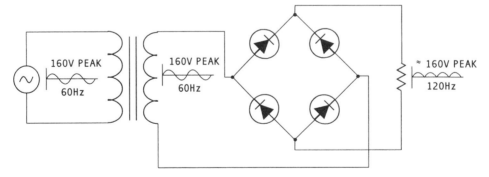

Figure 11-28. *A bridge rectifier.*

11-6B. Filtering

Because pulsating DC power is unacceptable for most electronic circuits, filtering is needed to remove this pulsating component while supplying a constant unvarying current.

A basic power supply filter is shown in *Figure 11-29*. This filter will remove any AC component from the rectified output, while permitting the DC. The filter consists of C_1, which filters most of the ripple, and R_S and C_2, acting as an AC voltage divider: Any ripple remaining after C_1 is dropped across R_S (due to its higher AC resistance over that of C_2's lower reactance to the ripple frequency). Replacing R_S with an inductor would have basically the same effect. However, RC filters work best with a low current drain, while LC filters will continue filtering even when the current drain is at a high level. Placing the capacitor at the input, as in the above example, will supply a higher final DC output, but with inferior voltage regulation. Placing an inductor at the input will supply a lower output voltage, but will furnish better voltage regulation.

A rectifier/filter variation that can be used for devices (such as some op-amps) that require both negative and positive power supply voltages (*Figure 11-30*) simply adds a second half-wave rectifier. This second half-wave rectifier passes the negative portion of the AC wave, and filters it through C_1.

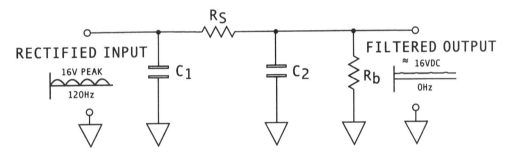

Figure 11-29. *A common power supply filter.*

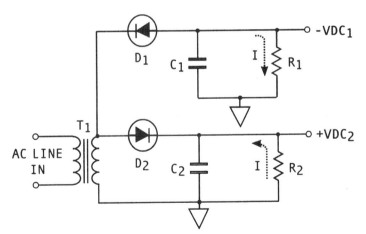

Figure 11-30. *A rectifier/filter that furnishes both negative and positive voltages.*

11-6C. Regulation

Most modern electronic circuits will only function reliably if they are provided with a supply voltage at a constant and unvarying amplitude. Considering that an unregulated P.S.'s output voltage will change when either the AC line voltage fluctuates or the resistance of the load varies, regulators are needed to prevent, or at least lessen, these changes. Regulators also assist the filters in smoothing the output voltage.

There are many different types of voltage regulators. The simple *Zener shunt voltage regulator* of *Figure 11-31* drops excess voltage across the resistor R_S as current increases through both the Zener and R_S. This occurs because the sum of the Zener voltage and the voltage drop across R_S must equal the input voltage ($V_{Rs} + V_Z = V_{IN}$). The Zener changes resistance as current varies to keep its Zener voltage (V_Z) constant which, considering it is parallel with the load, keeps the load voltage constant. In other words, as Zener current increases, Zener resistance decreases — if Zener current decreases, then Zener resistance increases, thus keeping the voltage drop across the Zener (and the load) constant ($V=IR$). V_{IN} must be slightly higher than V_Z for this Zener regulator to stay in regulation.

With the *series-pass transistor regulator* of *Figure 11-32*, if the voltage across the load should attempt to increase for any reason, the regulator's output voltage will remain relatively steady. Since the voltage across the load must equal the voltage drop across the series-pass transistor's (Q_1) E-B junction, subtracted from the voltage drop across the Zener (V_Z) — and this V_Z value cannot change — and considering that the base voltage is set by this Zener, then any increase in the regulator's output voltage will make the emitter that much more positive than the base. This is the same as making the base less positive, resulting in less B-E voltage, producing less emitter current and a subsequently larger voltage drop across the series pass transistor. This means a decrease in any attempted increase in voltage across the load.

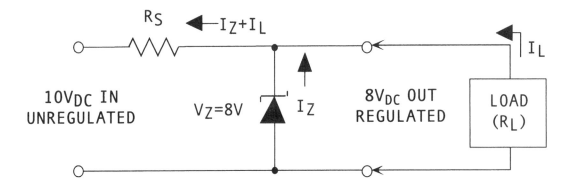

Figure 11-31. *A basic and low-cost Zener shunt regulator.*

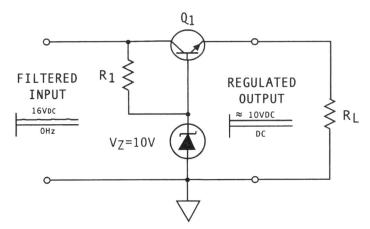

Figure 11-32. *A series-pass transistor regulator.*

The *series-pass regulator with feedback* block diagram (*Figure 11-33*) shows the major sections to this type of popular regulator. The circuit itself (*Figure 11-34*) consists of R_1, R_2, and R_3 forming a sample and adjust (voltage divider) circuit to set the desired output voltage. By sliding the wiper of the adjustable (pot) R_2 downward, the voltage on the base of Q_2 decreases, reducing its forward bias so that Q_2 will conduct less, causing its collector voltage, and the base voltage of Q_1, to increase. With the bias of Q_1 now increased and with Q_1 subsequently conducting harder, more current is sent through the load, increasing V_{OUT}. The Zener sets the reference voltage at the base of Q_2 by keeping the transistor's emitter clamped at this voltage, while R_5 fixes the idling current through the Zener. R_4 is both the collector resistor for Q_2 and the bias resistor for the base of Q_1.

Series-pass regulators with feedback are a large improvement over the simple Zener shunt regulator, as very little current is now wasted through the Zener shunt path. Series-pass regulators with feedback also allow, unlike both the Zener shunt and the basic series-pass regulator, a simple scheme to obtain a variable output voltage that can be changed anywhere from a little above the Zener's V_Z to just under the unregulated supply voltage. These feedback regulators also furnish improved regulation due to the error detector and its amplifier, since any change in voltage is amplified and fed into the base of the series-pass transistor.

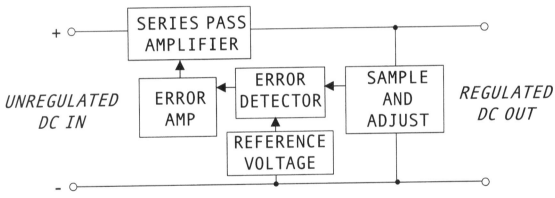

Figure 11-33. *A block diagram of a series-pass regulator with feedback.*

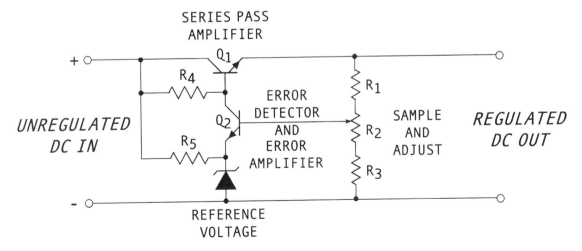

Figure 11-34. *The complete circuit of a series-pass regulator with feedback.*

Three-terminal regulators (or *3Ts, Figure 11-35*) are the most prevalent type of regulator because of low cost and simplicity. For any circuit needing a supply that must deliver three amperes or less (or higher currents with other external components), the 3T's are the regulators of choice. These IC regulators contain internal current limiting and thermal protection circuits (if internal power dissipation becomes too high, then the regulator automatically shuts down), and are lighter in weight and smaller in size, as well as much more efficient, than discrete regulators.

In the 3T circuit of *Figure 11-35* R_1 and R_2 set the fixed output voltage, while R_2 can be varied to adjust this output voltage over a certain range. Both positive and negative voltage output versions, with common values of 5, 8, 12, 15, and 24 volts, are available.

Another widespread type of IC regulator is the multipurpose *723 adjustable voltage regulator* of *Figure 11-36*. Operated alone it can supply up to 150 mA to a load, but with a supplementary series-pass transistor almost any desired load current can be accepted. With various external passive components, it is commonly utilized as a 2 to 37 volt series linear regulator; or as a current, shunt, or even switching regulator. The 723 contains on-board current limiting set by an external resistor.

Highly efficient *switching power supply regulators* are found in an increasing number of lower-powered equipment. They perform regulation by switching their output completely on and completely off while varying the resulting pulse's duty cycle or the number of pulses per second in step with the output load requirements. Most operate with switching speeds between 5 kHz and 500 kHz.

Due to the switching noise that can be produced, their use is still limited in certain communications gear. Switching power supplies demand serious shielding, filtering of all input and output leads, and short internal leads, to minimize RFI generation into surrounding space.

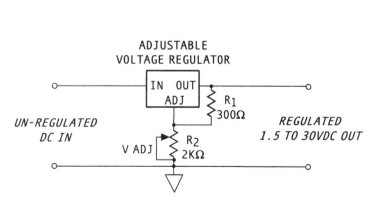

Figure 11-35. *The popular three-terminal regulator with adjustable voltage output.*

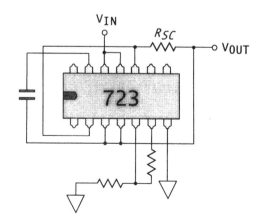

Figure 11-36. *A 723 IC voltage regulator.*

Figure 11-37 is one type of switching regulator. Q_1 is the switching pass transistor. D_1 protects Q_1 from the inductive kickback of L_1 (opening any inductive circuit causes a current pulse, producing damaging arcing in unprotected circuits due to:

$$t = \frac{L}{R}$$

when a charged RL circuit is opened, in this case by the switching transistor, the resistance is infinite, causing an extremely fast discharge time for the coil, giving off a very high spike of voltage). L_1 and C_1 filter the pulses coming out of Q_1 to DC, while the voltage divider of R_1 and R_2 program the desired output voltage.

The switching regulator functions as follows: When the load voltage decreases, should the load decrease in resistance or the input voltage to the regulator decline, then the comparator senses this change in output as a drop in voltage across R_2 below that of the Zener reference voltage, which then switches on the VCO, starting it oscillating and outputting pulses into Q_1. Q_1 then pulses on and off, outputting pulses into the filter network consisting of L_1 and C_1, which filters the pulses into a DC level. By increasing the pulse's duty cycle into the LC filter, the average DC output voltage increases. Much of this circuit, as well as similar switching schemes, can be obtained in IC form.

IC switching regulators allow the substantial reduction in the size of any regulator heat sinks needed over that of a similar linear regulator IC in a power supply, and typically can furnish over 80% efficiency levels at around 3A output currents (in other words, little power is wasted as internal heat), with very wide input voltage ranges of between 4 and 40V possible. Most include built-in current limiting and thermal protection.

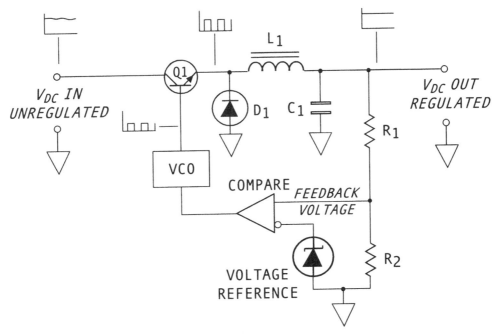

Figure 11-37. *An example of a switching regulator circuit.*

Figure 11-38 shows a common IC version of the discrete circuit of *Figure 11-37*. Such a regulator would be found in many switching power supplies, along with its characteristic support components arrangement. As discussed above, R_1 and R_2 program the output voltage value, with the feedback line between R_1 and R_2 supplying feedback to the regulator's internal comparator circuits (R_2 can be made adjustable to vary V_{OUT}). L_1 and C_{OUT} filter the output, while D_1 protects the IC from the inductive kickback of the discharging L_1, furnishing a return path for the load current when the IC switches off. C_1 is needed for regulator stability at higher current draws. If undesirable ripple amplitudes are still present at the output of this circuit, then a supplementary LC ripple filter is sometimes added after R_1 and R_2.

With this breed of regulators, it must be understood that when an output capacitor is found to be defective, it *must* be replaced with the exact same type, or higher ripple outputs may result, as well as possible regulator instability due to variations in the capacitor's *equivalent series resistance* (*ESR*, which is the value of its internal in-series resistance. The capacitor may also overheat due to the increased ESR and series impedance, causing it to fail). The diode, D_1, must as well be replaced with the same type, or instability and RFI may occur. This is caused by diode switching speed variations from diode to diode.

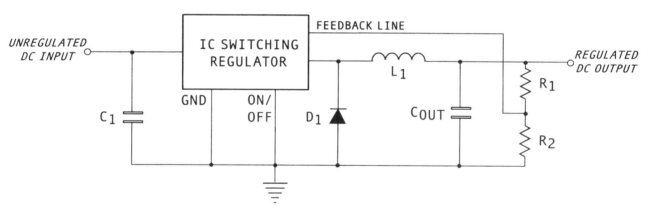

Figure 11-38. *A popular switching regulator IC with bias components.*

11-6D. *Switching Mode Power Supplies*

A complete switching mode power supply (SMPS) is shown in *Figure 11-39*. The MOV (*Metal Oxide Varistor*) shorts any high-amplitude transients to ground, giving limited protection from any lightning strikes or surges. D_1, D_2, D_3, and D_4 form a bridge rectifier, and rectifies the unregulated AC, while C_1 filters to DC. TH1 and TH2 are current inrush limiting thermistors (when cold the thermistors resistance is quite high, restricting damaging excessive current draw at startup. After the thermistors heat up, due to normal internal power dissipation, their resistance falls). Q_1 is the chopper transistor, which is switched on and off by the *pulse width modulator* (PWM). The PWM is turned on, varying the width of its output pulses, when the SMPS's output voltage falls lower than the *REF* voltage. R_1 is the Q_1 startup resistor. The *isolator*, which can be a simple *opto-isolator*, provides isolation between the low-voltage secondary and the higher-voltage primary. The chopped DC is sent through the high-frequency transformer, T_1, and is rectified by D_5, filtered by C_2, L_1, and C_3, and out of the SMPS as regulated DC.

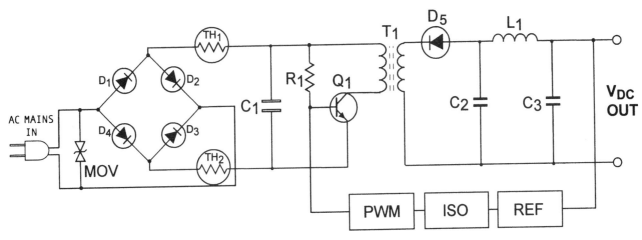

Figure 11-39. *A complete switch mode power supply.*

A few caveats when working with switching mode power supplies. Never run an SMPS without a load at the output, or it may become damaged or run improperly. Also, be careful were you place your probe's ground, since sections of the switching power supply will be floating, and should not be grounded at all. Plus, some sections of an SMPS may contain up to 800V, such as across the *chopper* transistor, creating a severe shock hazard, while the input capacitor may contain 300V or more (always discharge capacitors).

11-6E. Circuit Protection

If either the load or the power supply should have a failure, high currents could flow into the opposite device, destroying it before any fuse could open. There are two common circuits adopted for load and power supply protection.

Adding a *current-limiting* circuit (*Figure 11-40*) to the above discrete regulators results in protection for the power supply if the load should develop a short. Without current protection the power supply's series-pass transistor could be destroyed, even before the fuse opened, if any low-resistance path formed in the load. If current increases to the point where the voltage drop across R_6 is above 0.6V, the base-emitter junction of Q_3 is turned on, causing current to flow through R_4, making the base of Q_1 less positive, thus decreasing the collector current of Q_1. This action will lower the current within the power supply to a safe level.

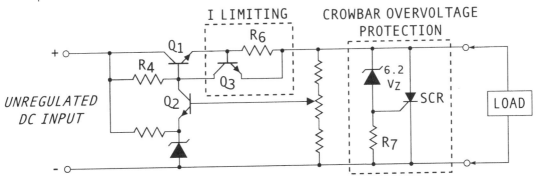

Figure 11-40. *A discrete regulator equipped with current-limiting and overvoltage protection.*

Overvoltage protection is used to safeguard the load if the power supply should malfunction. If the power supply voltage rises above its design value, which in *Figure 11-40* is above 6.2V, the 6.2V Zener will begin to conduct, causing a voltage drop across R_7, which latches the SCR on, shorting out the load from the power supply circuit through the SCR, causing the PS fuse to blow. It is often called a *crowbar* circuit because it completely shorts out the power supply.

11-7. Digital Signal Processing

In *digital signal processing* (DSP) the transmitted or received signal is enhanced by applying digital technology. Signal-to-noise and fidelity are improved, as is adjacent channel interference.

An audio (or IF) signal is placed into the DSP circuit IC and *sampled* (converted to digital pulses by an A/D converter). The onboard microprocessor then processes these signals using algorithms (a series of steps employed to solve a problem) to remove any noise, echo, and/or most of the interference. The processed output is then fed into a D/A converter to change the digital data of the converted audio signal back into an analog audio frequency. This recovered audio is then amplified and output to a speaker or headphones in a receiver, or sent to a modulator in a transmitter.

DSP is capable of processing both incoming or outgoing voice and digital signals, as well as video, with built-in DSP chips; or through add-on external units.

11-8. Miscellaneous Circuits

There are many other important circuits that contribute to the operation of a transmitter, receiver, or transceiver. These miscellaneous circuits are encountered in most communications gear, and a quick explanation of their operation is appropriate to aid in further understanding and troubleshooting radio equipment.

11-8A. Comparator Amplifier

Comparator amplifiers (*Figure 11-41*) exploit an op-amp in its open loop (no feedback) configuration to compare two input voltage levels. Due to an op-amp's very high gain without negative feedback, *any* difference in potential between its inverting and non-inverting input causes saturation — either to its maximum negative or maximum positive supply voltage, depending on which input goes high.

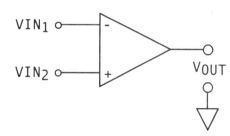

Figure 11-41. *An operational amplifier configured as a comparator.*

The comparator can be employed as a *zero-crossing detector.* When the input signal goes even slightly positive, the output voltage will switch to its full positive rail (which is near Vcc), and visa-versa for a negative crossing.

The comparitor is also used as a *level detector.* A reference voltage is placed at one input, while the voltage to be sampled is placed at the other. Whichever input contains the higher amplitude will saturate the output at that corresponding polarity, *comparing* the two voltages.

11-8B. Clippers, Slicers and Clampers

A *clipper* clips off one or both alternations, and is used to set a voltage to be no higher than to some design value. *Figure 11-42* shows a *diode clipper*, which only allows one alternation to pass while clipping off the top of the other alternation at 0.7V (if a silicon diode is utilized). In *Figure 11-43* an *active* clipper, which is basically an overdriven amplifier, will cut off the top of both peaks while placing the AC wave on a DC level (the clipped wave never crosses zero volts). However, only one peak may be clipped, depending on the bias applied to the transistor (to create either saturation *or* cutoff).

A *slicer* (*Figure 11-44*) slices off both *AC* alternations by using two Zener diodes consisting of any desired Zener voltage. The first alternation is passing through the reverse-biased first Zener, dropping 15V, while then passing through the second Zener in the forward direction, which drops another 0.7V. Since the two diodes are in series, and any load present is in parallel, a total of 15.7V is the maximum amplitude that is allowed to be passed from input to output in either the positive or negative alternations of the AC wave.

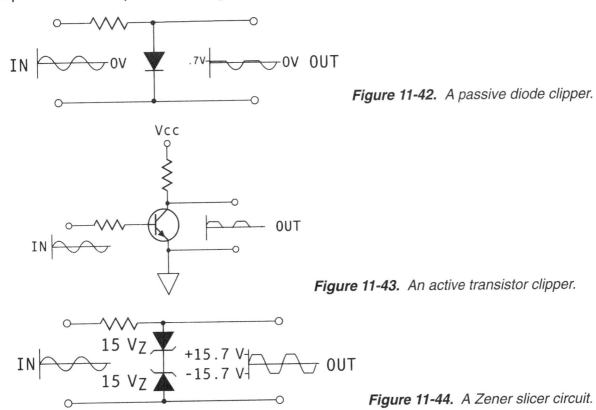

Figure 11-42. A passive diode clipper.

Figure 11-43. An active transistor clipper.

Figure 11-44. A Zener slicer circuit.

A *clamper* (or *DC restorer*, *Figure 11-45*) clamps the incoming AC to some positive or negative DC level. It converts an AC wave into a varying DC wave by *clamping* the output to zero volts or below (or above, if the diode is reversed). First, the capacitor is charged up by the input signal as it flows through the diode. Then, when the input signal drops and can no longer forward bias the diode, the capacitor discharges through the resistor, creating a DC level for the input signal at the output load. The capacitor must be of a relatively high value, or distortion of the output wave will occur.

Occasionally a clipper is considered to be a circuit that limits one alternation above a certain set amplitude, while a slicer is considered to be a clipper that limits both alternations above a certain set amplitude.

Slicers and clippers are commonly used as *limiters* in audio and FM IF circuits, or to reshape certain waveforms.

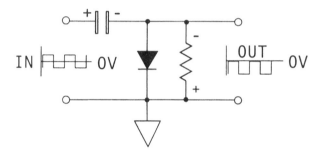

Figure 11-45. *A diode clamper circuit.*

11-8C. Voltage Multipliers

Voltage-doublers (*Figure 11-46*) are utilized when increased voltages are needed in low-current circuits. A peak AC voltage at the input is rectified and filtered into a 2 X V_{PEAK} DC voltage at the output.

The voltage-doubler works as follows: An incoming AC voltage forward biases the first diode, allowing current flow to charge the first capacitor. The second diode is reverse biased and cannot conduct. When the polarity of the AC reverses, the first diode becomes reverse biased and does not conduct, while the second diode is now forward biased and passes current, charging up the second capacitor to the opposite polarity to that of the first capacitor. The voltage is now taken from between the two oppositely charged capacitors. This output voltage is the difference between the two capacitors' potential difference. Doubling of the input voltage occurs, as well as the filtering of the AC into a DC, caused by the ability of the capacitors to store a charge (if the capacitors are large enough to supply the load with a relatively steady current).

Multipliers, in different configurations, are commonly available up to voltage quintuplers.

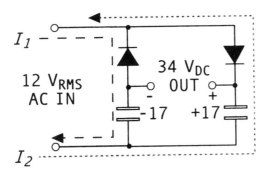

Figure 11-46. *An AC-to-DC voltage-doubler circuit.*

11-8D. Constant-Current Source

Certain circuits and test equipment need a constant-current source to function properly. The discrete constant-current source of *Figure 11-47* functions because there is a consistent 5.6V applied to the base of the transistor, supplied by the reverse-biased Zener. Consequently, a constant five volts is dropped across the emitter resistor and a constant 0.6V is also dropped across the base-emitter junction of the transistor (5.6 V_Z - 0.6 V_{BE} = 5V). As a result, the collector current will be a steady 5 mA

$$I_C = \frac{5V_E}{1\,K\Omega} = 5\,mA$$

through the load, no matter how much the load resistance may vary (up to a point, of course).

There are many other configurations to implement constant-current source circuits and current regulation, most of which are available in IC form or as an integral part of numerous other ICs (such as differential amplifiers).

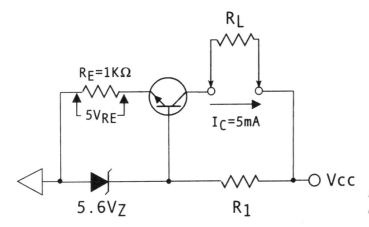

Figure 11-47. *A simple discrete constant-current source.*

11-8E. Link Coupling

Link coupling (Figure 11-48) can be placed between a transmitter and a remotely located antenna to transfer energy over long distances to the antenna. Link coupling consists of a few turns of wire connected, commonly, to a coaxial cable. This will minimize losses due to inductive reactances in long RF cable runs. Link coupling will lower harmonic transmission to the antenna, as the links have a low value of capacitive coupling between the coil output of the transmitter and the small turns of the link coils.

Link coupling is also adopted to couple test equipment, such as oscilloscopes and spectrum analyzers, to a transmitter — with almost no loading on the circuit under observation.

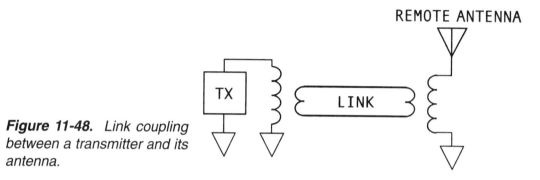

Figure 11-48. *Link coupling between a transmitter and its antenna.*

11-8F. Electronic Switches

Not only does the transistor perform as an amplifier of AC and DC signals, but it is also capable of switching DC, AC, and RF signals. *Figure 11-49* presents a simple circuit for switching RF currents. When the transistor's base is grounded through the manually operated mechanical switch, the NPN transistor does not conduct and does not allow any RF to pass to its output. However, when the mechanical switch is placed across the positive supply, base current instantly begins to flow, and the transistor quickly saturates, creating a very low emitter-to-collector resistance. This permits the RF signal to pass to the output. Transistor switches can be used to remotely switch RF currents when long wire or trace runs containing RF is undesirable.

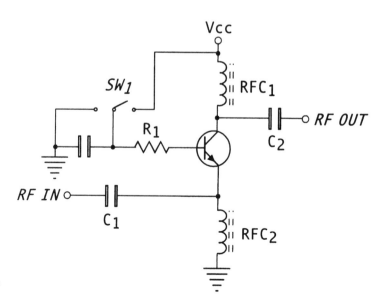

Figure 11-49. *A transistor RF switch.*

DC, especially for amplifier supply voltages, may also be switched. The transistor switch of *Figure 11-50* is activated when SW_1 closes, supplying a positive bias voltage to the base, turning on the transistor and allowing DC current to pass through the load.

Instead of directly placing a positive voltage on the transistor's base with a mechanical switch, a positive voltage can be sent to the base in response to other circuit conditions, such as when a transceiver changes from receive to transmit — the transmitter section now needs power, and not the receiver. The transistor switch can supply this power, as can the *diode switch*.

A very important function of diodes in RF is likewise for switching applications. Diodes are utilized to switch-in different values of crystals, circuits, or entire sections, into a signal path. *Figure 11-51* shows such a simple diode switch. The AC signal passes through C_1 and is stopped by D_1. This is because the mechanical switch (SW_1) is in the OFF position, supplying a negative, or reverse, DC bias to the diode. To forward bias the diode and allow the signal to pass, the switch is turned to the ON position, which places a positive voltage that forward biases the diode into conduction, allowing the AC signal to ride on this DC bias. C_2 blocks the DC bias and passes the AC signal. R_1 is a current-limiting resistor to ground for the DC bias voltage, while the RFC blocks the RF from entering the power source.

Using a diode instead of a simple mechanical switch likewise allows for remote switching of an RF signal without the added inductance and stray capacitance of a wire run to a distant panel switch.

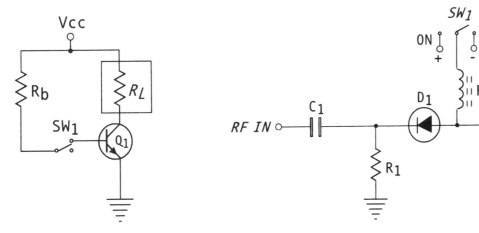

Figure 11-50. A transistor switch for DC.

Figure 11-51. A diode switch for RF.

11-8G. VOX

VOX, or *Voice Operated Transmission*, is used in certain radios to automatically transmit when listening on a transceiver, without having to physically press a PTT (*Push-To-Talk*) switch when the operator begins speaking. When the operator stops speaking, the transceiver will automatically switch back to receive. It is a hands-free way to communicate.

In a VOX circuit (*Figure 11-52*) some of the audio is tapped from one of the transmitter's audio amplifiers, amplified, rectified to pulsating DC by D_1, filtered to DC by R_1 and C_1, and then applied to Q_1. The DC turns on this switching transistor Q_1, energizing the relay to switch the transmitter on, the receiver off, and the antenna into the path of the transmitter.

VOX anti-trip circuits are utilized to prevent the output from the *receiver's* speakers from tripping the sensitive VOX circuit of above. It works by rectifying some of the received audio, which is then used to block the VOX from actuating. Only sound *not* coming from the receiver can cause VOX.

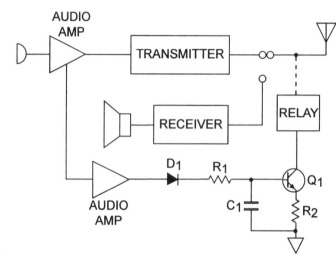

Figure 11-52. *A transceiver's VOX circuit.*

11-8H. Noise Blankers

Noise blankers are one way to remove high-powered RF noise impulses from the receiver's output. These noise impulses can be generated by auto ignitions, sparking motors, switching circuits, etc., and are very unpleasant to listen to.

The noise blanker of *Figure 11-53* is one such circuit to lessen the effects of such impulses. The IF is tapped and fed to the *noise amplifier,* which conducts and amplifies any signal *over* that as set by the *threshold control.* Any signal over this value is considered to be a noise impulse. The signal then goes through the phase shifter for a 180° phase shift, and back into the IF strip. This action cancels the impulse noise, since it is now 180° out-of-phase with the original noise impulse.

Another type of noise blanker is that of *Figure 11-54*, which splits the IF into two legs. One leg of the IF is sent into a *noise amplifier*, which amplifies any noise present, and on to a *noise detector* — which detects and activates the *pulse generator* when any impulse noises are present. The other leg of the split IF goes through a *delay line* — which delays the IF just enough to make up for the delay of the other leg, keeping both legs in phase — and then into a *blanking gate*. This blanking gate is activated by the pulse generator when impulse noise is present. The blanking gate "turns off" the noise impulse, quieting the receiver for the duration of the pulse.

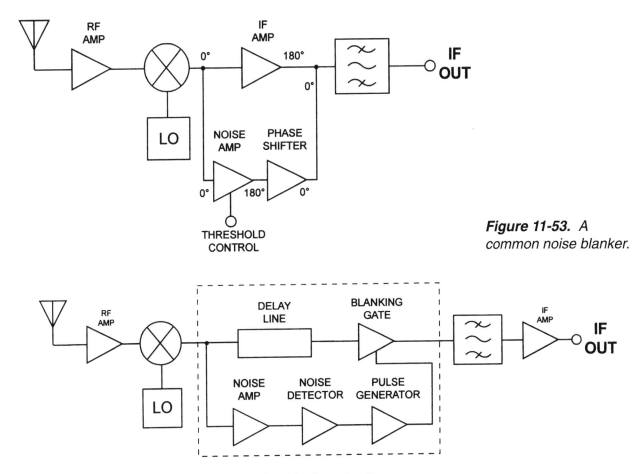

Figure 11-53. *A common noise blanker.*

Figure 11-54. *Another common noise blanker circuit.*

A *noise limiter* is an additional method for reducing impulse noise. It limits the amplitude of the noise pulses in a receiver so that they are not too objectionable, while the *noise blanker* actually silences the receiver during noise pulses. A noise limiter may employ any circuit that will limit a high-amplitude impulse, such as diodes that shunt the high-amplitude AM signal to ground.

FM receivers do not normally need noise blanking or limiting for impulse noise due to their built-in limiters, which have the same effect.

APPENDIX A

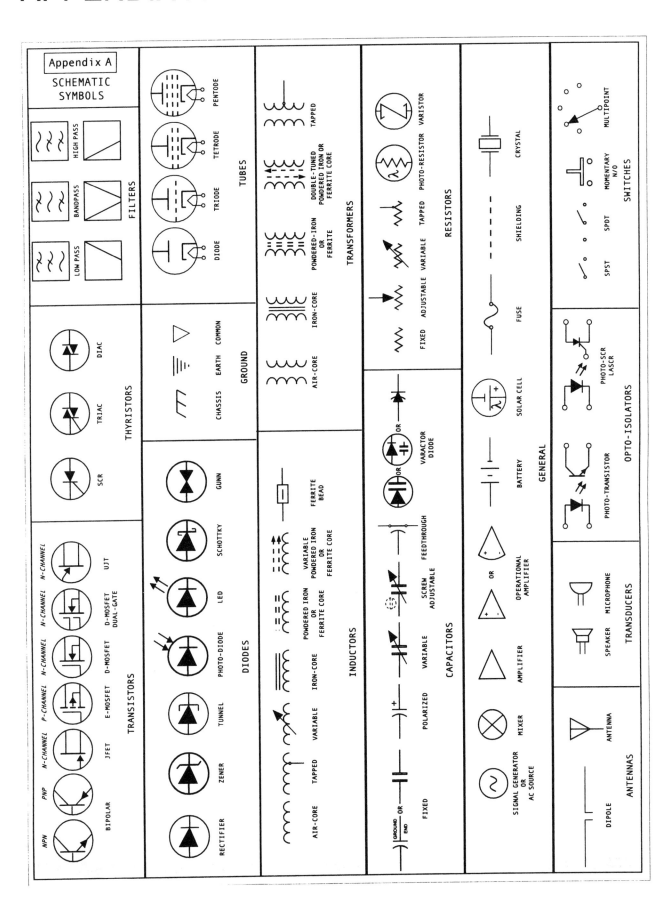

Appendix A

SCHEMATIC SYMBOLS

APPENDIX B

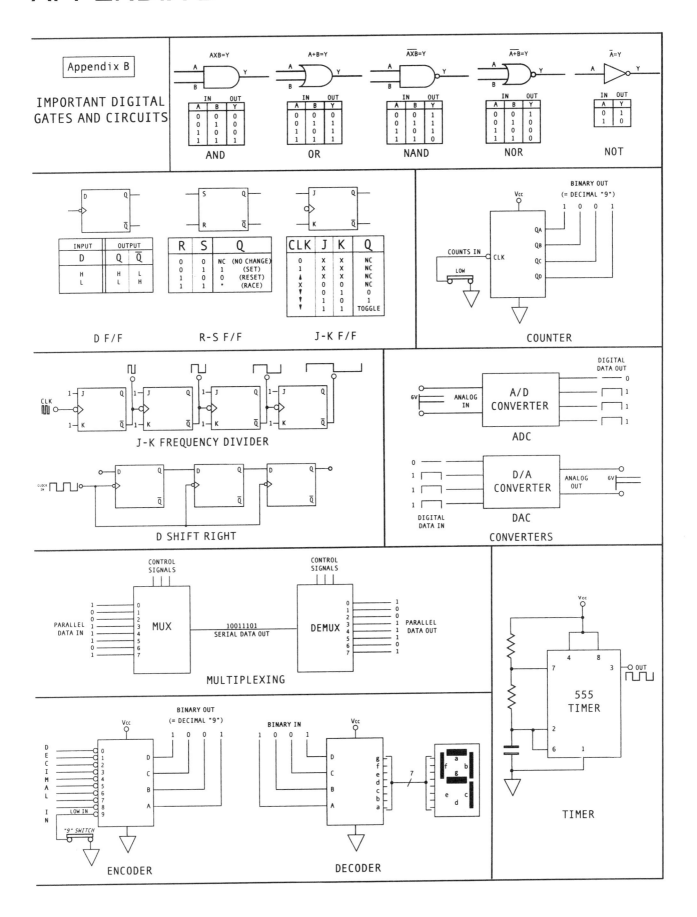

Appendix B

IMPORTANT DIGITAL
GATES AND CIRCUITS

AXB=Y

IN		OUT
A	B	Y
0	0	0
0	1	0
1	0	0
1	1	1

AND

A+B=Y

IN		OUT
A	B	Y
0	0	0
0	1	1
1	0	1
1	1	1

OR

$\overline{AXB}$=Y

IN		OUT
A	B	Y
0	0	1
0	1	1
1	0	1
1	1	0

NAND

$\overline{A+B}$=Y

IN		OUT
A	B	Y
0	0	1
0	1	0
1	0	0
1	1	0

NOR

$\overline{A}$=Y

IN	OUT
A	Y
0	1
1	0

NOT

INPUT	OUTPUT	
D	Q	$\overline{Q}$
H	H	L
L	L	H

D F/F

R	S	Q	
0	0	NC	(NO CHANGE)
0	1	1	(SET)
1	0	0	(RESET)
1	1	*	(RACE)

R-S F/F

CLK	J	K	Q
0	X	X	NC
1	X	X	NC
▲	X	X	NC
X	0	0	NC
▼	0	1	0
▼	1	0	1
▼	1	1	TOGGLE

J-K F/F

COUNTER

BINARY OUT
(= DECIMAL "9")
1 0 0 1

J-K FREQUENCY DIVIDER

D SHIFT RIGHT

ADC

DAC

CONVERTERS

MULTIPLEXING

TIMER

ENCODER

DECODER

APPENDIX C

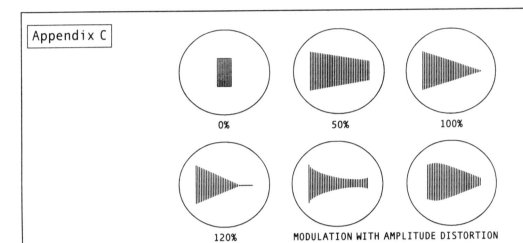

0%   50%   100%

120%   MODULATION WITH AMPLITUDE DISTORTION

**AM TRAPEZOIDAL
MODULATION PATTERNS**

SMD CAPACITOR
STANDARD TWO PLACE CODE

VALUES (PICO FARADS)					MULTIPLIER
A-1.0	H-2.0	b-3.5	f-5.0	X-7.5	0 = X 1.0
B-1.1	J-2.2	P-3.6	T-5.1	t-8.0	1 = X 10
C-1.2	K-2.4	Q-3.9	U-5.6	Y-8.2	2 = X 100
D-1.3	a-2.5	d-4.0	m-6.0	y-9.0	3 = X 1000
E-1.5	L-2.7	R-4.3	V-6.2	Z-9.1	4 = X 10000
F-1.6	M-3.0	e-4.5	W-6.8		5 = X 100000
G-1.8	N-3.3	S-4.7	n-7.0		ETC.

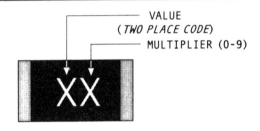

VALUE
(*TWO PLACE CODE*)

MULTIPLIER (0-9)

XX

SMD CAPACITOR
STANDARD SINGLE PLACE CODE

VALUES (PICO FARADS)				MULTIPLIER
A-1.0	J-2.0	T-3.9	4-7.5	ORANGE = X 1.0
B-1.1	K-2.2	V-4.3	7-8.2	BLACK = X 10
C-1.2	L-2.4	W-4.7	9-9.1	GREEN = X 100
D-1.3	N-2.7	X-5.1		BLUE = X 1000
E-1.5	0-3.0	Y-5.6		VIOLET = X 10000
H-1.6	R-3.3	Z-6.2		RED = X 100000
I-1.8	S-3.6	3-6.8		

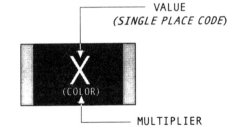

VALUE
(*SINGLE PLACE CODE*)

X
(COLOR)

MULTIPLIER

**READING SMD CHIP
CAPACITOR VALUES**

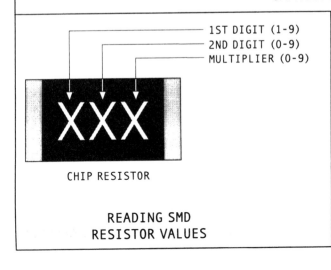

1ST DIGIT (1-9)
2ND DIGIT (0-9)
MULTIPLIER (0-9)

XXX

CHIP RESISTOR

**READING SMD
RESISTOR VALUES**

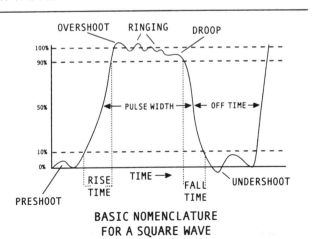

OVERSHOOT   RINGING   DROOP

100%
90%

PULSE WIDTH   OFF TIME

50%

10%

0%

RISE TIME   TIME →   FALL TIME   UNDERSHOOT

PRESHOOT

**BASIC NOMENCLATURE
FOR A SQUARE WAVE**

APPENDIX D

COMMON ELECTRONIC COMPONENTS

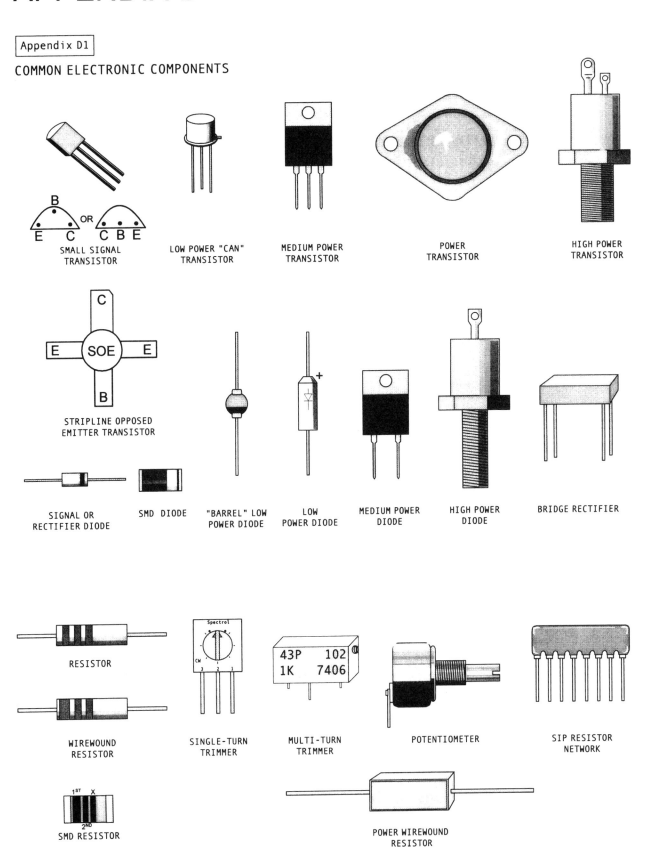

SMALL SIGNAL TRANSISTOR

LOW POWER "CAN" TRANSISTOR

MEDIUM POWER TRANSISTOR

POWER TRANSISTOR

HIGH POWER TRANSISTOR

STRIPLINE OPPOSED EMITTER TRANSISTOR

SIGNAL OR RECTIFIER DIODE

SMD DIODE

"BARREL" LOW POWER DIODE

LOW POWER DIODE

MEDIUM POWER DIODE

HIGH POWER DIODE

BRIDGE RECTIFIER

RESISTOR

WIREWOUND RESISTOR

SINGLE-TURN TRIMMER

MULTI-TURN TRIMMER

POTENTIOMETER

SIP RESISTOR NETWORK

SMD RESISTOR

POWER WIREWOUND RESISTOR

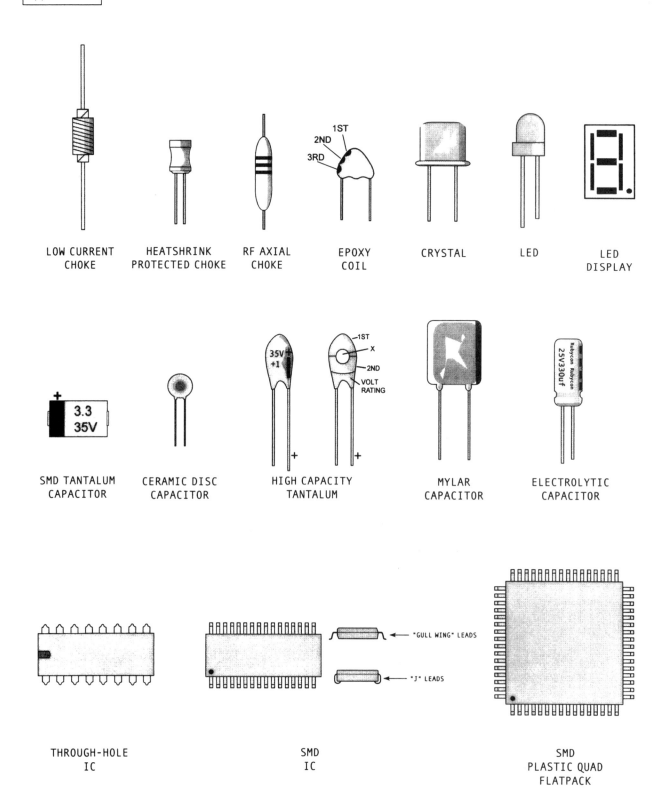

LOW CURRENT
CHOKE

HEATSHRINK
PROTECTED CHOKE

RF AXIAL
CHOKE

1ST
2ND
3RD

EPOXY
COIL

CRYSTAL

LED

LED
DISPLAY

SMD TANTALUM
CAPACITOR

CERAMIC DISC
CAPACITOR

35V
+1

1ST
X
2ND
VOLT
RATING

HIGH CAPACITY
TANTALUM

MYLAR
CAPACITOR

ELECTROLYTIC
CAPACITOR

THROUGH-HOLE
IC

SMD
IC

"GULL WING" LEADS

"J" LEADS

SMD
PLASTIC QUAD
FLATPACK

APPENDIX E

IMPORTANT REFERENCE TABLES

FREQUENCY RANGES

FREQUENCY	CLASSIFICATION
3 TO 30 kHz	VERY LOW FREQUENCIES (VLF)
30 TO 300 kHz	LOW FREQUENCIES (LF)
300 TO 3000 kHz	MEDIUM FREQUENCIES (MF)
3 TO 30 MHz	HIGH FREQUENCIES (HF)
30 TO 300 MHz	VERY HIGH FREQUENCIES (VHF)
300 TO 3000 MHz	ULTRA HIGH FREQUENCY (UHF)
3 TO 30 GHz	SUPER HIGH FREQUENCIES (SHF)
30 TO 300 GHz	EXTREMELY HIGH FREQUENCIES (EHF)

THE PREFIX AND ITS VALUE

NAME	FACTOR
PICO	10^{-12}
NANO	10^{-9}
MICRO	10^{-6}
MILLI	10^{-3}
KILO	10^{3}
MEGA	10^{6}
GIGA	10^{9}
TERA	10^{12}

SILICON TRANSISTOR'S V_{BE}

TRANSISTOR TYPE	CUTTOFF V	SATURATION V
LOW POWER	0.6	0.8
MEDIUM POWER	0.6	1.5
HIGH POWER	0.5	2.0

RESISTOR COLOR CODES

COLOR	VALUE	MULTIPLIER	TOLERANCE (%)
BLACK	0	1	±20
BROWN	1	10	± 1
RED	2	100	± 2
ORANGE	3	1 000	± 3
YELLOW	4	10 000	—
GREEN	5	100 000	± 5
BLUE	6	1 000 000	—
VIOLET	7	10 000 00	—
GRAY	8	0.01	—
WHITE	9	0.1	—
GOLD	—	0.1	± 5
SILVER	—	0.01	±10
NO COLOR	—	—	±20

APPENDIX F

Important formulas

$$V = IR$$

$$I = \frac{V}{R}$$

$$R = \frac{V}{I}$$

$$P_{TOTAL} = P_1 + P_2 + P_N$$

$$P = IE$$

$$P = I^2 R$$

$$P = \frac{E^2}{R}$$

Total impedance of a circuit: $Z_{TOTAL} = \dfrac{V_{APPLIED}}{I_{TOTAL}}$

Resonant frequency of an LC circuit: $fr = \dfrac{1}{2\pi\sqrt{LC}}$

$$f = \frac{1}{t}; \ t = \frac{1}{f}; \ \lambda = \frac{C}{f}; \ f = \frac{C}{\lambda}$$

(Where ƒ is the frequency in Hertz, t is the time in seconds, λ is the wavelength in meters, C is the speed of light in meters)

RMS (Effective Value)=.707 x V_{Peak}

V_{Peak}=1.414 x RMS

$V_{Average}$=.636 x V_{Peak}

To find instantaneous voltage anywhere on a sine wave, if degree of rotation is known:
$V_{INSTANTANEOUS} = SINE\ OF\ DEGREE\ OF\ ROTATION\ x\ V_{Peak}$

Time that one wave leads or lags a second wave: $t_L = \dfrac{\theta}{360} x \dfrac{1}{f}$

kWh = kW x Hours

Voltage divider formula: $V_{R1} = \dfrac{R_1}{R_T} x\ V_T$

Current divider formula: $I_{R1} = \dfrac{R_2}{R_1 + R_2} x\ I_T$

For transformer turns ratio: $\dfrac{N_P}{N_S}$

Matching transformer turns ratio needed for impedance match: $\sqrt{\dfrac{Z_P}{Z_S}}$

Series inductances: $L_T = L_1 + L_2 + L_N$

Parallel inductances: $\dfrac{1}{L_T} = \dfrac{1}{L_1} + \dfrac{1}{L_2} + \dfrac{1}{L_N}$

Series inductive reactance: $X_{L\,TOTAL} = X_{L1} + X_{L2} + X_{LN}$

Parallel inductive reactance: $\dfrac{1}{XL_T} = \dfrac{1}{XL_1} + \dfrac{1}{XL_2} + \dfrac{1}{XL_N}$

Inductive reactance: $X_L = 2\pi f L$

Total current in a parallel resistive/reactive circuit: $I_T = \sqrt{I_R{}^2 + I_L{}^2}$

Series capacitance: $\dfrac{1}{C_T} = \dfrac{1}{C_1} + \dfrac{1}{C_2} + \dfrac{1}{C_3} + ETC$ (Equivalent to thickening the dielectric)

Parallel capacitance: $C_T = C_1 + C_2 + C_n$

Series capacitive reactance: $X_{CT} = X_{C1} + X_{C2} + X_{CN}$

Total parallel capacitive reactance: $\dfrac{1}{X_{CT}} = \dfrac{1}{X_{C1}} + \dfrac{1}{X_{C2}} + \dfrac{1}{X_{CN}}$

Reactance of a capacitor at a desired frequency: $X_C = \dfrac{1}{2\pi f C}$ (Applies only to sine waves)

$$Q = \frac{fr}{BW\ at\ -3dB}\ \text{ or }\ \frac{X_L}{\text{Re}}$$

Q in a series resonant circuit: $Q = \dfrac{V_{OUT}}{V_{IN}}$

Q in a parallel resonant circuit: $Q = \dfrac{Z_T}{X_L}$

Time it takes for current to rise to 63.3% of its steady-state value in a resistive/reactive circuit:

$t = \dfrac{L}{R}$ and $t = RC$

Reactive power: $SIN\theta(VI)$ (Measured in VARS (Volt Amp Reactive))

Apparent power: EI (Measured in VA (Volt Amps))

Real power (RMS) in watts = I^2R

The power factor: $\dfrac{I^2R}{IE}$ or $\dfrac{POWER_{REAL}}{POWER_{APPARENT}}$ (PF will always be less than 1)

PEP (AM) = $4\ x\ P_{CARRIER}$

Common-emitter amplifier gain calculations: $A_V = \dfrac{V_{OUT}}{V_{IN}}$ $\quad \beta = \dfrac{I_C}{I_B}$ $\quad A_P = \dfrac{P_{OUT}}{P_{IN}}$

(With β as current gain; A_V as voltage gain; A_P as power gain)

Transistor output power: $P_{OUT} = I_C \times V_C$

FET amplifier gain: $G_m \times Z_{OUT}$

AC current gain in a CE amplifier: b $(h_{fe}) = DI_C/DI_B$

DC current gain in a CE amplifier: $\beta_{DC}(h_{FE}) = \dfrac{I_C}{I_B}$

Input resistance for CC amplifiers: $R_{IN} = h_{fe} \times R_L$ $\quad$ or $\quad$ b $\times R_L$

Amplifier efficiency: $\dfrac{RF_{POWER\ OUT}}{DC_{POWER\ IN}} X\ 100$

For power supply regulation: $\% REGULATION = \dfrac{V_{NO\ LOAD} - V_{FULL\ LOAD}}{V_{FULL\ LOAD}} X\ 100$

$$dB = 10 \ LOG \frac{P_2}{P_1}$$

$$dBm = 10 \ LOG \frac{P_2}{1mW}$$

$$dB = 20 \ LOG \frac{\frac{V_2}{\sqrt{Z_2}}}{\frac{V_1}{\sqrt{Z_1}}} \quad \text{or, if } Z_1 \text{ equals } Z_2; \ 20 \ LOG \frac{V_2}{V_1}$$

To measure total AM power (with modulation): $P_T = \left(I_C \sqrt{(1 + \frac{m^2}{2})} \right)^2 \bullet Z_{OUT}$

I_c = Unmodulated carrier current

m = Modulation Index

Z_{OUT} = Z_{OUT} of final amplifier

FM BW: $2N \cdot f_{m\,max}$

N = # of significant side bands

$f_{m\,max}$ = Maximum modulating frequency

FM modulation index (m): $\dfrac{f_{DEVIATION}}{f_{AUDIO}}$

FM deviation ratio: $\dfrac{f_{DEVIATION\ MAX}}{f_{AUDIO\ MAX}}$

ERP: $POWER_{TX\ OUT}$ x $(Antilog\dfrac{ANT\ GAIN\ IN\ dB, LESS\ LINE\ LOSSES}{10})$

RF power density: $\dfrac{W}{CM^2}$ (watts per centimeter squared)

Simple signal-to-noise-ratio measurement: $\dfrac{V_{SIGNAL}}{V_{NOISE}}$

SINAD: $\dfrac{SIGNAL + NOISE + DISTORTION}{NOISE + DISTORTION}$

Measuring SWR with a wattmeter: $1 + (\dfrac{P_{REFLECTED}}{P_{FORWARD}}) \div 1 - (\dfrac{P_{REFLECTED}}{P_{FORWARD}})$

Damping factor: $\dfrac{Z_L}{Z_{OUT}}$. *(With Z_L the Z of a speaker, Z_{OUT} the Z_{OUT} of an amplifier)*

Common Mode Rejection Ratio: $CMRR = \dfrac{A_D}{A_{CM}}$ (*With A_D being the difference gain, or 180° out-of-phase signals, and A_{CM} the common mode gain, or 0° in-phase signals*)

Duty cycle: $(\%) = \dfrac{PULSE\,WIDTH}{PERIOD}\ X\ 100$

BW $= \dfrac{fr}{Q}$

GLOSSARY

1 dB Compression Point — The point at which the output of a device is 1 dB less than it should be if the output linearly followed the input. This is the point an amplifier is considered to be at maximum amplification.

Absorption Trap — A type of wave trap that is inductively coupled to the output coil of an amplifier. It consists of a resonant tank circuit, tuned to the undesired signal, which absorbs the undesired frequency's energy.

Absorption Wave Meter — A tuned circuit with a tunable capacitor and a rectified meter. The coil of the tank circuit is loosely coupled to another tank or RF producing circuit. When the variable capacitor of the meter is varied a peak in the meter will indicate that the wave meter has coupled energy from the circuit under test, and thus that both are at the same resonant frequency. The resonant frequency is read from a dial or chart. Can be utilized to detect parasitics; measure the frequency of an oscillator; or as a general purpose RF indicator.

Active Device — A component that generates or amplifies a current or voltage. Common examples would be transistors and vacuum tubes.

Active Region — The region of only 0.2V where a semiconductor is capable of amplifying, which is between saturation (0.7V) and cutoff (0.5V). Between these two V_{BE} values the I_B, and thus the I_C, is controlled.

AGC Saturation Point — AGC voltages are linear only up to the " AGC saturation point ", or where the AGC voltage increases and becomes nonlinear at its AGC knee.

Alpha Cutoff Frequency (f_{ab}) — The frequency at which a common-base amplifier increases until the gain reaches 70.7% of its initial frequency of 1 kHz.

Amplifier Efficiency — The percent of RF output power compared to DC input power (this DC input power to a stage comprises the RF output power, as well as the heat dissipated in the active device (its plate or collector), and the device's LC and bias circuits).

Amplitude Shift Keying (ASK) — A early form of digital modulation using discrete amplitudes to denote bits (0 or 1). Normally not practiced due to the noise sensitivity of AM systems.

Angle-Of-Radiation — The angle of the transmitting antenna's radiation in relation to the earth's surface.

Apparent Power — The power that appears to be consumed in a reactive circuit, and is calculated by voltage multiplied by current (VI). Measured in VA, or Volt Amps.

Application Specific Integrated Circuit (ASIC) — A custom-designed and fabricated IC.

Attenuator — Either fixed or adjustable (*step attenuators*) apparatus for reducing signal amplitude, while maintaining the proper input and output impedances of any device attached to its ports.

Autotransformer — A single-winding transformer with only 3 leads. Cannot be used for isolation purposes.

Autodyne Detector — A CW detector that employs a BFO and product detector.

Automatic Level Control (ALC) — Controls the gain of a transmitter, controlling the drive to the final amplifier or some earlier stage. It can be used in AM, FM, and especially SSB transmitters. ALC will reduce excessively high amplitude output levels, but amplify medium and low-level signals normally.

Automatic Noise Limiter (ANL) — A circuit, usually employing diodes, placed just after the detector to cut off any noise spikes that would reach the audio section of the receiver.

Autopatch — A device for connecting a transceiver to the telephone lines by remote control.

Axial Leads — Leads on a component placed in-line but opposite one another — the ordinary arrangement for most resistors and signal diodes.

Backplane — An area where everything is connected; a common connection point (a *rat's nest*).

Balanced Amplifier — Another term for a push-pull amplifier.

Ballast Resistor — Some power BJTs have a very low value of internal emitter resistor incorporated within their package, making them less susceptible to thermal runaway.

Balun — A wideband transformer that is capable of matching balanced line (twin lead, a dipole antenna) to unbalanced line (coaxial cable), as well as altering RF impedances.

Bandgap Reference — Low-voltage device used as a temperature-stable voltage reference (similar to Zeners).

Bandpass — The frequency limits between *half-power points* of a signal or filter. For example, a certain channel may have a bandpass of between 92.788 MHz and 92.800 MHz. Compare *bandwidth*.

Bandspreading — The technique, either mechanically or electronically achieved, that allows the tuning of a receiver across a wide range, yet permits accuracy when tuning in a narrow-frequency channel. A *planetary drive* is an example of mechanical bandspreading.

Bandwidth — The amount of frequency spectrum, in hertz, utilized by a filter or channel. For example, a certain channel has a *bandpass* between 92.788 MHz and 92.800 MHz: it thus has a *bandwidth* of 92.800 MHz minus 92.788 MHz, or *12,000 Hz.*

Baseband Signal — The modulation signal, whether voice, music, or digital information.

Battery Ampere Hours — The length of time a battery can supply a certain amount of current. For example, a rating of 120 ampere-hours means a battery may deliver 120 amps for 1 hour, or 10 amps for 12 hours. This simple correlation does not hold up for high or fast current drains.

Beam Width — The width of the major lobe of an antenna's output electromagnetic wave, in degrees, between the half-power points.

Bifilar Winding — A special transformer winding technique: The primary and secondary windings are wound adjacent to each other, allowing for almost unity coupling. There is, however, high capacitance between the primary and the secondary, which allows harmonics through with little attenuation.

Bilateral Amplifier — A single amplifier that amplifies in two directions. Usually used for circuit sharing transceivers in transmit and receive. There are commonly two transistors: Q1 may be turned on during transmit while Q2 is biased off; on receive Q2 may the turned on while Q1 is cut-off.

Binary Coded Decimal (BCD) — A digital code that differs from binary in that a nibble represents only a single decimal. For example, decimal 12 to binary equals $12_{10} = 1100_2$; decimal 12 to BCD equals $12_{10} = (0001_2) (0010_2)$

Bit — Binary digit. A "0" or a "1".

Blanketing — A very strong signal that destroys reception for all receivers in a certain band. The offending signal may be heard all across, or only at certain discrete points, within the band.

Blocking Dynamic Range — The difference, in decibels, between the noise floor and the desired signal that will create 1 dB of gain compression. This denotes the signal level, over the noise floor, that initiates undesirable desensitization (causing decreased sensitivity in a receiver).

Blocking Interference — A receiver that is overloaded by a very strong received signal, causing a reduction or total lack of output.

BNC Connector — A high-frequency connector good up to 2 GHz. Both 50 and 75Ω versions are available. Used for test equipment and handheld vertical antennas in low-power applications.

Broadband Power Amplifiers — Amplifiers employed in modern low- and medium-power transmitters that need not be tuned to transmit. Simply turning the tuning dial to change frequencies, or switching the band switch, is all that is needed. With this type of amplifier spurious signals and harmonics within this wide bandpass are not attenuated. However, the transmitters are far easier and faster to use.

Buffer Amplifier — Typically, a common-collector amplifier, which naturally has a high input impedance and low output impedance. The buffer is placed between an oscillator or amplifier and the load. Buffering prevents the load from drawing too much current from the previous circuit, which would load that circuit down:

$$I = \frac{V}{R}$$

The buffer still allows the same signal voltage to be applied to the load.

Buffer Register — Temporarily stores a word during data processing.

Butterworth Characteristic — A description of a filter's type of response. The Butterworth has a very flat response with no peak within its bandpass.

Bypass Capacitor — A capacitor utilized to shunt AC around a component.

Byte — 8 bits (1 character).

C/N Ratio — The amount, in decibels, between the signal's carrier and the noise amplitude. Can be measured with a spectrum analyzer.

Capacitance and Inductor Multipliers — Circuits that can create artificially high values of capacitance and inductances, at low frequencies, by the use of op-amps and low-value capacitors. A capacitance multiplier may be used at the output of a power supply for filtering, and can comprise a transistor and a capacitor (the beta of the transistor multiplies the capacitance of the capacitor, or bXC=Cnew).

Capacitor — A component constructed basically of two conducters isolated from one another by a dielectric. Can store electrical energy. To read a capacitor: If marked with, for example, 2R8, the capacitor equals 2.8 pF or uF (the pF or uF depends on the capacitor's size, as well as the whim of the manufacturer); if marked with three numbers, such as 213, then its value would be 21,000 pF, or .021 uF (some capacitors may be marked 151K, which equals 15 X 10, or 150 pF — with K being a tolerance value).

Carrier — The frequency used to carry voice or data modulation from the transmitter to the receiver.

Carrier Balance — SSB carrier nulling in the balanced modulator.

Carrier Shift — In AM transmitters, the unequal increase or decrease in amplitude of either the negative or positive alternation of the modulated wave, which should always remain symmetrical. Positive carrier shift, which results in an increase of average AM output power, can be caused by parasitics, improper tuning of the finals, bad neutralization, or excess audio drive. Negative carrier shift, which results in a decrease in the average AM output power, can be caused by any distortion in the audio modulation, overmodulation, bad tuning of the resonant tanks, or bad regulation. Carrier shift may be monitored by a circuit built into many AM commercial transmitters.

Cascade Amplifier — A string of amplifiers in which each amplifier is connected in series to the following amplifier. This arrangement multiplies the gain of each amplifier by the next amplifier (if amplifier A has a gain of 100, and amplifier B has a gain of 50, then the total gain will be 100 X 50, or a total gain of 5000).

Ceramic Filter — A narrow bandwidth bandpass filter. Smaller and cheaper, but with a lower Q, and thus a wider bandwidth, than a crystal filter. Can be in a two-terminal or three-terminal package, with the third terminal placed at ground potential. The input frequency is inserted at the center terminal, with the output taken from the outer terminal of the three-terminal device. In a receiver the ceramic (or crystal) filter will generally be located between the mixer and the first IF amplifier, and sets the selectivity of the IF.

Channel — A segment of the frequency spectrum that includes a carrier frequency with its attendant sidebands which are standardized by common agreement; or a bandpass taken up by one modulated signal.

Characteristic Impedance — Also called *surge impedance*. The AC resistance of a transmission line as seen by the generator (such as a transmitter) and the equivalent value of the termination needed to make that line *flat* (no reflected waves). This value does not change with the frequency of the signal or length of line, and is a function of the physical size, spacing, and properties of the transmission line.

Chebyshef Characteristic — A description of a filter's type of response. The Chebyshef has a certain amount of ripple within its passband but has a sharp cutoff response.

Chip Capacitors — Leadless monolithic capacitors built up of numerous layers.

Choke Action — As the frequency increases in an inductor, inductive reactance will also increase. A device referred to as a *choke* uses this action to provide extremely high impedance to AC or RF, while letting DC pass unattenuated.

Circuit Sharing — Modern circuit design that gives joint use of certain circuits in a transceiver, such as the oscillators, filters, power supplies, frequency synthesizers, and amplifiers, to both the receiver and transmitter sections.

Class E and G Amplifiers — Both are similar in that they use 4 BJTs, one low-voltage and one high-voltage power supply, and function as push-pull amplifiers. Only 2 of the BJTs function under low power requirements, while the other 2 switch in, along with a high-voltage power supply, when high power is needed. Can be used in power audio amplifiers.

CMOS Logic — A logic arrangement employing Complementary Metal-Oxide Semiconductors. Uses both P and N channel MOSFETs with very low current consumption. However, they are susceptible to static discharge damage and are slower than TTL logic. Newer designs exploit Zener static protection. Power supply can be anywhere between 3 to 15 volts, which can also be unregulated.

Combiner — A circuit used to mix different signals in a linear manner, leaving them unaffected; there are no new signals created. More sophisticated combiners can be used to combine the signals of several different power amplifiers into a single high-power RF output.

Common-Mode Conducted Signals — Wires that behave as a single conductor, with all signals in phase, acting as a receptor or radiator for undesired signals. Generally has a return through ground. Can be cured by using chokes and ferrite beads. (See *differential-mode conducted signals*)

Communication-Service Monitor — A consolidated test unit that may be used to service FM, AM, and SSB in the field. May contain a signal generator, frequency counter, modulated FM and/or AM generator, audio generator, FM deviation meter, oscilloscope, and spectrum analyzer.

Compandoring — In SSB, compressing speech amplitudes at the transmitter and expanding them at the receiver. Some SSB compandored transmitters send a 3 kHz pilot tone to permit rapid tuning of a mobile receiver.

Complex Wave — A wave made up of a fundamental (sinusoidal) wave and its harmonics. Square and triangle waves are common examples. These complex waves will be distorted by reactive components, due to the different capacitive or inductive reactances and phase angles affecting each of the harmonics forming the wave.

Compression — Loud sounds are made quieter, quiet sounds are made louder in an amplifier. Used to lower bandwidth requirements and remove any chance of overmodulation in a transmitter.

Conductive Pad — Some transistors use conductive pads between their bodies and their heatsinks, which look like the normal insulating washers — but are electrically and thermally conductive, and can replace thermal grease in many applications. Some pads are made of rubber and are not electrically conductive.

Connector, RF — A device used to obtain a permanent or temporary connection for transferring RF energy, at the proper impedance, from one circuit or cable to the next. See specific connector.

Converter (DC-to-DC Converter) — A circuit utilized to transform one DC voltage value into another DC voltage value. This is accomplished by "chopping" the DC input voltage so that it can be converted into another value by a built-in transformer. An output filter converts this chopped wave back into direct current.

Copper Losses — Losses in lines or coils caused by I^2R losses, skin effects, and crystallization (wire damage due to high temperature or bending by wind, causing small, high-resistance cracks).

Coupler (or Antenna-Coupler) — An external device attached at the output of a transmitter to match the antenna's input impedance to the transmitter's output impedance.

Coupling Capacitor — Couples AC to another stage, while blocking DC.

Cross Modulation (XMOD) — The unwanted transfer of intelligence from an adjacent strong channel to the desired weak channel. The desired signal is modulated by the undesired signal in the nonlinearities of an amplifier in the receiver's front end, causing amplitude-modulated interference, which can be heard in addition to the desired signal. The amplifier may only be partially nonlinear, or may be driven into nonlinear operation by the presence of a strong local AM broadcast station. Can be reduced by traps and filters at the receiver's antenna input in order to filter out the unwanted frequency. To measure cross-modulation, the modulation of the channel of interest must be turned off, or its own modulation will obscure the XMOD.

Crosstalk — Undesired electromagnetic or electrostatic transfer of signals between systems, channels, or sections of a system.

Crystal — A thin piece of quartz cut to a precise thickness to produce resonance at a specific frequency. Crystals come in many different packages, with the most popular being the "tin can" and SMD types.

Crystal Oven — Employed in extremely accurate oscillators to stabilize the temperature of a crystal, which eliminates temperature-induced crystal frequency drift. Consists of a small oven chamber maintained at a slightly elevated temperature over that of ambiance by a heating element and internal thermostat.

Cutoff Frequency — The points on a signal's selectivity curve that are 0.707, or 3 dB (the half power points), below the signal's peak amplitude. Regularly used to measure the bandwidth of a tuned circuit.

Damping Factor — A numerical value of an amplifier and speaker impedance combination:

$$\frac{Z_{SPEAKER}}{Z_{POWER\ AMP}}$$

The lower the output impedance of a power amplifier and the higher the speaker input impedance, the cleaner and sharper the sound output of an audio amplifier.

Darlington Pair — Twin transistors directly coupled, with a very high input impedance and a very high gain ($\beta \times \beta$).

Decoding — An electronic digital function of converting from binary to decimal.

Degenerative Feedback — 180° out-of-phase feedback from an amplifier's output back into its input. This action lowers both its gain and its distortion levels. Can be designed into a circuit by utilizing feedback-producing components. Or it can be undesirable feedback if it occurs due to the interelectrode capacitances of the active amplifying device. Also called *negative feedback*.

Delta Markers (◊) — Movable markers on a measuring device's CRT that can indicate a signal's E_{min}/E_{max} or f_{min}/f_{max} .

Desensitization (Desensing) — A close unwanted frequency can create an overloading in a receiver's first RF amplifier or mixer, causing a loss of sensitivity to the desired signal. Also called blocking. Caused by the strong signal reducing the gain of the amps, which may also cause distortion (since the amplifier may be thrown into a nonlinear part of its operation). The offending signal may not actually be heard unless the receiver is tuned near this signal.

Device-Under-Test (DUT) — The device currently under examination for faults, alignment, or repair.

DIAC — A *thyrister* employed as triggers for TRIACs to equalize the ON voltage for each AC alternation.

Diddlesticks — Plastic RF alignment tools.

Die Form — An IC chip used without a package to conserve space.

Differential-Mode Conducted Signals — Signals that are 180 degrees out-of-phase with each other on two-conductor wire. The majority of desired signals are in differential mode. Undesired signals in differential mode can be cured by high-pass filters which shunt the undesired signals to the out-of-phase return line. (See *common-mode conducted signals*)

Digital Signal Processing (DSP) — An incoming signal is converted to digital pulses, manipulated, then converted back to an analog waveform. Used in digital state-of-the art adjustable filters. DSP improves S/N ratios and fidelity.

Diode Noise — In a power supply rectifier, noise is created when the diode or diodes begin to conduct, causing a brief but steep wave front that creates RFI. This RFI shows up as a "rush" sound on transmit or an irritating receiver noise. Usually removed by suppression capacitors or additional lead filtering.

Diode-Transistor Array — A linear IC with a number of diodes or transistors inside the IC package (similar to a resistor array).

Diplexer — A device that allows more than one transmitter to use the same antenna. It is also used as a type of bandpass filter in some FM radios.

Direct Conversion — In a receiver, converting the modulated RF signal directly into an audio frequency, which is then detected and fed to audio amplifiers; no IF is utilized.

Directional Coupler — A device used to sample the power traveling down a transmission line in one direction.

Disc Operating System (DOS) — A system program that informs a computer how to operate the disc drive, set up files, erase files, find files, etc.

DMOS — Method of producing a high-power, high-current, high-gain, high-frequency RF power MOSFET.

Double-Tuned Transformer — A transformer that contains two slugs; one tunes the primary, the other tunes the secondary.

Downlink — The frequency utilized by a satellite to send signals to earth.

Driven Guard Shield — A noise and anti-stray capacity shield used in some sensitive circuits and special probes. Comprised of a shield that is actively powered from a unity gain, low-impedance reproduction of the true input signal, which drives zero volts and zero currents between the signal and the shield (ground potential).

Driver Stage — The stage that drives the power amplifier. A *pre-driver* drives the driver.

Dual-Tone Squelch System (DTSS) — A selective calling system in which squelch is opened in a receiver only when a specific three-digit code is received.

Dummy Load — A load for transmitters that simulates an antenna, but does not cause radiation. Used for testing and adjustment.

Duplexer — A device that allows one antenna to be operated for both transmitting and receiving.

Duty Cycle — Ratio of a square wave to its ON time and OFF time.

Dynamic Instability — Undesired changes in a transmitter's frequency.

Dynamic Range — The measurement, in decibels, between the faintest and loudest sounds reproduced without significant noise or distortion in a system.

Eddy Currents — Induced currents in the core of a transformer with resulting I^2R (power) losses given up as heat. The higher the frequency the greater these losses become. Eddy currents can be reduced by laminating the core with thin sheets instead of one solid core, or with the use of ferrites or air cores.

Effective Radiated Power (ERP) — The actual power delivered to the antenna times the power gain of the major lobe of the antenna.

Electromagnetic Compatibility (EMC) — Electronic devices are tested to insure that they do not emit excessive interfering signals, nor are they excessively susceptible to another device's signals. Test may run all the way from 60 Hz, or line frequency, up to the CPU clock harmonics of 1 GHz and above. Entrances to any electronic enclosure must be screened (RF resistant). All wires and cables into and out of a device must not radiate emissions nor must these wires and cables allow EM fields into the enclosure of the device.

Electromagnetic Interference (EMI) — Interference to any electronic device caused by a producer of electromagnetic energy, such as transmitters, computers, electric motors, relays, etc.

Electronics Industries Association (EIA) — Sets equipment and other standards in the electronics field.

EMC Test Antenna — Antenna used to both radiate and detect RFI during EMC tests. They must have a very wide bandwidth (10 kHz-1 GHz). The EMC test antenna may be attached to a Class A DC 1 GHz transmitter, or a frequency generator and linear amplifier, to test a device for EMI susceptibility. The antenna may also be attached to special test receivers which are used to make measurements of a single frequency at a time, or to a spectrum analyzer to check multiple frequencies, testing for emissions.

EMI Current Probes — Probes optimized to measure RF currents on conductors. They are basically toroidal transformers.

EMI Injection Probes — Toroidal probes used to inject high-frequency currents into wires and cables for EMC susceptibility testing.

Emissions — Whether an electronic device generates excessive EM fields.

Encoding — An electronic digital function of converting from decimal to binary.

Engineering Change Order (ECO) — A formal order to change components or specifications on a manufactured device.

Epoxy-Dipped Tantalum Capacitors — Miniature electrolytics that have a very long shelf life. They are polarized.

Exciter — The carrier oscillator for a transmitter, or the driven element of a directional antenna.

Fading — The intermittent decline of a received signal in a receiver. Can be caused by out-of-phase signals reaching the receiver's antenna due to multipath reception, or changing conditions in the ionosphere. Can also occur if the RF signal of interest bounces off metal lightpoles, building pylons, signs (or anything metalic), and reaches the receiver at a different time than the direct signal. Multipath problems are especially pronounced against metal objects when the frequency is high.

Fan-Out — Maximum number of digital gates one gate can drive.

Faraday Shield — Also referred to as an *electrostatic shield*. A shield placed between two coils, such as the primary and secondary of an air-core transformer in an RF tank circuit, that drastically reduces the natural capacitance that occurs between two conductors separated by a dielectric (in this case, air). Harmonic and spurious signal frequency transfer between coils are almost eliminated, and are passed to ground. The shielding allows the resonant frequency to be coupled to the secondary through transformer action, with no adverse effects on the collapsing and expanding *magnetic* fields that cut the conductors of the secondary.

Fast Fourier Transform (FFT) — An algorithm for transforming data from the time domain to the frequency domain. Will transform a time domain signal on an oscilloscope, that is properly equipped with FFT processing, into a frequency domain display of the signal. It is perfect for displaying harmonics that would otherwise not be seen in the time domain.

FCC Type Acceptance — A type acceptance number given by the FCC to a manufacturer for a device that meets certain frequency, bandwidth, and stability requirements. If a nonoriginal or nonapproved part is used, type acceptance could be nullified, making the device illegal.

F-Connector — Common RF connector used in TV receiver connections and very low-power applications.

Federal Communications Commission (FCC) — The United States federal regulatory agency that governs communications, and is responsible for radio spectrum allocations and the licensing of individuals and radio stations.

Feedback Amplifier — An audio amplifier that exploits a negative feedback loop to supply degenerative feedback in order to minimize distortion. However, this will lower overall stage gain. Distortion in the feedback loop may be quite high.

Feedforward Amplifiers — Wideband amplifiers that have low distortion products with high stability and high gain.

Feedthrough Capacitor — A capacitor that allows DC and low-frequency AC to pass through an enclosure, while grounding noise. Commonly used at the input to a shield or metal cabinet containing a transmitter, receiver, or power supply. Also called an *EMI filter*.

Fencing — Internal copper shielding within a transmitter or receiver.

Field Strength Meter — A piece of test equipment that determines the strength of a signal at a certain distance from a transmitter's antenna, usually in mV. With high-powered transmitters, field strength readings are usually taken starting at one mile and extending out to 20 miles. When using a field strength meter: If the power of a transmitter increases from 1 kilowatt to 4 kilowatts, the field strength meter will indicate a doubling of the mVs/m, due to I^2R and V^2/R (since doubling the current or voltage means four times the power).

Field Strength Antenna — A measurement, generally in μV/meter, of the electromagnetic energy radiating from an antenna.

Final Power Amplifier (FPA) — The last power amplifier before the antenna.

Flatpack — A very small and thin integrated circuit package with RF and DC pin connectors exiting straight from the sides of the package. For surface-mount (SMD) or microwave microstrip circuit applications.

Flip-Flops — Digital two-state memory elements.

Flywheel Effect — The ability of a tuned circuit to complete a sinusoidal wave if only a pulse is received. The capacitor and inductor of a tank circuit exchange energy back and forth, causing oscillation at the tank's resonant frequency. Just as long as a pulse is received at set times to reenforce and replace the power lost in the unavoidable resistances present, the circuit will continue to oscillate, forming nearly pure sinusoids.

FM (or *Phase*) Noise — The short-term frequency deviations in the desired output frequency. Commonly expressed in dBc (*decibels below the carrier*).

Four-Layer Diode — Semiconductor device found in overvoltage indicators, relaxation oscillators, and time delay circuits. Four-layer diodes are turned on by a trigger pulse, and can be utilized for switching on SCRs.

Frequency Shift — An unwanted shift in oscillator frequency. Usually caused by a low C/L ratio, with the resulting low Q, in the tuned circuit of the oscillator; or by loading down of the oscillator; or by poor power supply regulation.

Frequency Synthesis (FS) — A technique for generating extremely accurate multiple frequencies from one (but generally more) crystals.

Frequency Tolerance — The amount, in a percentage, that the carrier wave is allowed to deviate from its center frequency (Ex: ± .005% of the carrier frequency). AM and FM broadcast stations must be within 0.002 percent of their assigned frequency. AM = plus or minus 20 Hz (0.00002 times 1 MHz). FM = plus or minus 2 kHz (0.00002 times 100 MHz). Fixed land transmitters at between 50 to 450 MHz must be within 0.0005%. (Do not confuse percent with a simple decimal multiplication. For example: Half of 100 is 50%, or 0.5 times 100; with 0.5 equaling 50% divided by 100%).

Frequency Translation — Changing one frequency to a higher or lower frequency, while still retaining the original modulation information.

Full Duplex — Both ends of the communication channel can communicate with each other during the same time frame.

Fundamental Frequency — The frequency at which a pure sine wave's energy is located. Except in frequency multipliers (harmonic generators), it is the frequency that the tuned circuits of RF devices are tuned to.

Fundamental Overload — Interference to an electronic device, such as a radio or television receiver, caused by the fundamental frequency of a transmitter reaching the receiver's amplifiers and producing undesired results. The receiver is incapable, through careless or faulty design, of rejecting out-of-channel signals.

Fuse — A part used to protect a circuit by opening when excessive current begins to flow. Must be placed only on the hot side of a circuit, and never on the ground return. Fuses come in many different packages, with the most common being the glass tube type with metal end caps. Some PCB consumer fuses are cartridge type that look like green 1/4-watt resistors. These are designated with an "F" or a "PR" (protector) on the PCB. "IC protectors" are miniature fuses in TO92 plastic cases but have only two leads. They have a very rapid response time, and may be designated with ICP, PR, or F.

Gain Block — A high-gain amplifier in a single package. Usually viable from DC to at least 3 GHz. It is very stable and is used in narrow and wideband linear amplifier applications. Normally a MMIC in GASFET for 50W use. Needs very little external biasing components. A gain block can also refer to a single amplifier stage or cascaded stages in any device.

Gain Compression — The point in an amplifier or receiver in which increasing the input signal does not linearly increase the output signal. When 1 dB of gain compression occurs, the receiver or amplifier is considered at its maximum input signal capability, and is now desensitized (the 1 dB compression point is the point in which an amplifier's gain decreases by 1 dB over its small-signal value).

GASFET — A gallium arsenide field-effect transistor used at UHF and above.

Gate Turnoff Thyristors (GTOs) — Thyristors that can be shut down with a gate pulse, unlike normal thyristors.

Gemini Header — A mounting plate that consists of two matched FETs or BJTs on a common flange.

Grid Dip Meter — Basically an oscillator with plug-in coils for gross adjustment and a variable capacitor for fine tuning. When the frequency of the grid dip meter is adjusted to the resonant frequency of a circuit's tank, the energy of the meter is coupled to the tank, lowering the amplitude of the oscillator's oscillations, thus causing a dip in the meter. Can be utilized to measure the approximate resonant frequency of any tuned circuit.

Ground Fault — Loss of ground, caused ordinarily by a severed ground strap or wire, or a bad ground connection. Also refers to any alternate return path to ground for current, such as through the human body.

Ground Fault Circuit Interrupter (GFCI) — A circuit that prevents a person from getting electrocuted by sensing that the hot and neutral wires are carrying equal currents. If the neutral is carrying less current, then the device trips. When this occurs, someone may have touched the hot *and* the ground (ground fault), causing the current to go directly to ground through the person's body, and not to the neutral wire as designed. A GFCI must normally be hooked to a 3-wire system. Used in damp environments.

Group Delay — The relative delay of certain frequencies over that of other frequencies.

Guaranteed Minimum Value (GMV) — A capacitor rating that guarantees that its printed value of capacitance will at least be at that level.

Gyrator — An active audio filter.

Half-Duplex — Each end of the channel can communicate with the other, but not at the same time.

Handy-Talkie — A walkie-talkie. Customarily referred to simply as an HT.

Hardware — The physical computer equipment, such as the CPU or the hard drive.

Harmonic Distortion Analyzer — A test instrument used to measure total harmonic distortion (THD). Use by: 1. Injecting an audio frequency into the receiver input at the proper amplitude. 2. Set analyzer's selector to *broadband*, then feed total signal into input, while setting meter to 100%. 3. Switch in the *fundamental rejection filter* and take reading (meter is calibrated in % THD).

Harmonic Generator Frequency Interference — Any harmonic of the fundamental frequency generated from a transmitter, frequently causing interference, such as from the master oscillator or frequency doublers or triplers. Any of these can leak from the transmitter's shields or cabinet, creating interference.

Harmonic Suppression — The degree to which harmonics at the output of a transmitter are attenuated.

Harmonics — The frequencies generated by a distorted sinewave at multiples of its fundamental. The harmonic's frequencies can be at two, three, four or more times that of the fundamental. If it could be attained in practice, a pure sine wave would have no harmonics, but would consist merely of its unadulterated fundamental frequency. Due to different transmission characteristics during different parts of the day, a harmonic may sometimes travel further than the fundamental — such as a 14 MHz second harmonic of 7 MHz. A strong signal can drive a receiver's RF amplifier into nonlinearity, causing the signal to be heard at harmonicaly related frequencies of 1F, 2F, 3F, etcetera.

Heat Spreader — A thin plate of copper that extends beyond a device, and is between a device and its heat sink, in order to help dissipate the device's heat over a wide area.

Helical Resonators — Basically a helically wound *resonant* transmission line surrounded by a conductive shield. Can be operated between 50 MHz and 500 MHz. They are employed as highly selective filters at the input to high-frequency receivers. Helical resonators have a flat response and narrow bandpass (approximately 2-3 hundred kHz; an LC tuned filter may have a bandpass of 2 MHz) when used with 4 to 6 resonators in a receiver's front end. The resonator can be tuned by an air dielectric variable capacitor or by tuning screws.

Helix — A single layer coil.

HEXFET — A low-drive current, low-resistance, high-gain, temperature-stable metal-oxide semiconductor field-effect transistor, used typically in switching power supplies and audio amplifiers.

Hook-Up Wire — Wire utilized to connect circuits or equipment.

Hot Chassis — Some chassis can be "hot" in that they are connected to one side of the AC line (without a transformer present for isolation) which, when touched, may cause a danger-ous electrocution hazard. This is due to the difference in potential between the hot chassis and ground. (A hot chassis must not be grounded — but can be AC bypassed by a low-value capacitor to ground).

House Wiring — Mains wiring available in a residence: green = Ground (the round plug on the socket or plug); white = Neutral (the tall slot); red or black = Hot, 120 volts (the short slot).

Hum — A low-frequency AM sound, noise, or interference, that can be at 60 Hz, or a harmonic of this line frequency, such as 120 or 180 Hz, that may be riding on the signal of interest, such as an RF carrier, or on the DC supply voltage. Can be caused by an unshielded power transformer, or poor filtering of the power supply. Hum, as viewed on a spectrum analyzer, would be sidebands that are very close (60 or 120 Hz) to the signal of interest (requiring a very narrow RBW filter setting to view).

Hybrid Amplifiers, RF Linear — First used in the CATV industry, but can now be found in some radio circuits. They are entire Class A RF amplifier circuits in one small package, with 17-35 dB of gain, frequency ranges of 1-500 MHz, output power levels of up to 0.4 watts, and with internally-matched input impedance and output impedances of 50Ω, and have a wide bandwidth with low distortion. They are sometimes referred to as RF gain blocks.

Hybrid Combiner — A device that allows two signal generators to inject their signals into a receiver without adversely affecting each other.

Hysteresis Loss — Losses in a transformer's core created by friction of the shifting mag-netic domains in the ferromagnetic core, caused by the alternating current signal, producing heat and a resulting power loss. Air-core coils have almost no hysteresis loss.

I/Q Signals — I is the input or output with 0 degrees phase shift, while Q is the input or output with 90 degrees phase shift. (Q = Quadrature; while I = In-phase).

IF Notch Filter — A tunable wavetrap used in a receiver's IF stages.

Image Suppression — In cellular test systems it is the amount of suppression, in dB, between the wanted and unwanted sideband during an SSB test (if the USB is desired, then the LSB is undesired, and must be suppressed).

IMD Dynamic Range — Is the difference, in decibels, between the noise floor and the power of two uniform input signals that creates a third-order product that is 3 dB over the noise floor. Causes false signals.

Incidental AM (IAM) — Unwanted AM of an FM signal.

Incidental FM (IFM) — Unwanted FM of an AM signal.

Inrush Current — The large transient currents that occur on equipment startup, caused by the currunt inrush through the power supply transformer, unheated tube filaments, uncharged filter capacitors, etc. In a device or power supply with inrush current limiting protection thermistors, they must be given a chance to cool off after device shutdown before repowering the circuit, or temporary excessive currents will flow, possibly blowing the fuse or damaging other components.

Instability — Tendency for an amplifier to oscillate. An unconditionally stable amplifier will not oscillate.

Intermodulation Distortion (IMD) — IMDs are created by the beating of two or more frequencies in a nonlinear device, such as a transistor or a tube, causing related spurious signals. Can only be created when two or more frequencies are involved. The IMD frequencies can be at almost any frequency, since not only can the fundamental frequencies mix, but so can their harmonics.

International Telecommunications Union (ITU) — A coalition of 154 member countries that negotiate worldwide RF spectrum allocations.

Inverter — A circuit used to convert DC into AC.

Isolator — A device constructed of ferrites, generally used at microwave frequencies, that allows a signal to pass in one direction, but not in the reverse direction. It is a one-way device that can be employed to prevent reflected waves or externally generated frequencies from entering the finals of a transmitter.

Jitter — The rapid fluctuation of a signal in frequency or amplitude.

Junction Tetrode Transistor — A four-terminal high-frequency transistor with two base leads: one grounded, the other at 0.4V. This reduces base resistance and increasing its upper frequency limit.

Kludge — Rube Goldberg-type device or setup.

Knife-Edge Diffraction — When an electromagnetic wave reaches the top of a barrier (such as a mountain), it has the tendency to bend slightly, causing reception in an area where it would not normally be possible.

Lead Acid Cells — A common high-current backup battery that is rechargeable, with a voltage per cell of 2.1V and a specific gravity of 1.27 at full charge.

Leakage Reactance — A transformer response caused by a certain amount of magnetic flux being lost from the primary to the secondary of a transformer. This will cause the voltage out of the transformer, under load, to be less than expected.

Lecher Lines — Used to measure wavelength, and thus frequency. They consist of a transmission line with adjustable sliding contacts. When these contacts reach the standing waves, the distance is measured between the voltage peaks or valleys, thus obtaining its wavelength.

Line Flattener — A device similar to a *transmatch*, but cannot match end-fed wire antennas.

Linear Mixing — Linear mixing creates one frequency that rides on another, usually creating hum. Unlike nonlinear mixing, no extra frequencies are produced.

Liquid Crystal Display (LCD) — When electricity is applied to liquid crystals the molecules of the crystals become opaque, forming letters and numbers on a small display panel.

Litz Wire — Specially woven insulated multistrand wire capable of high-frequency operation with low losses. Oscillator circuits will have less drift when Litz wire is used for the inductor, along with fiberglass (instead of ceramic) forms.

Loading Coils — Electrically increases the length of an antenna to make it appear longer than it actually is, decreasing its resonant frequency. However, they will also narrow the effective bandwidth of an antenna. (Adding a capacitor in series with an antenna will increase its resonant frequency, or electrically shorten the antenna).

Local Area Network (LAN) — A connection, by wire or radio, of computers over a small area, commonly confined to a single building or group of buildings, to share data or peripheral equipment.

Locking — Locking two oscillators together refers to electronically synchronizing the frequency of one oscillator to the frequency of another.

Logarithmic Taper — The resistance change of a potentiometer that varies logarithmicaly to simulate the response of the human ear, in contrast to linear tapers. Operated as audio potentiometers.

Low Noise Amplifier (LNA) — Amplifiers, usually in the VHF, UHF, and above regions, designed to add little noise to the input signal. They are used in the RF amplifiers of quality receivers.

LSA Diode — A device similar to the Gunn diode, but can output several times the Gunn's power.

Maggie — A magnetron vacuum tube.

Magnetostriction — When placed under a magnetic field, nickel constricts, while iron expands. If alternating magnetism is used, these materials will vibrate.

Mains Voltage — EMF supplied by interior building electrical wires. Also known as 120/240 voltage; 117/234 voltage; or 110/220 voltage.

Mean Time Between Failure (MTBF) — The approximated average time, typically in hours, before a device is expected to fail in normal operation.

Metglass — A special transformer core made from "iron-glass", which has very small hysterysis losses.

Microcontroller — A circuit, placed on a single chip, that contains a CPU, RAM, an EPROM, I/O ports, timers, and an interrupt controller. Programmed to perform a special function.

Microphone — A transducer that converts sound waves into current. There are many types: *Ceramic and crystal* mikes have a flat frequency response, require no power supply, have a high-impedance output, but have low output levels, with some hum pickup; *dynamic* mikes have a moving coil with low-impedance output, require no power supply, have good frequency response and high output levels, with low hum pickup and large dynamic range; *ribbon* mikes have a wide frequency response, high output level, with no power supply necessary, have low-impedance output, a large dynamic range, and low hum pickup; *condenser mikes* use a "voice" variable capacitor, have low output levels, and many need external power supplies.

Microphonics — Physical vibration causing noise modulation of a receiver or transmitter. Created by loose components, especially in the oscillator regions, such as crystals or capacitors. More common in tube equipment due to their sensitive electrodes.

Microprocessor — A CPU on a single chip. Frequently called the MPU (microprocessor unit).

Nicrowave Integrated Circuits (M.I.C.) — IC's with components integral to a small stripline or microstrip sections.

Modulation — Combining a lower frequency, the modulation, with a higher frequency, the carrier. The modulation frequency may be voice or data.

Modulus — The modulus of a digital counter is the number of counting states before it begins to repeat itself.

Monolithic — A component formed from a single material, with etching performed on or in this substrate material. A regular IC is monolithic.

Monolithic Microwave IC (MMIC) — High-frequency integrated circuits for microwave communications.

Monolithic-Crystal Filter — A small, low-cost, multielectrode crystal filter used for frequencies between 5 MHz to 300 MHz.

Motorboating — Unwanted self-oscillations in an audio amplifier causing a motorboat type "putt-putt" noise. May occur intermittently.

Multipath Fading — Also called *phase cancellation*, occurs if the RF signal of interest bounces off metal lightpoles, building pylons, or anything metal — as well as the ionosphere — and reaches the receiver at a different time than the direct signal, which produces an out-of-phase reception condition, causing fading. Multipath problems are especially pronounced against metal objects when the frequency is very high.

Multiple-Path Reception — A receiver receiving both ground wave and sky wave propagation, or by one-hop and two-hop paths, or by sky wave and surface wave. The two signals are alternately in-phase and out-of-phase with each other, which creates *fading*.

Multiplexers — A device with many inputs but only one output. By using control signals, any input can be steered to the single output.

Multivibrators (MV) — Circuits that utilize two transistors that alternate between saturation and cutoff. MV's are pulse generators of square or rectangular waves and are composed of four basic types. A. STABLE MV: Sends pulses only when triggered; B. ASTABLE MV (Free running): Sends pulses constantly without a need for triggering; C. BISTABLE MV (A flip-flop): One input causes one 180° phase-shifted pulse out; D. MONOSTABLE MV (A one-shot): Causes one output pulse of any preset length for each input pulse.

N-Connector — A 50Ω RF low-reflection connector good up to a maximum of 18 GHz. A 75Ω version is available, but the two are not compatible. Used for test equipment hookups and for coaxial and chassis connections.

Near-Field Probes — Specialized probes that can be attached to spectrum analyzers to detect cable and PCB RF emissions.

Network — A circuit grouping of electronic components. An example is the RF RC PI network that can transform one impedance to another by modifying the reactive component and changing the circuit resistance to a desired value. Also refers to a group of interconnected receiving, transmitting, or transceiving stations.

Network Analysis — A testing technique requiring the feeding of a known signal into the DUT and comparing this signal with the output signal.

No Trouble Found (NTF) — Comment found notated if a device-under-repair is found to be apparently functioning properly.

Noise Bridge — A device used to find the impedance of an antenna's input. Uses the concept of a balanced bridge.

Noise Figure (NF) — The degree that a receiver approaches the quietness of a theoretical receiver. The higher the noise figure, the noisier the receiver. In the real world, 4 dB is superior for a HF receiver. Since under 30 MHz atmospheric noise is quite high, VHF-and-above receivers have better noise figures to take advantage of the low atmospheric noises. Noise figure (NF) is basically the amount of noise, in dB, that is added by the receiver. Noise figure = Input S/N minus the output S/N of the amplifier.

Noise Floor — The level of all the combined noise (mostly thermal) of an electronic device or component. Below this a signal cannot easily be detected; and normally limits the possible amplification of any such signal.

Non-Return To Zero (NRZ) — A method of decreasing bandwidth requirements for data transmission by lowering the number of times the signal changes states, by forming any serial string of incoming "1"s as an unchanging HIGH state — until a "0" occurs, in which case the level drops to a LOW until the next HIGH occurs. (See *RZ*).

Notch Filter — A sharp bandstop filter. Some are readily tunable by the operator to reject an unwanted signal within the receiver's IF passband by tuning an adjustable capacitor, which adjusts the notch frequency within this IF passband.

Octave — A single octave is twice a given frequency or half a given frequency. If a 1 MHz signal is used as an example, then one octave above 1 MHz would be 2 MHz (2x1 MHz), while two octaves above 1 MHz would be 4 MHz (2x2x1 MHz), three octaves would be 8MHz (or 2x2x2x1 MHz), etc.

Optoelectronic Devices — Devices that either generate or control current by the application of light energy, or devices that generate light by the application of current. Examples are *photoresistors* (photocells), ordinarily made of CdS (Cadmium Sulfide), that vary resistance (between 300Ω to 200 MΩ) when light is applied or removed. Have slow response times, a memory effect and a nonlinear response to changes, but are very sensitive to light variations. *Photodiodes* are used as photoresistors. *Phototransistors* are operated as high-sensitivity photodiodes, but have higher current-carrying capabilities, but at slower response times. *Photovoltaic*, or solar cells, when exposed to light, generate a voltage. When a photon hits the cell an electron can be dislodged from the atom's valence, thus causing electron flow. They have low sensitivity, low open circuit voltage, and low closed circuit current. *LEDs generate light energy* when a voltage is applied to their leads. Typical values of current used by a LED would be 50 mA with a forward voltage of 1.6V.

Optoisolator — A device that utilizes an LED and a photodiode or phototransistor (an *optical coupler*) to completely isolate one circuit from another, with no noise feedback or loading on the first circuit. The signal is fed into the input, converted to light by the LED, sent through a small gap, and focused into a light-sensitive device that converts the light beam back into a signal voltage.

Original Equipment Manufacturer (OEM) — Used when referring to original replacement parts manufacturer for the device under repair.

Pad Attenuator — A circuit that reduces the output amplitude of an input signal, while maintaining the input and output impedances of the circuits over their complete frequency range.

Padder Capacitor — A trimmer capacitor for small levels of low-end frequency adjustment in a superheterodyne's local oscillator.

Padder Trimmer — A trimmer capacitor for small levels of high-end frequency adjustment in a superheterodyne's local oscillator.

Parameter — The electronic characteristics of a device.

Parametric Amplifier — Also known as para-amps. High-frequency amplifiers (primarily 30 GHz and above) with a traveling wave tube or varactor operated as the para-amp's main element. Very low noise, which can be further reduced by cryogenic cooling.

Parametric Instability — Instability in a high-frequency transistor amplifier, which can cause the outbreak of oscillations caused by the collector-to-base nonlinear capacitance creating low-frequency modulation of the amplifier's output frequency. High-value bypass capacitors can suppress these oscillations.

Parts-Per-Million — A ratio measurement of the purity or accuracy of a device. For example, a certain frequency counter is specified as being accurate to within ± 0.1 PPM. It reads a transmitter's output as 156.52 MHz. Thus, the reading of the counter will differ by a maximum of ±15.652 Hz from the transmitter's actual frequency, or:

$$\frac{0.1}{10^6} \; 156.52 \; MHz = 15.652 \; Hz$$

Peak Envelope Power (PEP) — The peak power of an AM or SSB envelope with 100% modulation applied to the carrier:

$$PEP = V_{RMS} \bullet I_{RMS} = \frac{V^2{}_{RMS}}{R}$$

Peak Inverse Voltage (PIV) — The maximum reverse voltage a diode can sustain before breakdown occurs.

Peak Limiter — A circuit that limits a signal's output amplitude to some preset maximum level.

Peak Limiting Amplifier — An amplifier used in AM broadcast to linearly limit the high-amplitude peaks of music and speech to prevent distortion and overmodulation.

Periodic Waveform — A wave that does not change from one cycle to the next. It can be a sine, square, sawtooth, etc., waveform. The sine wave is the only periodic waveform that cannot be distorted by RC, LC, or RL networks.

Permeability Tuning — Adjusting the frequency of a resonant tank by the movement of a coil's slug.

Personal Communication System (PCS) — A loosely defined standard for digital cellular and other advanced personal communication technologies.

Phantoms — RFIs that occur when two transmitters are ON at the same time, and beat together in any nonlinear impedance, with the resulting sum or difference frequency causing a beat note, or other interference, in any local receivers.

Phase Modulation (PM) — A technique of producing indirect FM.

Phase Noise — A measure of an oscillator's short-term frequency fluctuations.

PL Tone — A Private Line tone to break squelch of a suitably equipped receiver.

Point-To-Point Wiring — An old method of circuit wiring employing wire to connect components, instead of utilizing the more modern printed-circuit board traces to conduct current. Still used in some high-current applications and in wire-wrap circuits.

Pole — In reference to filters, a single pole is an increase in signal attenuation of 6 dB as the frequency doubles or is cut in half (one octave) at the filter's resonant frequency on the frequency response curve. Pole also refers to the number of crystals in a filter: a two-pole filter has two crystals. The more poles, the sharper the filter's skirt and the faster the drop on the response curve.

Port — The point or terminal in a circuit or device where a signal may be injected or extracted.

Potentiometer — An adjustable device that controls voltage. Also an old term occasionally used to describe voltmeters.

Potted Circuit — A circuit that is encased in a nonconductive compound for weather and vibration resistance.

Potted Inductor — An inductor encased in a nonconductive compound and surrounded by its own shield.

Power Divider — A resistive network that splits power placed at its input port equally among more than one output device, and without any phase changes.

Power Down Control Voltage (PD) — Some RF amplifier ICs have a PD pin: 0 volts at this pin cuts off power output from the IC; while 5 volts may allow full power out. Saves battery power.

Power Factor (PF) — In a capacitor, it is the percentage of loss and leakage, or 1 equals all loss and no capacitance; while 0 equals no loss and all capacitance.

Power Supply Unit (PSU) — A term for a power supply circuit.

Power Transistors — Transistors built to dissipate high power levels. Can be of the bipolar junction or FET type. Virtually all RF power FETs are of the E-MOSFETs N-Channel type.

Preforms — Small circles of solder placed under a component before soldering.

Premixer — A technique of obtaining a stable higher *variable* frequency by the nonlinear mixing of a stable crystal oscillator's frequency with that of a variable frequency oscillator's.

Pulse Amplifier — An amplifier optimized for fast and complex pulse RF waveforms.

Q-Point — The area, in an amplifier, where the DC operating bias is set when no signal is present. Depending on the location of the Q-point, an amplifier can operate Class A, AB, B, or C. Also called the quiescent, or resting, point.

Quadrature Modulator (Vector or I/Q Modulator) — A RFIC used for modulation, which can modulate a carrier with data. It is capable of creating most forms of analog and digital modulation: AM, PM, FM, and SSB all can be implemented by the quadrature modulator.

Quadrature Phase — Two sine waves that are 90° out of phase with each other.

Quasi-Complementary Symmetry — A complementary symmetry amplifier that employs two NPN Darlington pairs instead of one NPN and one PNP transistor. The two NPN Darlington transistors are lower in cost and easier to match than the transistors in a true complementary symmetry amplifier.

Radial Leads — Leads on a component in parallel with one another. The common lead arrangement for capacitors.

Radials — Also referred to as a *ground plane* or *counterpoise*. Used when insufficient conditions exist for grounding a quarter-wavelength vertical antenna. Consists of four or more quarter-wavelength wires laid on or in the ground, to simulate a proper earth ground.

Radiation Resistance — The AC resistance of an antenna at its feedpoint.

Radio Frequency Interference (RFI) — Interference to any electronic device caused by a producer of RF, such as a transmitter, a computer, or the LO in a receiver.

Rails — The long thin plastic packaging for storing ICs, or the maximum negative and positive voltages in a comparator.

Random Access Memory (RAM) — Memory that can be read and written from.

Reactive Power — An indication of the *apparent* power used in the reactive component of a circuit. Calculated by SINq (VI). Measured in VARS, or Volt Amps Reactive.

Reactor — A very high Q-coil utilized principally for its high reactance.

Read Only Memory (ROM) — Memory that can be read from, but not written to.

Real Power — The only true power consumed in a reactive circuit is the power dissipated in the resistance, while the reactive components do not consume power, they store it. Real power is calculated by I^2R.

Receiver Birdies (or *Crossovers*) — Undesirable mixer/LO harmonics that cause spurious beat notes, and vary in frequency during tuning of the receiver. They are minimized by double conversion. Birdies sound like squawks and beeps, and are audible to the listener as "phantom signals".

Receiver Signal-Strength Meter — A indicator of the relative or real power reaching a receiver. The meter usually obtains its signal from the AGC circuits or from the RSSI output from a RFIC.

Reflected Impedance — In transformer action, the impedance of the secondary load is reflected back into the primary winding; any load on the secondary will be reflected back to the primary. Calculated by the turns ratio of the transformer times the impedance of the secondary load.

Reflectometer — Also called an *SWR bridge*. Employed to measure the standing-wave ratio of a transmission line.

Regenerative Feedback — Amplifier feedback that is in-phase from its output back into its input. This feedback can create increased amplification or oscillations, either desirable or undesirable. Also called *positive feedback*.

Register — A data storage circuit.

Relative Field-Strength Meter (RFS) — A field-strength meter that employs relative, and not absolute, readings.

Reluctance — Magnetic resistance.

Repeater — A mechanism that retransmits a signal that it receives from a transmitter, generally concurrently. Used to extend the range of a low-powered transmitter or to lengthen the line-of-site of a low-lying antenna.

Resistor Array — Multiple resistors in a single IC DIP (Dual Inline Package) or SIP (Single Inline Package).

Resonator — A device that, at its natural resonant frequency, vibrates, or oscillates, and will resonate at no other injected frequency. Adopted instead of standard LC circuits at higher frequencies for filtering. A crystal is a resonator.

Return-To-Zero (RZ) — In the digital RZ format for data transmission, each "1" bit in an incoming serial data stream changes the RZ's signal state to a HIGH pulse, while any incoming "0" bits are held continuously LOW until an incoming "1" again occurs. (See *NRZ*).

Return Loss — Is the measurement, usually expressed in decibels, of the loss in strength of the RF signal returned (the *reflected power*) from an end termination or discontinuity in a transmission line.

RF — The part of the electromagnetic spectrum considered to be between approximately 50 kHz up to 3 GHz.

RF Connectors — Same as RFX connectors but good up to 11 GHz, and includes SMA, SMC, SMB, etc.

RF Integrated Circuits (RFICs) — Integrated circuits optimized and built specifically for RF use. Passive components for RFICs are usually, but not always, off-chip due to their large footprint if on-chip. Most RFICs are now powered from a single voltage supply.

RF Leakage — The amount of RF energy that leaks from a connector, cable, component, or circuit. Expressed in decibels.

RFX Connectors — Can be a general term for BNC, TNC, N, and PL-259 connectors — usually all below 1 GHz.

Rheostat — An adjustable resistor that controls current.

Ringing — Damped sine wave oscillations occurring when a tank circuit is hit by a pulse.

S/N Ratio — The ratio, in dBs, of the signal voltage or power, to the noise voltage or power.

Sampling Scopes — Oscilloscopes that sample waveforms at repetitive time slices, then reassemble the waveform and display it on the CRT. Since only discrete samples are made, this can extend an oscilloscope's frequency response by 100 or more (for repetitive signals only).

Schmitt Trigger — A circuit that can receive an incoming ragged pulse and convert it into a clean square wave.

Selective Calling — Tone-coded signals are sent to receivers from a transmitter, opening the squelch of only a specific receiver, thus opening a communications channel.

Selective Fading — Fading that strikes at certain frequencies, commonly caused by multipath transmission or single, double-hop transmissions reaching the receiver's antenna out-of-phase at times, in-phase at others.

Selectivity — The ability of a receiver to choose one signal out of many while rejecting all others. The receiver's bandwidth must be narrow enough, and with sharp enough skirts, but be wide enough to not attenuate the sidebands of the incoming signals. Generally the filters in the IF control this. Placing filters in series (cascading) narrows bandwidth, thus increasing selectivity.

Selectivity Factor — A filter measurement that describes the steepness of its skirt, or

$$\frac{(BW) - 3dB}{(BW) - 60dB}$$

Selenium Rectifier — A line bridge rectifier constructed of selenium and finned aluminum plates. Survives transients well.

Self-Resonance — A point is reached when a specific frequency causes a component to become resonant due to its distributed and stray capacitive and inductive reactances. Especially a problem with inductors: RFCs using single-layer coils may, at a certain frequency, reach either a parallel or series resonance. By winding the RF choke into separate flat wire "pies", these resonant peaks may be avoided.

Sensitivity — The ability of a receiver to pick up low-amplitude signals. A function of the gain, as well as the internal noise, of the receiver. Commonly expressed as the minimum signal that is 10 dB above the noise. Sensitivity is greatly influenced by the gain of the IF amplifiers and the noise generated in the RF and mixer sections. The best communication receivers can receive signals as low as 0.2 μV, while low-cost AM consumer receivers may not be able to receive anything less than 75 μVs.

Shielding — A device fabricated to protect a sensitive circuit or component from stray electric or magnetic fields. At low frequency, copper is a poor magnetic shield, while soft iron is a great magnetic shield. Copper is a very good RF shield however, while soft iron will work (but not as effectively). The lower the frequency, the thicker the electromagnetic shield must be to be effective. Shields around low-frequency transformers must be of a solid ferromagnetic material to encase the magnetic fields and protect the rest of the circuit from stray magnetic leakage.

Shot Noise — Random noise created by electron movement in a semiconductor or vacuum tube, resulting in a "hissing" sound.

Shrink-Fit Tubing — Plastic tubing placed over a wire or connection that, when heated, shrinks to tightly conform to the conductor. Performs the function of insulation.

Sideband Cutting — The higher audio frequencies become distorted due to the bandwidth of the IF filters being too narrow, and attenuating the higher sideband frequencies in a receiver.

Signal Analysis — A testing technique requiring the monitoring of a device's input and output signals during normal operation.

Signal Diode — A low-power diode.

Signal Monitors — Low-cost *spectrum analyzers* which monitor a narrow band of signals, usually those in a receiver's IF.

Signal Tracer — A piece of test equipment, operated in *signal tracing*, that detects the audio modulation on the injected RF, and amplifies this audio so it can be heard by the operator.

Simplex Operation — One side of a communications channel can only communicate with the other, but never visa-versa. In voice communications, it refers to the alternating communications between two transceivers on the same channel, but not at the same time, without the use of a repeater.

Single-Ended — A one-transistor amplifier; or an amplifier grounded at one end.

Skin Effect — A consequence of high frequencies, creating an increase in a conductor's effective AC resistance. The RF begins to travel closer to the surface as the frequency is raised, until finally conduction takes place only on the conductor's surface. This causes a higher value of AC resistance than that measurable by a normal ohmmeter. The skin effect also produces a decrease in the Q of a coil.

Skip Zone — Also called the *zone of silence*. An area where no reception is possible. It occurs at the end of ground wave propagation, but before the sky wave's return to earth.

SMA Connector — A common microwave connector good up to 24 GHz. Generally used for semi-rigid cables and components that will be semi-permanently connected.

Small Outline Integrated Circuit Package (SOIC) — A *SMD* IC used in *surface-mount technology* applications.

Software — A computer's detailed digital programs for performing a desired function.

Spaghetti — Hollow insulating sleeves for uninsulated wire.

Spectrum — The entire range of the electromagnetic waves from the long wavelength sound waves, to the infared, to visible light, to ultraviolet, to X-rays. The part of the electro-magnetic spectrum of concern in normal wireless communications is between approximately 20 kHz (VLF) up to 300 GHz (EHF).

Speech Compression — A form of AGC in the audio stages of an AM and SSB transmitter. Speech compression maintains the average power output at an increased level by amplifying low-amplitude audio more than higher-amplitude audio.

Speech Processor — A circuit that controls the amplitude and/or frequencies of a modulat-ing speech input to a transmitter's modulator. Suppresses overmodulation caused by exces-sive speech volume, as well as excessive bandwidth created by high-frequency audio.

Splatter — A collection of spurious sidebands caused by overmodulation.

Spread — Refers to a tolerance variation in a device, such as in low-cost FETs, which may be rated as having a I_{DSS} of 10 mA, but may be actually anywhere between 7-15 mA.

Spurious Emissions (*Spurious Radiation* or *"Spurs"*) — Interfering emissions that are outside of the design bandpass of a transmitter.

Square Wave — A periodic wave, made up of odd harmonics, with very rapid transistions forming two equal, but opposite, ON states. If a square wave is not exactly square (the duty cycle being more or less than 50%), then some even harmonics will be present.

Step Attenuator — A device that allows adjustable attenuation from its input to its output, in discrete dB calibrated steps, while maintaining impedances (input and output impedances are customarily 50Ω). Utilized to lower the output power of transmitters and frequency gen-erators to safe levels for insertion into a spectrum analyzer, frequency counter, etc.

String — A group of amplifiers in series. Specifically, a group of frequency multiplier stages are called a "string", such as the local oscillator string that increases the low frequency of the LO to a higher level.

Surface-Mount Devices (SMD) — IC's and discrete components operated in surface-mount technology (SMT) applications. They are soldered to the surface of the PCB, and do not have leads. Transistors, capacitors, inductors, resistors, crystals, etc., all come in SMD form.

Surface-Mount Technology (SMT) — A method of using special IC's and discrete compo-nents that are not soldered in through-holes in the PCB board, but are soldered to the board's surface.

Susceptibility — Whether an electronic device is adversly susceptible to another device's EM fields.

Sweep Generator — A frequency generator that can sweep through a band of frequencies, with user-variable sweep widths and output amplitudes. Employed to test the frequency response of amplifiers and filters in conjunction with an oscilloscope.

Swinging Chokes — Inductors that have a thin airgap in their core so that their inductance decreases with an increase in current. Used to economically improve regulation in certain power supplies.

T/R Subsystem — An add-on unit that contains a power amplifier for transmission and a low-noise amplifier for receiving, along with T/R switches. It is used to increase the power of an existing transceiver, as well as its receiver sensitivity.

Telemetry — The transmission of measurement signals remotely, such as pressure, positions, frequency, temperature, speed, etc.

Television Interference (TVI) — Interference to a television caused by a producer of RF, such as a transmitter, computer, or the LO in a receiver.

Temperature-Compensated Crystal Oscillator (TCXO) — A crystal oscillator that is highly stable, and employs a temperature sensor (a *thermistor*), which is used to generate a correction voltage to the compensation network (a *varactor*), which then attempts to overcome the drifting effects of the varying temperature. Commonly utilized in PLL circuits. Not as stable as the OCXO (Oven-Controlled Crystal Oscillator).

Temperature-Compensated Zeners — Zeners that do not have large changes in Zener voltage (V_z) with large changes in temperature.

Termination — A load used to terminate an output circuit, customarily in its characteristic impedance, to avert damage or unexpected operation. Can consist of an antenna, resistor, or circuit. A *resistive termination* is constructed specifically to prevent reflected waves.

Thermal Noise — Noise created in an amplifier by the random movement of electrons (*white noise*).

Thermal Runaway — A condition created in a semiconductor by increasing heat causing increased current, which produces further heat until, unchecked, the common result is destruction of the semiconductor device.

Thermistor — A temperature-sensitive resistor that changes its resistance, in a nonlinear fashion, as the ambient temperature changes. Thermistors may have either a positive or a negative temperature coefficient, and are sometimes used to prevent thermal runaway in amplifiers (they are usually placed in physical contact with one of the amplifier's transistors).

Thermocouple — A device constructed of two dissimilar metals that produces electricity by the application of heat to these metals, causing a difference in potential.

Thin Small Outline Package (TSOP) — An IC employed in *surface-mount technology* applications.

Thyristors — Solid-state electronically controlled switches. They may also be used to control the amount of power reaching a load. SCRs, TRIACs, and DIACs are common examples of thyristors.

TNC Connector — An RF connector good up to 12 GHz. Similar to the BNC, except threaded connections are employed instead of a bayonet.

Toner — A code generator to break the CTCSS or DPL on a receiver for testing. Use by placing the output of the toner into the EXT MOD input of the frequency generator, and attach the frequency generator to the radio receiver. The code synthesizer (tone gen) must modulate a frequency generator's frequency when being fed to a CTCSS-equipped receiver. When using a tone generator to break the DPL (digital private line) squelch, the signal generator's FM deviation must first be adjusted by switching in the tone gen's 134 Hz square wave test frequency, which is usually mode 2. With the toner connected to the signal generator's external modulation input port, adjust the FM deviation (from ±0.5 kHz) to ±1 kHz. Some signal generators have an inverted (180 degrees phase shifted) output during external modulation—if so, press the INVERT key on the toner to correct. Usually the proper DPL code sequence can be found in the receiver's manual, and then entered into the toner. This is normally a three-digit octal DPL code number. DPL generation (mode 1): Select "DPL function"; "Enter" three-digit octal code; press "Send" key. To turn off the DPL signal generation, press the CLEAR button, which sends a 134 Hz 250 mS burst OFF code, and then ceases sending. There are 83 such octal toner codes.

Tracking — The ability, in a superheterodyne receiver, for the local oscillator to maintain the intermediate frequency (IF); or for the RF and IF tank circuits to mutually remain onfrequency when utilizing ganged capacitors.

Transfer Oscillator — A device used to measure the frequency of a received signal by attaching a frequency generator to a frequency counter. While a receiver is receiving a signal, the generator is tuned until zero beat occurs in the receiver. This is the exact frequency of the received signal.

Transformer (XFMR) — An electronic component designed to transfer its power from its primary winding to its secondary winding by magnetic coupling, usually at a different voltage and current than that input into the primary winding. However, since the inductive reactance of the primary of a transformer limits the current flow, attaching DC to a transformer will allow heavy currents to flow; low-frequency AC would have same effect in a non-line-frequency transformer, while high frequency in any line transformer would increase inductive reactance and other losses.

Transient Intermodulation — Distortion caused by an amplifier's inability to respond to an input signal's fast changing amplitude variations.

Transient Voltage Suppressors (TVS) — Devices, typically Zener diodes, maximized for use as transient voltage suppressors in sensitive circuits. Some circuits have transient protection at turn-on by using transient suppressors such as spark gaps, vacuum encapsulated gaps, or solid-state suppressors. These send the transient to ground by tripping when a certain maximum voltage is reached.

Transistor-Transistor Logic (TTL or T²L Logic) — A digital logic circuit consisting of transistors. Its power supply must be a well regulated +5 volts.

Transit-Time Problem — In a tube, as an electron is "boiled" off of the cathode, it begins to travel toward the control grid, heading toward the plate. Before this electron can pass the control grid, at higher frequencies, the control grid has already begun to turn negative, and now repels the electron back toward the cathode. Lowers gain and causes *transit-time noise*. Transistors are also susceptible (See *Frequency limitations*).

Transition Region — In a transistor, the region between cutoff and saturation.

Translator — A type of repeater operated to retransmit TV signals into inaccessible areas.

Transmitter Noise — Noise generated in and transmitted by a transmitter due to the normal noise generated within its circuits.

Transponders — Satellite-based transponders are broadband repeaters that can retransmit many channels at once, while non-satellite transponders automatically broadcast an ID signal in response to another transmitter's interrogation signal.

Trap Antennas — A type of multi-band antenna capable of operating on different bands due to the presence of several resonant tuned tanks along its length. Trap antennas can radiate harmonics.

Trimmer — An adjustable fine-tuning resistor, capacitor, or inductor of small values and sizes.

Trunked RF Systems — Consist of a base station that can communicate with a large amount of portable units. Generally in the 806 to 866 MHz range, with 5 to 20 channels available and serving up to 2500 radio units. Trunked radio (UHF): one type may use 5, 10, 15, or 20 channels, plus a control station channel, which all mobiles monitor. When the PTT switch is placed on transmit, a tone is sent over the control channel. This tone identifies which fleet it belongs to. The burst is decoded by the control station's microprocessor. The control station scans all channels for a clear one. It then transmits a burst of digital data, which tunes all of that fleet's radios to the clear channel. A digital burst of information is then sent to the transmitter initiator to go ahead. As soon as the calling mobile hangs up the microphone, an all-clear tone is sent to all of that fleet's mobiles to automatically tune back to the control channel. If all channels are busy, the control station sends back a busy (red) light; when it clears, the green light is turned on automatically. Another type of trunked system does not use a control channel station. Each mobile has a built-in microprocessor and ROM with plug-in channel modules. Each mobile constantly scans all of the channels until a digital burst is received on one of these channels. It then locks on to this channel.

Twisted Pair — A transmission line consisting of two insulated wires twisted together. Employed for short runs in low-frequency applications to reduce low-frequency magnetic hum pickup and EMI-producing radiated fields. Characteristic impedance is typically about 100Ω. There are two types; shielded and unshielded.

Two-Tone 3rd-Order Intermodulation Distortion — The total amount of distortion, in dB relative to the desired waveform, of the output signal waveform that exists when two simultaneous input frequencies are applied.

UHF Connector — Also referred to as a PL-259 connector. Used with medium to low RF power levels at 200 MHz and below, with 50Ω impedances. A very common and mechanically strong connector.

Unijunction Transistor — The UJT is a three-terminal, single-junction electronic device operated as a switch in UJT relaxation oscillators, which produce certain types of nonsinusoidal waveforms, as well as a trigger in SCR/TRIAC circuits; and as a switch. A *PUT* is a programmable UJT.

Unit-Under-Test (UUT) — The unit currently under examination for faults, alignment, or repair.

Uplink — The frequency employed to send a signal up to a satellite.

Vacuum Relays — Relays that are in a vacuum so that they may survive high voltage and current.

Vacuum-Dielectric Capacitors — Special capacitors that can be of the fixed or variable type, and have a very high W.V. (Working Voltage), low capacitance, and are good up to 1 GHz. Used mainly in high-power transmitters.

Varactor (or *Varicap*) — A diode optimized to vary its internal capacitance with a change in its reverse bias voltage. Used in some oscillators in the frequency determining networks to electrically change their resonant frequency, and in receivers which use automatic varactor tuning of an LC filter to obtain a wide tuning range — without the operator manually adjusting for each band.

Variable Crystal Oscillator (VXO) — A crystal oscillator that uses a variable capacitor or inductor in series or parallel with the crystal to obtain very limited frequency adjustment capabilities.

Variac — An adjustable output voltage transformer. A variac does not provide isolation.

Varifilter — A type of adjustable notch filter which works by mixing down the IF and sending it through two filters whose frequency is slightly different. Tuning the mixing oscillator adjusts the notch-out.

Varistor — Consists of the *MOV* (Metal Oxide Varistor) and the *ZNR* (Zinc Oxide Resistor), and are used to protect circuits against transients: When a spike of voltage reaches a certain level, the varistor's resistance goes down, shunting the transient to ground. (Most quality electronic equipment have MOV diodes built-in for surge protections. With a bad surge an MOV may short, causing the fuse to continually blow in the main's junction box.)

Vias — *Through-hole vias* are plated hollow connections placed between tracks on opposite sides of a double-sided (tracks on both sides) PCB, or completely through a multi-layer PCB from top to bottom. *Blind vias* connect one exposed layer of a multi-layer board to a hidden lower layer.

Virtual Ground — A point in a circuit that appears to be at ground, but is not actually attached to ground. For example, since the inverting input to an op-amp has such an extremely high input impedance, very little current (virtually zero) flows into the op-amp, making the input a virtual ground because zero current equals zero volts, which equals ground potential.

VMOS — An older method of producing a high-power, high-current, high-gain, high-frequency RF power MOSFET. A VMOS single-layer device amplifier can reach 200W input. Replaced by *DMOS*.

Voltage-Controlled Crystal Oscillator (VCXO) — A crystal VCO. It has great frequency stability but limited tuning range.

Voltage-Dependent Resistor (VDR) — A device that decreases resistance continuously as the voltage rises. Can be used in current regulation circuits.

Wall Wort — An AC wall adapter, or *battery eliminator*. Many times these AC-to-DC converters may not be internally regulated, but may depend on the equipment they are powering for their regulation.

Wave Analyzer — Essentially a frequency-selective voltmeter.

Wave-Trap — Also called a bandstop filter. Designed to eliminate a certain frequency from entering a receiver or exiting a transmitter.

Wireless Local Area Network (WLAN) — A network that allows the wireless connectivity of each computer in a network to each other or to commonly shared peripherals.

Word — A group of bits filling a single storage location in a computer. May consists of 8 bits, 16 bits, 32 bits, etc.

Zero Beat — When two signals are mixed together in a nonlinear fashion, both the sum and difference of the two signals are created. If the difference frequency is sent to a speaker, then as the two original signals are brought closer together in frequency, an audio tone can be heard. Since the difference between these two signals decreases as they become closer in frequency, a point is reached where there will be no difference in frequency between the two signals. This is a null point, where no audio tone can be heard at all. This null point is referred to as "zero beat".

Zone Of Silence — See *skip zone*.

BIBLIOGRAPHY
Recommended Reading

The following are some of the finest books in the electronics field for technicians. A few may be out of print, while others may be published in newer editions than those shown below.

The ARRL Handbook for Radio Amateurs

Schetgen, Robert (Editor)

ARRL, 1995

ISBN 0-87259-172-7

Published yearly. Contains beginning, intermediate and advanced RF troubleshooting, construction, theory, and design information.

Basic Electronics

Grob, Bernard

McGraw Hill, 1984

ISBN 0-07-024928-8

A very informative book for those beginning in electronics.

Basic Electronics

The U.S. Navy

Dover Publications, 1973

ISBN 0-486-21076-6

Not withstanding the title, it is a great beginning, intermediate, and advanced guide to RF theory and troubleshooting.

The Circuit Designer's Companion

Williams, Tim

Newnes, 1991

ISBN 0-7506-1756-X

For those interested in the proper design of analog circuits without the need for immense mathematical knowledge.

Communication Electronics

Frenzel, Louis

McGraw-Hill, 1995

ISBN 0-02-801842-7

One of the finest books on practical and current RF circuits and theory.

Computer Technician's Handbook

Margolis, Art

McGraw-Hill, 1990

ISBN 0-8306-3279-4

No RF, but plenty of useful digital troubleshooting information for any *technician.*

Electronic Circuits and Applications

Grob, Bernard

McGraw-Hill, 1982

ISBN 0-07-024931-8

Covers RF, digital, power supplies, audio circuits, etc.

Electronic Communication

Shrader, Robert

Glencoe, 1991

ISBN 0-07-057157-0

A must-have book for anyone in radio communications.

Electronic Communications Systems

Dungan, Frank

Breton Publishers, 1987

ISBN 0-534-07698-X

Overview of communications in general, including RF, microwaves, telephones, satellites, etc.

The Illustrated Dictionary of Electronics

Turner, Rufus

Tab Books, 1982

ISBN 0-8306-1366-8

Fantastic reference.

Lenk's RF Handbook

Lenk, John

McGraw-Hill, 1993

ISBN 0-8306-4560-8

Covers the field of RF troubleshooting and operation.

Motorola RF Application Reports

Authors: Various

Motorola Literature Distribution, 1995

ISBN: None (Catalogue number HB215/D)

A impressive book on HF, VHF, and UHF amplifier design.

Radio Handbook

Orr, William

Howard W. Sams Publishing, 1995

ISBN 0-672-22424-0

A extraordinary book on two-way communication.

Radio Operator's License Q&A Manual

Kaufman, Milton

Hayden Books, Div. of Howard W. Sams & Co., 1987

ISBN 0-8104-0666-7

One of the great books on RF. Targeted to those interested in obtaining the prestigious FCC GROL license.

Troubleshooting Analog Circuits

Pease, Robert A.

Butterworth-Heinemann, 1993

ISBN 0-7506-9499-8

A notable book on troubleshooting low-frequency linear circuits.

INDEX

Surface-Mount Technology for PC Boards
James K. Hollomon, Jr.

The race to adopt surface-mount technology, or SMT as it is known, has been described as the latest revolution in electronics. This book is intended for the working engineer or manager, the student or the interested layman, who would like to learn to deal effectively with the many trade-offs required to produce high manufacturing yields, low test costs, and manufacturable designs using SMT. The valuable information presented in *Surface-Mount Technology for PC Boards* includes the benefits and limitations of SMT, SMT and FPT components, manufacturing methods, reliability and quality assurance, and practical applications.

Digital Electronics
Stephen Kamichik

Although the field of digital electronics emerged years ago, there has never been a definitive guide to its theories, principles, and practices — until now. *Digital Electronics* is written as a textbook for a first course in digital electronics, but its applications are varied.

Useful as a guide for independent study, the book also serves as a review for practicing technicians and engineers. And because *Digital Electronics* does not assume prior knowledge of the field, the hobbyist can gain insight about digital electronics.

Some of the topics covered include analog circuits, logic gates, flip-flops, and counters. In addition, a problem set appears at the end of each chapter to test the reader's understanding and comprehension of the materials presented. Detailed instructions are provided so that the readers can build the circuits described in this book to verify their operation.

Professional Reference
510 pages - Paperback - 7 x 10"
ISBN: 0-7906-1060-4 - Sams: 61060
$26.95 ($36.95 Canada) - July 1995

Electronic Theory
150 pages - Paperback - 7-3/8 x 9-1/4"
ISBN: 0-7906-1075-2 - Sams: 61075
$16.95 ($22.99 Canada) - February 1996

CALL 1-800-428-7267 TODAY FOR THE NAME OF
YOUR NEAREST PROMPT PUBLICATIONS DISTRIBUTOR

 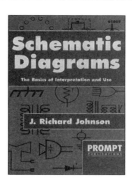

The Microcontroller Beginner's Handbook
Lawrence A. Duarte

Schematic Diagrams
J. Richard Johnson

Microcontrollers are found everywhere — microwaves, coffee makers, telephones, cars, toys, TVs, washers and dryers. This book will bring information to the reader on how to understand, repair, or design a device incorporating a microcontroller. *The Microcontroller Beginner's Handbook* examines many important elements of microcontroller use, including such industrial considerations as price vs. performance and firmware. A wide variety of third-party development tools is also covered, both hardware and software, with emphasis placed on new project design. This book not only teaches readers with a basic knowledge of electronics how to design microcontroller projects, it greatly enhances the reader's ability to repair such devices. Lawrence A. Duarte is an electrical engineer for Display Devices, Inc. In this capacity, and as a consultant for other companies in the Denver area, he designs microcontroller applications.

Step by step, *Schematic Diagrams* shows the reader how to recognize schematic symbols and determine their uses and functions in diagrams. Readers will also learn how to design, maintain, and repair electronic equipment as this book takes them logically through the fundamentals of schematic diagrams. Subjects covered include component symbols and diagram formation, functional sequence and block diagrams, power supplies, audio system diagrams, interpreting television receiver diagrams, and computer diagrams. *Schematic Diagrams* is an invaluable instructional tool for students and hobbyists, and an excellent guide for technicians.

Electronic Theory
240 pages - Paperback - 7-3/8 x 9-1/4"
ISBN: 0-7906-1083-3 - Sams: 61083
$18.95 ($25.95 Canada) - July 1996

Electronic Theory
196 pages - Paperback - 6 x 9"
ISBN: 0-7906-1059-0 - Sams: 61059
$16.95 ($22.99 Canada) - October 1994

CALL 1-800-428-7267 TODAY FOR THE NAME OF YOUR NEAREST PROMPT PUBLICATIONS DISTRIBUTOR